Ronald Larsen

An Introduction to the Theory of Multipliers

Springer-Verlag Berlin Heidelberg New York 1971

Professor RONALD LARSEN
Department of Mathematics, Wesleyan University
Middletown, U.S.A.

Geschäftsführende Herausgeber

Prof. Dr. B. Eckmann
Eidgenössische Technische Hochschule Zürich

Prof. Dr. B. L. van der Waerden
Mathematisches Institut der Universität Zürich

AMS Subject Classifications (1970): 43-02, 43A22, 42-02, 42A18, 46-02, 46J99

ISBN 3-540-05120-1 Springer-Verlag Berlin · Heidelberg · New York
ISBN 0-387-05120-1 Springer-Verlag New York · Heidelberg · Berlin

Library of Congress Catalog Card Number 78-134023. Printed in Germany. Type-setting and printing: Universitätsdruckerei Stürtz, Würzburg

To the memory of my mother

Preface

When I first considered writing a book about multipliers, it was my intention to produce a moderate sized monograph which covered the theory as a whole and which would be accessible and readable to anyone with a basic knowledge of functional and harmonic analysis. I soon realized, however, that such a goal could not be attained. This realization is apparent in the preface to the preliminary version of the present work which was published in the Springer Lecture Notes in Mathematics, Volume 105, and is even more acute now, after the revision, expansion and emendation of that manuscript needed to produce the present volume.

Consequently, as before, the treatment given in the following pages is eclectric rather than definitive. The choice and presentation of the topics is certainly not unique, and reflects both my personal preferences and inadequacies, as well as the necessity of restricting the book to a reasonable size. Throughout I have given special emphasis to the functional analytic aspects of the characterization problem for multipliers, and have, generally, only presented the commutative version of the theory. I have also, hopefully, provided too many details for the reader rather than too few.

Many interesting and important results have had to be omitted. Some of these topics have been discussed briefly in the notes to be found at the end of each chapter, while for many others I have merely suggested appropriate references. I have also attempted to indicate in the notes some sources for the material developed in the chapter in question. However, no attempt has been made to list all the sources of any one given result. Such an attempt seemed to me to be unnecessary, nor, given the copious amount of material on multipliers which exists in the literature, could it have been accomplished without some errors and omissions. I do wish to apologize though to those authors whose work has been intentionally or unintentionally omitted from these and other bibliographical references in the book. In no case should the lack of a reference for a specific result be construed as a claim to originality on my part. Indeed, the vast majority of the results presented are due to other mathematicians, although I may be responsible for an occasional novelty of proof.

At the suggestion of the publisher, I have collected in a number of appendices at the end of the book most of the more or less standard results from various areas of mathematics which are employed without proof in the body of the work. Hopefully these appendices will facilitate the reading of the material. References in the text to results in the appendices are either to the paragraph in which a particular result or concept is to be found, or explicitly to the result in question. For example, Plancherel's Theorem is Theorem F.8.2, that is, the second theorem in the eighth paragraph of Appendix F.

In general, I have tried to indicate the location of sources outside of the present text as accurately as possible. For this reason, whenever possible, such citations refer to a paragraph or theorem number in the appropriate reference, rather than to a page number. Thus, for example, references for a proof of the Fourier Inversion Formula are given as: Gaudry [6], II.4.1, II.5.2 and Rudin [5], 1.5.1, that is, the proof appears in sections 4.1 and 5.2 of Chapter II of Gaudry [6] and in section 5.1 of Chapter 1 of Rudin [5]. In some instances a reference could not be accurately given in this manner, and in these cases we have indicated the source in the usual fashion by page numbers.

After the completion of the manuscript a number of additional papers related to the theory of multipliers, which I had either overlooked or of which I had not been previously aware, came to my attention. Although it was impossible to include mention of these articles in the text proper, they are cited in footnotes scattered throughout the book and are listed in the bibliography.

A word is perhaps in order about the paucity of references to *Abstract Harmonic Analysis, II*, by E. Hewitt and K. A. Ross [Grundlehren der mathematischen Wissenschaften, Band 152]. This is due to the fact that this excellent volume was not available to me at the time of writing the manuscript. However, it certainly should be added as one of the chief sources for the material from harmonic analysis used in this book.

I am particularly grateful to Robert B. Burckel for his contributions to the form of the present volume. He painstakingly read the previous version of the manuscript which appeared in the Springer Lecture Notes in Mathematics series, and suggested many corrections, additions and improvements. A goodly portion of whatever clarity and precision is to be found in the present treatment is due to him, and I am deeply in his debt for his assistance. Those errors which do appear in the text are, of course, my own responsibility.

I wish to thank Miss Rosemarie Stampfel for her truly outstanding job of typing the manuscript. Her skill and experience transformed what could have been an onerous task into a pleasant one.

Thanks are also due to the University of California, Santa Cruz, for financial support during the final stages of preparation of the manuscript, and to Springer-Verlag for their understanding, cooperation and efficiency in producing the book.

Finally, I wish to express my appreciation to my wife, Joan, for her editorial help preparing the manuscript for publication, and especially for her patience and understanding during my frequent black moods while the manuscript was being written.

Durham, Connecticut
January, 1971

RONALD LARSEN

Table of Contents

Symbol Index

"Vil du være med så heng på!"
Tyrihans

Chapter 0

Prologue: The Multipliers for $L_1(G)$

0.0. Introduction. The concept of a multiplier first appears in harmonic analysis in connection with the theory of summability for Fourier series, namely, in attempts to describe those sequences $\{c_n\}$ for which $\sum_n c_n a_n e^{int}$ is always the Fourier series of a periodic integrable function whenever $\sum_n a_n e^{int}$ is such a Fourier series. Subsequently the notion has been employed in many other areas of harmonic analysis, such as the study of properties of the Fourier transformation and its extensions, the investigation of homomorphisms of group algebras, the characterization of group algebras, the study of factorization problems in various topological algebras, the characterization of convolution in terms of its mapping properties, the study of the conjugate Fourier series and of lacunary Fourier series. Multipliers have also appeared in a variety of other contexts, among which we mention the general theory of Banach algebras, representation theory for Banach algebras and locally compact groups, the study of Banach modules, the theory of singular integrals and fractional integration, interpolation theory, stochastic processes, the theory of semigroups of operators, operational calculus, partial differential equations, the study of certain approximation problems and the question of the existence of invariant means.

Our main concern will not be with these various applications of the theory of multipliers. Rather we shall address ourselves primarily to the task of characterizing multipliers. The meaningful presentation of this problem in the context of various topological algebras and linear spaces, and its investigation, together with some few examples of applications, makes up the content of the succeeding chapters.

0.1. Multipliers for $L_1(G)$. One way of obtaining some insight into the multiplier problem is to examine the situation for the group algebra of a locally compact Abelian group. Before doing this we should remark on the fact that we have yet to define what we mean by a "multiplier". This equivocation is intended since, as will be evident shortly, there is often more than one equally valid definition. The appropriate choice of a definition will often depend on the context in which we are considering multipliers.

If G is a locally compact Abelian group then we shall denote by $L_1(G)$ the linear space of equivalence classes of complex valued functions which are absolutely integrable with respect to Haar measure λ on G. With the norm

$$\|f\|_1 = \int_G |f(t)|\, d\lambda(t) \qquad (f \in L_1(G))$$

$L_1(G)$ is a Banach space, and a Banach algebra with the usual convolution multiplication

$$f * g(t) = \int_G f(t s^{-1})\, g(s)\, d\lambda(s) \qquad (f, g \in L_1(G)).$$

$M(G)$ will denote the Banach space of bounded regular complex valued Borel measures on G where the norm of $\mu \in M(G)$ is $\|\mu\| = |\mu|(G)$. $|\mu|$ denotes the total variation of the measure μ. $M(G)$ is also a Banach algebra with identity under the multiplication

$$\mu * \nu(E) = \int_G \mu(E s^{-1})\, d\nu(s) \qquad (\mu, \nu \in M(G)).$$

$L_1(G)$ is a closed ideal in $M(G)$ (F.1). We denote the dual group of G by $\hat{G}$ (B.2). If $f \in L_1(G)$ and $\mu \in M(G)$ then the Fourier and Fourier-Stieltjes transforms are defined as

$$\hat{f}(\gamma) = \int_G (t^{-1}, \gamma)\, f(t)\, d\lambda(t) \qquad (\gamma \in \hat{G}),$$

and

$$\hat{\mu}(\gamma) = \int_G (t^{-1}, \gamma)\, d\mu(t) \qquad (\gamma \in \hat{G}),$$

where $(\cdot,\cdot)$ denotes the usual pairing between G and $\hat{G}$ (B.2). These transforms are homomorphisms, that is, $(f * g)\hat{} = \hat{f}\hat{g}$ and $(\mu * \nu)\hat{} = \hat{\mu}\hat{\nu}$ (F.5 and F.6). The collection of all Fourier or Fourier-Stieltjes transforms will be denoted by $L_1(G)\hat{}$ or $M(G)\hat{}$, respectively. If $s \in G$ then τ_s will denote the *translation operator* defined by $(\tau_s f)(t) = f(t s^{-1})$. It is obvious that each translation operator defines a linear isometry of $L_1(G)$ onto $L_1(G)$. $L_\infty(G)$ will denote the Banach algebra of equivalence classes of essentially bounded measurable functions on G under pointwise multiplication and the usual essential supremum norm $\|\cdot\|_\infty$ (D.9).

The main results on the multipliers of $L_1(G)$ are contained in the following theorem.

Theorem 0.1.1. *Let G be a locally compact Abelian group and suppose $T: L_1(G) \to L_1(G)$ is a continuous linear transformation. Then the following are equivalent:*

(i) *T commutes with the translation operators, that is, $T\tau_s = \tau_s T$ for each $s \in G$.*

(ii) *$T(f*g)=Tf*g$ for each $f, g\in L_1(G)$.*

(iii) *There exists a unique function φ defined on $\hat{G}$ such that $(Tf)\hat{} = \varphi\hat{f}$ for each $f\in L_1(G)$.*

(iv) *There exists a unique measure $\mu\in M(G)$ such that $(Tf)\hat{} = \hat{\mu}\hat{f}$ for each $f\in L_1(G)$.*

(v) *There exists a unique measure $\mu\in M(G)$ such that $Tf=f*\mu$ for each $f\in L_1(G)$.*

Proof. Suppose $T\tau_s=\tau_s T$ for each $s\in G$. Let $k\in L_\infty(G)$. Then the mapping $f\to\int_G Tf(t)\,k(t^{-1})\,d\lambda(t)$ defines a bounded linear functional on $L_1(G)$ since

$$\left|\int_G Tf(t)\,k(t^{-1})\,d\lambda(t)\right|\leqq \|k\|_\infty\,\|Tf\|_1\leqq \|k\|_\infty\,\|T\|\,\|f\|_1,$$

where $\|T\|$ denotes the usual operator norm of T. Consequently, (Theorem D.9.2), there exists a function $\kappa\in L_\infty(G)$ such that

$$\int_G Tf(t)\,k(t^{-1})\,d\lambda(t)=\int_G f(t)\,\kappa(t^{-1})\,d\lambda(t) \qquad (f\in L_1(G)).$$

If $f, g\in L_1(G)$ then we have

$$\begin{aligned}\int_G Tf*g(t)\,k(t^{-1})\,d\lambda(t)&=\int_G\Big[\int_G Tf(t\,s^{-1})\,g(s)\,d\lambda(s)\Big]\,k(t^{-1})\,d\lambda(t)\\ &=\int_G\Big[\int_G(\tau_s T)\,f(t)\,g(s)\,d\lambda(s)\Big]\,k(t^{-1})\,d\lambda(t)\\ &=\int_G g(s)\Big[\int_G(T\tau_s)\,f(t)\,k(t^{-1})\,d\lambda(t)\Big]\,d\lambda(s)\\ &=\int_G g(s)\Big[\int_G \tau_s f(t)\,\kappa(t^{-1})\,d\lambda(t)\Big]\,d\lambda(s)\\ &=\int_G \kappa(t^{-1})\Big[\int_G f(t\,s^{-1})\,g(s)\,d\lambda(s)\Big]\,d\lambda(t)\\ &=\int_G T(f*g)(t)\,k(t^{-1})\,d\lambda(t).\end{aligned}$$

Since this remains valid for each $k\in L_\infty(G)$, we conclude via a consequence of the Hahn-Banach Theorem (D.6b) that $Tf*g=T(f*g)$ for all $f, g\in L_1(G)$.

Thus (i) implies (ii).

Suppose that $Tf*g=T(f*g)$ for all $f, g\in L_1(G)$. Then since $L_1(G)$ is commutative (F.1) it follows immediately that $Tf*g=f*Tg$ for all $f, g\in L_1(G)$. In particular, then for each $f, g\in L_1(G)$, we have $(Tf)\hat{}\,\hat{g}=\hat{f}(Tg)\hat{}$. For each $\gamma\in\hat{G}$ choose a $g\in L_1(G)$ such that $\hat{g}(\gamma)\neq 0$ (Theorem E.2.2) and define $\varphi(\gamma)=(Tg)\hat{}(\gamma)/\hat{g}(\gamma)$. The equation $(Tf)\hat{}\,\hat{g}=\hat{f}(Tg)\hat{}$

shows that the definition of φ is independent of the choice of g. For φ so defined it is apparent that $(Tf)\hat{} = \varphi \hat{f}$ for each $f \in L_1(G)$. If ψ is a second function on $\hat{G}$ such that $(Tf)\hat{} = \psi \hat{f}$ for each $f \in L_1(G)$ then the equation $(\varphi - \psi)\hat{f} = 0$ for each $f \in L_1(G)$ reveals that $\varphi = \psi$.

Thus (ii) implies (iii).

Suppose now that $(Tf)\hat{} = \varphi \hat{f}$ for each $f \in L_1(G)$, that is, $\varphi \hat{f} \in L_1(G)\hat{}$ whenever $\hat{f} \in L_1(G)\hat{}$. If we set $\|\hat{f}\| = \|f\|_1$, then $L_1(G)\hat{}$ becomes a Banach space (F.10) and $S\hat{f} = \varphi \hat{f}$ defines a linear mapping from $L_1(G)\hat{}$ to $L_1(G)\hat{}$. Suppose $f_n; f, g \in L_1(G)$ are such that $\lim_n \|\hat{f}_n - \hat{f}\| = 0$ and $\lim_n \|\varphi \hat{f}_n - \hat{g}\| = 0$. Then for each $\gamma \in \hat{G}$ we have, since $\|\hat{f}\|_\infty \leqq \|f\|_1$ for each $f \in L_1(G)$ (F.5), that

$$\hat{g}(\gamma) = \lim_n \varphi(\gamma) \hat{f}_n(\gamma) = \varphi(\gamma) \lim_n \hat{f}_n(\gamma) = \varphi(\gamma) \hat{f}(\gamma).$$

Hence S is a closed mapping. From the Closed Graph Theorem (Theorem D.6.1) we conclude that S is continuous, that is, there exists a constant K such that $\|\varphi \hat{f}\| = \|S\hat{f}\| \leqq K \|\hat{f}\|$ for each $f \in L_1(G)$.

Let $\gamma_1, \gamma_2, \ldots, \gamma_n \in \hat{G}$, $\varepsilon > 0$ and choose $f \in L_1(G)$ such that $\|\hat{f}\| = \|f\|_1 < 1 + \varepsilon$ and $\hat{f}(\gamma_i) = 1$, $i = 1, 2, \ldots, n$ (F.7e). If $c_1, c_2, \ldots, c_n$ are any complex numbers and $\hat{g} = \varphi \hat{f}$ then we see that

$$\begin{aligned}
\left| \sum_{i=1}^n c_i \varphi(\gamma_i) \right| &= \left| \sum_{i=1}^n c_i \hat{g}(\gamma_i) \right| \\
&= \left| \int_G \left[\sum_{i=1}^n c_i (t^{-1}, \gamma_i) \right] g(t)\, d\lambda(t) \right| \\
&\leqq \|g\|_1 \left\| \sum_{i=1}^n c_i (\cdot, \gamma_i^{-1}) \right\|_\infty \\
&= \|\varphi \hat{f}\| \left\| \sum_{i=1}^n c_i (\cdot, \gamma_i^{-1}) \right\|_\infty \\
&= K \|\hat{f}\| \left\| \sum_{i=1}^n c_i (\cdot, \gamma_i^{-1}) \right\|_\infty \\
&< K(1+\varepsilon) \left\| \sum_{i=1}^n c_i (\cdot, \gamma_i^{-1}) \right\|_\infty .
\end{aligned}$$

Since ε is arbitrary it follows that for any $\gamma_1, \gamma_2, \ldots, \gamma_n \in \hat{G}$ and complex numbers $c_1, c_2, \ldots, c_n$ we have

$$\left| \sum_{i=1}^n c_i \varphi(\gamma_i) \right| \leqq K \left\| \sum_{i=1}^n c_i (\cdot, \gamma_i^{-1}) \right\|_\infty .$$

Moreover, φ is continuous. This is evident because $\varphi\hat{f}$ is continuous for each $f \in L_1(G)$ and for each open set U in $\hat{G}$ with compact closure there exists an $f \in L_1(G)$ such that $\hat{f}$ is a constant on U (F.7e). Appealing now to a well-known characterization of Fourier-Stieltjes transforms (Theorem F.9.1) we conclude that there exists a $\mu \in M(G)$ such that $\varphi = \hat{\mu}$. Clearly μ is unique (F.7b).

Hence (iii) implies (iv).

The implication (iv) implies (v) is obvious since $L_1(G)$ is an ideal in $M(G)$.

Finally, if $Tf = f * \mu$ for each $f \in L_1(G)$ then $(T\tau_s)f = \tau_s f * \mu = \tau_s(f * \mu) = (\tau_s T)f$ for each $s \in G$. Hence $T\tau_s = \tau_s T$.

Therefore (v) implies (i) and the proof of the theorem is complete. □

A *multiplier* for $L_1(G)$ is a continuous linear operator on $L_1(G)$ which satisfies any one, and hence all, of the five characterizations given in the preceding theorem. It is evident that the multipliers form a linear space and that each translation operator is a multiplier.

Generally one starts with the definition that a multiplier is either a continuous linear operator T which commutes with the translation operators or one such that $Tf * g = T(f * g)$. A third definition is also frequently used. Namely, one calls a function φ on $\hat{G}$ a multiplier for $L_1(G)$ if $\varphi\hat{f} \in L_1(G)\hat{}$ whenever $f \in L_1(G)\hat{}$. If one defines $Tf \in L_1(G)$ by the equation $(Tf)\hat{} = \varphi\hat{f}$ then, since $L_1(G)$ is semi-simple (F.7b), it follows that T is a well-defined linear transformation from $L_1(G)$ to $L_1(G)$. Moreover, if $\lim_n \|f_n - f\|_1 = 0$, and $\lim_n \|Tf_n - g\|_1 = 0$ then it is evident that $(Tf)\hat{} = \varphi\hat{f} = \hat{g}$, from which we conclude that T is a closed transformation. An application of the Closed Graph Theorem (Theorem D.6.1) then reveals that the transformation T is continuous. From the definition of T it is obvious that $Tf * g = T(f * g)$ and hence T is a multiplier in the sense previously indicated. Thus the third definition of a multiplier for $L_1(G)$ is equivalent to the other definitions.

It should also be noted that the proof of the implication (iii) implies (iv) actually establishes more than the theorem asserts. In particular, it shows that if φ is any function on $\hat{G}$ for which $\varphi\hat{f} \in L_1(G)\hat{}$ whenever $\hat{f} \in L_1(G)\hat{}$ then the bounded linear transformation T on $L_1(G)$ defined by φ is of the form $Tf = f * \mu$ for some unique $\mu \in M(G)$.

The structural richness of the group algebras is primarily responsible for the diverse equivalent formulations of the notion of a multiplier. However, if one wishes to study multipliers for more general spaces, then certain of the definitions may no longer be meaningful, and some may be more appropriate than others. For example, if A is a Banach algebra then neither the concept of an operator which commutes with translations may be meaningful nor, in the noncommutative case, is the machinery of the Gelfand representation theory (E.2) available. Here it seems natural to define a multiplier as a continuous linear operator T from A to A such that $(Tx)y = x(Ty)$ for all x, y in A. If A is semi-simple and commutative then we could also define a multiplier T by the equation $(Tx)\hat{} = \varphi\hat{x}$ for each x in A where φ is some function on the regular maximal ideal space of A such that $\varphi\hat{x} \in \hat{A}$ whenever $\hat{x} \in \hat{A}$. $\hat{x}$ of course denotes the Gelfand transform of x. Clearly the former definition is more general and hence is preferable.

On the other hand, if A is a topological linear space of functions or measures on a locally compact Abelian group then the definition of a multiplier as a linear operator which commutes with translations is evidently the most natural one. In this context it is also natural to extend the notion of a multiplier to include linear transformations between two different spaces which commute with translations.

In either context one generally attempts to obtain characterizations of the multipliers similar to (iii) and (v) established for $L_1(G)$ in Theorem 0.1.1. A prototype of still another kind of characterization is given by the following corollary to the previous theorem.

Corollary 0.1.1. *Let G be a locally compact Abelian group. Then the space of multipliers for $L_1(G)$ is isometrically isomorphic to $M(G)$.*

Proof. From Theorem 0.1.1 it is evident that the equation $Tf = f*\mu$ defines a bijective isomorphism between the multipliers for $L_1(G)$ and $M(G)$, and that if T corresponds to μ under this mapping then $\|T\| \leqq \|\mu\|$.

Conversely, for any $\gamma_1, \gamma_2, \ldots, \gamma_n$ in $\hat{G}$, complex numbers $c_1, c_2, \ldots, c_n$ and $\varepsilon > 0$, if we choose $f \in L_1(G)$ as in the proof of (iv) of Theorem 0.1.1 we see that

$$\left|\sum_{i=1}^{n} c_i \hat{\mu}(\gamma_i)\right| = \left|\sum_{i=1}^{n} c_i (f*\mu)\hat{}(\gamma_i)\right| = \left|\sum_{i=1}^{n} c_i (Tf)\hat{}(\gamma_i)\right|$$
$$\leqq \|T\|(1+\varepsilon)\left\|\sum_{i=1}^{n} c_i(\cdot, \gamma_i^{-1})\right\|_\infty .$$

Hence

$$\left|\sum_{i=1}^{n} c_i \hat{\mu}(\gamma_i)\right| \leqq \|T\| \left\|\sum_{i=1}^{n} c_i(\cdot, \gamma_i^{-1})\right\|_\infty ,$$

from which it follows immediately (Theorem F.9.2) that $\|\mu\| \leqq \|T\|$.

Therefore the mapping is an isometry. □

Characterizations such as the one given in this corollary will be established at several points in the following chapters. However, at times the mapping will only be a continuous linear mapping from the linear space of multipliers to the dual space of a certain Banach space. Such an interpretation of Corollary 0.1.1 is clearly valid as $M(G)$ is the dual space of the Banach space of continuous complex valued functions on G which vanish at infinity (Theorem D.9.4).

It should be noted that the assumption of continuity in the definition of a multiplier is often redundant and can be deduced from other portions of the definition. This possibility will be studied more thoroughly in the sequel.

Our main purpose then is to investigate multipliers in various contexts and attempt to develop characterization theorems analogous to the results for group algebras. The appropriate characterization will, naturally, depend on the space under consideration. The development breaks into two main areas, the multipliers for commutative Banach algebras and those for topological linear spaces. Of the latter we shall study the L_p-spaces in greatest detail.

The theory of multipliers can also be discussed for noncommutative topological algebras. We shall not investigate this material in any consistent manner but instead shall restrict our attention mainly to the commutative theory.

We shall conclude this section with the statement and proof of two further properties of the multipliers for $L_1(G)$. Similar results will also appear in our subsequent investigations.

If $g \in L_1(G)$ then we shall denote by T_g the multiplier for $L_1(G)$ defined by $T_g f = f * g$ for each $f \in L_1(G)$. A net of continuous linear operators $\{T_\alpha\}$ on $L_1(G)$ converges to the continuous linear operator T in the strong operator topology if $\lim_\alpha \|T_\alpha f - Tf\|_1 = 0$ for each $f \in L_1(G)$.

Theorem 0.1.2. *Let G be a locally compact Abelian group. Then:*

(i) *$\{T_g | g \in L_1(G)\}$ is strong operator dense in the space of all multipliers for $L_1(G)$.*

(ii) *The space of finite linear combinations of translation operators is strong operator dense in the space of all multipliers for $L_1(G)$.*

Proof. Let $\{u_\alpha\} \subset L_1(G)$ be an approximate identity, that is,

$$\lim_\alpha \|f * u_\alpha - f\|_1 = 0$$

for each $f \in L_1(G)$ (F.7d). Then for each $\mu \in M(G)$ we have $\{u_\alpha * \mu\} \subset L_1(G)$ since $L_1(G)$ is an ideal in $M(G)$. If $f \in L_1(G)$ we see that

$$\|f * \mu - f * u_\alpha * \mu\|_1 \leqq \|f - f * u_\alpha\|_1 \, \|\mu\| .$$

Consequently, $\lim_\alpha \|f * \mu - f * u_\alpha * \mu\|_1 = 0$ for each $f \in L_1(G)$.

Therefore $\{T_g | g \in L_1(G)\}$ is strong operator dense in the multipliers of $L_1(G)$.

To prove part two of the theorem it suffices to show, by the Hahn-Banach Theorem, that if F is a strong operator continuous linear functional on the space of operators on $L_1(G)$ which vanishes on the space of finite linear combinations of translation operators then it vanishes at every multiplier for $L_1(G)$. For each operator T on $L_1(G)$ a strong operator continuous linear functional F has the form

$$F(T)=\sum_{i=1}^{n}\int_G Tf_i(t)\,k_i(t^{-1})\,d\lambda(t)$$

where $f_i\in L_1(G)$ and $k_i\in L_\infty(G)$, $i=1,2,\ldots,n$ (Theorem D.8.1.).

Thus if F vanishes on the translation operators and T is a multiplier then

$$\begin{aligned} F(T)&=\sum_{i=1}^{n}\int_G Tf_i(t)\,k_i(t^{-1})\,d\lambda(t)\\ &=\sum_{i=1}^{n}\int_G f_i*\mu(t)\,k_i(t^{-1})\,d\lambda(t)\\ &=\sum_{i=1}^{n}\int_G\Big[\int_G f_i(t\,s^{-1})\,k_i(t^{-1})\,d\lambda(t)\Big]\,d\mu(s)\\ &=\int_G\Big[\sum_{i=1}^{n}\int_G \tau_s f_i(t)\,k_i(t^{-1})\,d\lambda(t)\Big]\,d\mu(s)\\ &=\int_G F(\tau_s)\,d\mu(s)=0. \end{aligned}$$

This completes the proof. □

0.2. Notation. In this section we wish to set some notation which will be used throughout the book.

The topological groups of the complex and real numbers with the usual topologies will be denoted by C and R. Z will denote the discrete group of the integers and Γ the compact group of complex numbers of unit modulus. If G is any locally compact Abelian topological group then $\hat{G}$ will denote the dual group of G, that is, the group of continuous characters on G. The pairing between elements of G and $\hat{G}$ will always be indicated by $(\cdot,\cdot)$. Thus by the Pontryagin-van Kampen Duality Theorem (Theorem B.2.1) we have that $(\cdot,\gamma)$ defines a continuous character on G for each $\gamma\in\hat{G}$, and $(t,\cdot)$ defines a continuous character on $\hat{G}$ for each $t\in G$. Moreover, all continuous characters are of this form. A topological group will always be assumed to be Hausdorff. e will generally denote the identity element of G.

All topological linear spaces X will be taken over the complex numbers, and the space of continuous linear functionals on X, that is, the dual space of X, will be denoted by X^*. If A is a commutative Banach algebra then $\hat{A}$ will denote the space of Gelfand transforms of A. If $B \subset A$ then $\hat{B}$ is the set of Gelfand transforms of elements in B.

If S is a locally compact Hausdorff topological space then $C(S)$, $C_o(S)$ and $C_c(S)$ will denote, respectively, the normed algebras under pointwise operations and the usual supremum norm, $\|f\|_\infty = \sup_{t \in S} |f(t)|$, of all continuous complex valued functions on S which are bounded, vanish at infinity, or have compact support. $C^R(S)$, $C_o^R(S)$ and $C_c^R(S)$ will denote the algebras of real parts of the functions in $C(S)$, $C_o(S)$ or $C_c(S)$ respectively.

If v is a nonnegative regular Borel measure on S then $L_p(v)$, $1 \leqq p < \infty$, will denote the Banach space of equivalence classes of complex valued functions whose p-th powers are absolutely integrable with respect to v under the norm

$$\|f\|_p = \Big(\int_G |f(t)|^p \, dv(t)\Big)^{1/p}.$$

The Banach space of equivalence classes of v-essentially bounded measurable complex valued functions with the norm

$$\|f\|_\infty = \operatorname{ess\,sup}_{t \in S} |f(t)|$$

will be denoted by $L_\infty(v)$. $L_p^+(v)$, $1 \leqq p \leqq \infty$, will denote the subset of $L_p(v)$ consisting of those functions which are nonnegative almost everywhere, while $L_p^R(v)$, $1 \leqq p \leqq \infty$, will denote the real parts of the elements of $L_p(v)$. The real and imaginary parts of a complex valued function f will be denoted by $\mathrm{Re}(f)$ and $\mathrm{Im}(f)$, respectively. If E is a Borel measurable subset of S then χ_E will denote the characteristic function of E. A simple function is a finite linear combination of characteristic functions. The linear space of integrable simple functions will be denoted by $\mathcal{S}(v)$. If S is a locally compact Abelian topological group G and v is Haar measure on S then we shall write $L_p(G)$, $1 \leqq p \leqq \infty$, for the spaces $L_p(v)$. Haar measure on G will always be denoted by λ or $d\lambda$, and on $\hat{G}$ by η or $d\eta$.

A general discussion of the preceding concepts, as well as a survey of the material necessary for an understanding of the text as a whole, is presented in the appendices to this volume.

The symbol $\square$ is used to indicate the completion of a proof.

0.3. Notes. The multipliers for the group algebra $L_1(G)$ have been investigated by Edwards [2], Helson [1], and Wendel [1, 2]. Their results are all contained in Theorem 0.1.1. Proofs of the theorem are also to be found in Brainerd and Edwards [1_{I}], Edwards [14_{II}], 16.3.3, Gaudry [6], V.1.3, Gulick, Liu and van Rooij [1_{I}] and Rudin [5], 3.8. Compact and weakly compact multipliers for $L_1(G)$ have been characterized by Akemann [1], Gaudry [3] and Kitchen [2]. Such multipliers T are shown to be of the form $Tf = f * \mu$ where $\mu \in L_1(G)$.

The characterization of the multipliers for a more general, but closely related, class of spaces, the weighted Lebesgue spaces, has been accomplished by Gaudry [4]. In this connection see also Edwards [4].

The second portion of Theorem 0.1.2 is due to Wendel [2]. He also characterizes the isometric multipliers of $L_1(G)$ as scalar multiples of the translation operators by complex numbers of modulus one. The work in Brainerd and Edwards [1,] and Wendel [1, 2] is valid for nonAbelian as well as Abelian groups.

Helson [1] and Wendel [1, 2] were actually concerned with the isomorphism problem for group algebras and not specifically with the description of multipliers. Some of the results they obtained by employing their characterizations of multipliers are summarized in the following theorem.

Theorem 0.3.1. *Let G_1 and G_2 be locally compact groups and suppose $S: L_1(G_1) \to L_1(G_2)$ is a surjective algebra isomorphism. Then G_1 and G_2 are topologically isomorphic whenever any of the following conditions is satisfied:*

(i) $\|Sf\|_1 \leqq \|f\|_1$ *for each* $f \in L_1(G_1)$.

(ii) G_1 *and* G_2 *are Abelian, either* $\hat{G}_1$ *or* $\hat{G}_2$ *is connected and* $\|S\| < 2$.

It is of course easily seen that $L_1(G_1)$ and $L_1(G_2)$ are algebraically and isometrically isomorphic whenever G_1 and G_2 are topologically isomorphic (Wendel [1]). Also in the preceding theorem the mapping S is an isometry if condition (i) is satisfied (Wendel [2]).

Subsequently, many authors have considered the problem of describing norm decreasing homomorphisms and isomorphisms of group algebras and algebras of measures which contain the group algebras, and the related question of the topological isomorphism of the underlying groups. Among these we refer the reader to Beurling and Helson [1], Cohen [2], Edwards [9], Forelli [1], Glicksberg [1], Greenleaf [1, 2, 4], Johnson [3], Kawada [1], Parrott [1], Rigelhof [1], Rudin [6] and Strichartz [1, 2].

Some of these results will be discussed in subsequent chapters. Here we wish to mention only two further results. The first comes from Strichartz [1].

Theorem 0.3.2. *Let G_1 and G_2 be locally compact groups and suppose $A_i \subset M(G_i)$ is a subalgebra such that $L_1(G_i) \subset A_i$, $i = 1, 2$. If $S: A_1 \to A_2$ is an isometric surjective algebra isomorphism then G_1 and G_2 are topologically isomorphic.*

In particular, the theorem is valid when $A_i = M(G_i)$, $i=1,2$, a result which was proved independently in Johnson [3]. In this connection see also Glicksberg [1], Greenleaf [1, 2] and Rigelhof [1] where the situation for norm decreasing maps is considered.

Secondly we wish to state the theorem of Kawada [1] which we shall have need of in a later chapter. A mapping $S: L_1(G_1) \to L_1(G_2)$ is called *bipositive* provided $Sf \geqq 0$ almost everywhere if and only if $f \geqq 0$ almost everywhere.

Theorem 0.3.3. *Let G_1 and G_2 be locally compact groups and suppose $S: L_1(G_1) \to L_1(G_2)$ is a surjective bipositive algebra isomorphism. Then G_1 and G_2 are topologically isomorphic.*

A generalization of Theorems 0.3.2 and 0.3.3 will also be discussed in 3.7.

We state one further result connected with isomorphisms of group algebras to which we shall refer in the notes to the succeeding chapter. The result is taken from Rudin [5], 4.6.4.

Theorem 0.3.4. *Let G_1 and G_2 be locally compact Abelian groups. If $S: M(G_1) \to M(G_2)$ is a surjective algebra isomorphism then S maps $L_1(G_1)$ onto $L_1(G_2)$. Conversely, if $S: L_1(G_1) \to L_1(G_2)$ is a surjective algebra isomorphism then S has a unique extension to an algebra isomorphism of $M(G_1)$ onto $M(G_2)$.*

We refer the reader to Rudin [5], 4, for a general discussion of the homomorphisms of group algebras.

Another possible definition of a multiplier, at least for topological linear spaces of functions, measures or distributions on a locally compact group G, has been studied by Edwards [6]. Namely, a linear transformation T from such a topological linear space to itself is a multiplier provided that it is the limit in an appropriate topology of finite linear combinations of translation operators. Some additional continuity restrictions may also be imposed on T. From Theorem 0.1.2 the appropriate topology for $L_1(G)$ is the strong operator topology. We shall not discuss this approach to multipliers but refer the interested reader to Edwards [6].

Similarly, there is a wide range of topics in connection with the study of multipliers which we shall not be able to discuss at any great length, neither in the text proper nor in the notes to each chapter. We do however wish to present some indication of references for these subjects so that the interested reader may investigate them further on his own. These topics and references are as follows: singular integrals and fractional integration: Benedek, Calderón and Panzone [1], Breuer and Cordes [1], Calderón [1–3], Calderón and Zygmund [1–4], Coifman [1], Cordes [1], Cotlar [1], Duren and Shields [1], Edwards [14_{II}], 12.8, 12.9, 16.4.8, Hardy and Littlewood [1, 4], Hirschman [1], Hörmander [1], Igari [1],

Jones [1, 2], Littlewood and Paley [1], Marcinkiewicz [1], Merlo [1], Mihlin [1–4], Muckenhoupt [1], Peetre [1–3], M. Riesz [2], J. Schwartz [1], Sobolev [1], Stein [1, 3],[1] Stein and G. Weiss [1, 2], Stein and Zygmund [1], Sunouchi [1], Taibleson [1, 2],[2] Thorin [1], Waterman [1, 2], G. Weiss [1] and Zygmund [2–4], [6], XII.8, XII.9 and XVI.1; conjugate functions: Calderón [1–3], Coifman [1], Cotlar [1], Edwards [14_{II}], 12.8, 12.9, 13.9 and 16.4.8, Hardy [1], Hardy and Littlewood [2], Jones [1], Kolmogorov [1], M. Riesz [2], Stein [2], Stein and G. Weiss [2], Tamarkin [1], G. Weiss [1] and Zygmund [3], [6], VII.1, VII.2 and XVI.1; partial differential equations: Calderón [3], Calderón and Zygmund [1], Gohberg [1], Hörmander [1], Jones [1, 2], Lax [1] and Peetre [1–3];[3] interpolation theorems: Edwards [14_{II}], 13.8, 13.9, 13.10, 13.11 and 16.5, Gaudry [5], Peetre [1–3], Stein [2] and Zygmund [6], XII.4 and XII.5; Banach modules: Comisky [1], Gulick, Liu and van Rooij [1], Kitchen [1], Liu, van Rooij and Wang [1], Máté [8] and Rieffel [2, 4]; representation theory: Máté [8] and Rieffel [2]; operational calculus: Boehme [1, 2], Edwards [3], Rowlands [1] and Weston [1–5]; invariant means: Gilbert [1], Leptin [1] and Máté [7].[4]

Besides the previous references we also wish to mention a collection of papers which treat the problem of describing the multipliers for various spaces. Many of these papers contain results which we shall discuss in some detail, while others contain results of a more specialized nature which we have not elected to include in the main exposition of the theory. In either case we shall not make explicit reference to these works again, preferring instead to base our development on alternative sources. These papers include some of the oldest ones dealing with the subject of multipliers. The papers are: Bochner [1], Fekete [1], Goes [1], Hille [1], Iltis [1], Kaczmarz [1], Kaczmarz and Marcinkiewicz [1], Kaczmarz and Stein [1], Kaczmarz and Steinhaus [1], Karamata [1, 2], Karamata and Tomić [1], Katayama [1], Littlewood and Paley [1], Marcinkiewicz [1], Mazurkiewicz [1], Orlicz [1], M. Riesz [1], Sidon [1, 2], Skvortsova [1], Stein and Zygmund [1], Steinhaus [1], Sunouchi [1], Tomić [1], Verblunsky [1], Young [1, 2] and Zygmund [6], IV.11 and V.5.

As general references to the study of multipliers we refer the reader to Edwards [14_{II}], 16, Gaudry [6], V, VI and VII, and Hewitt and Ross [2], 35 and 36. The latter reference contains an historical survey on many aspects of the subject.

Finally, we note that a number of authors use the term "centralizer" instead of "multiplier". This is true, for example, of Johnson [1, 2, 4], Kellogg [1] and Wendel [2]. However "multiplier" seems to be the older and more common term and for this reason we prefer it to that of "centralizer".

---

[1] See also Stein and Wainger [1]. — [2] See also Taibleson [3]. — [3] See also Telner [1]. — [4] C^*-algebras: Andersen [1], Busby [1, 2], Dauns [1], Eymard [2] and Flanders [1]. Segal algebras: Cigler [1] and Reiter [1, 2].

Chapter 1

The General Theory of Multipliers

1.0. Introduction. Our purpose in this chapter is to present a development of much of the theory of multipliers for Banach algebras. It is neither exhaustive of the material nor is the development the most general one that could be made. Instead we have emphasized the problem of characterizing the multipliers of various abstract Banach algebras.

For the most part we have restricted our attention to commutative algebras thereby availing ourselves of the machinery of the Gelfand representation theory. However, a number of the results are valid as they stand for noncommutative Banach algebras, while others have valid analogs in the noncommutative case. Some additional comments on the noncommutative results can be found in the notes at the end of the chapter.

The paradigm for many of the theorems is the group algebra $L_1(G)$ and its multipliers. We have tried to indicate these connections whenever appropriate.

1.1. Elementary Theory of Multipliers. If A is a Banach algebra then a mapping $T: A \to A$ is a *multiplier* of A if $x(Ty)=(Tx)y$ for all $x, y \in A$. For arbitrary Banach algebras essentially nothing is known concerning multipliers, but for algebras without order a considerable number of results are readily deduced. A Banach algebra A is *without order* if for all $x \in A$, $xA=\{0\}$ implies $x=0$, or, for all $x \in A$, $Ax=\{0\}$ implies $x=0$. Obviously if A has an identity it is without order. $E(A)$ will denote the Banach algebra of all continuous linear operators from A to A, and $M(A)$ the collection of all the multipliers of A. Clearly $M(A)$ is a linear space and $M(A) \neq \{0\}$ since the identity operator $I \in E(A)$ belongs to $M(A)$. Note that no assumptions of linearity or continuity are made in the definition of a multiplier. Indeed these properties are, in many instances, consequences of the definition, as is seen from the following theorem.

Theorem 1.1.1. *Let A be a Banach algebra without order. Then $M(A)$ is a closed commutative subalgebra of $E(A)$ which contains the identity operator of $E(A)$.*

Proof. If $T \in M(A)$, $x, y, z \in A$ and $a, b \in C$, then

$$x[T(a y+b z)]=(T x)(a y+b z)=a(T x) y+b(T x) z$$
$$=x[a T y+b T z].$$

Since A is without order we conclude that $T(a y+b z)=a T y=b T z$, that is, T is linear.

Moreover, if $y, z \in A$ and $\{y_n\} \subset A$ is a sequence such that $\lim_n \|y_n-y\|=0$ and $\lim_n \|T y_n-z\|=0$, then for each $x \in A$,

$$\|x z-x(T y)\| \leqq \|x\| \|z-T y_n\|+\|(T x) y_n-(T x) y\|$$
$$\leqq \|x\| \|z-T y_n\|+\|T x\| \|y_n-y\|.$$

Thus $x z=x(T y)$, and, as before, this implies that $z=T y$. Hence appealing to the Closed Graph Theorem (Theorem D.6.1) we conclude that T is continuous.

Therefore $M(A) \subset E(A)$.

If $\{T_n\} \subset M(A)$, $T \in E(A)$ and $\lim_n \|T_n-T\|=0$ then for $x, y \in A$,

$$\|x(T y)-(T x) y\| \leqq \|x(T y)-x(T_n y)\|+\|(T_n x) y-(T x) y\|$$
$$\leqq 2\|x\| \|y\| \|T_n-T\|,$$

and so $x(T y)=(T x) y$. Thus $M(A)$ is closed in $E(A)$.

Now if $T \in M(A)$ and $x, y, z \in A$ then $z[x(T y)]=z[(T x) y]=[(T z) x] y$ $=z[T(x y)]$, and hence $x(T y)=(T x) y=T(x y)$ for all $x, y \in A$. Consequently if $T, S \in M(A)$, then $[(T S) x] y=[T(S x)] y=T[(S x) y]=$ $(T S)(x y)=T[x(S y)]=x[(T S) y]$. Furthermore $x[(S T) y]=(S x)(T y)=$ $[T(S x)] y=[(T S) x] y=x[(T S) y]$. Thus $(S T) y=(T S) y$.

Therefore $M(A)$ is a commutative subalgebra of $E(A)$. ▯

The arguments given above assume that $x A=\{0\}$ implies $x=0$. It is apparent that with minor changes in the proof the result is also valid when $A x=\{0\}$ implies $x=0$.

Thus we see that when A is without order, $M(A)$ is a commutative Banach algebra with identity of continuous linear operators under the usual operator norm. It is also the case that $M(A)$ is complete in the *strong operator topology*, that is, in the topology on $E(A)$ in which a net $\{T_\alpha\}$ converges to T if and only if $\lim_\alpha \|T_\alpha x-T x\|=0$ for each $x \in A$ (D.7).

Theorem 1.1.2. *Let A be a Banach algebra without order. Then $M(A)$ is complete in the strong operator topology.*

Proof. Suppose $\{T_\alpha\} \subset M(A)$ is a Cauchy net in the strong operator topology. Then for each $x \in A$, $\{T_\alpha x\}$ is a Cauchy net in A and hence there exists $Tx \in A$ such that $\lim_\alpha \|T_\alpha x - Tx\| = 0$. If $x, y \in A$, then

$$\begin{aligned}\|x(Ty)-(Tx)y\| &\leqq \|x(Ty)-x(T_\alpha y)\| + \|(T_\alpha x)y-(Tx)y\| \\ &\leqq \|x\|\,\|Ty-T_\alpha y\| + \|y\|\,\|T_\alpha x - Tx\|,\end{aligned}$$

and so $x(Ty)=(Tx)y$.

Therefore $T \in M(A)$ and $M(A)$ is complete in the strong operator topology. ▯

In the proof of Theorem 1.1.1 we proved and utilized the fact that if $T \in M(A)$ then $T(xy)=x(Ty)=(Tx)y$ for all $x, y \in A$. This is an extremely useful observation and we shall make repeated use of it. For example, it enables us to prove the following result.

Theorem 1.1.3. *Let A be a Banach algebra without order and $T \in M(A)$. Then the following are equivalent:*

(i) *T is bijective.*

(ii) *T^{-1} exists and $T^{-1} \in M(A)$.*

Proof. Obviously (ii) implies (i), and if (i) holds then clearly T^{-1} exists and $T^{-1} \in E(A)$. Moreover, if $x, y \in A$ then

$$\begin{aligned}(T^{-1}x)y &= T^{-1}T[(T^{-1}x)y] = T^{-1}[(TT^{-1}x)y] = T^{-1}(xy) \\ &= T^{-1}[x(TT^{-1}y)] = T^{-1}T[x(T^{-1}y)] = x(T^{-1}y).\end{aligned}$$

Hence $T^{-1} \in M(A)$. ▯

If for any Banach algebra B with identity e we define the *spectrum* of an element $x \in B$ as $\sigma_B(x) = \{\lambda \mid \lambda \in C, (\lambda e - x)^{-1}$ does not exist$\}$ then we obtain immediately the following corollary.

Corollary 1.1.1. *Let A be a Banach algebra without order and $T \in M(A)$. Then $\sigma_{M(A)}(T) = \sigma_{E(A)}(T)$.*

Let $x \in A$ and define the *left and right multiplication operators* L_x and R_x by $L_x y = xy$ and $R_x y = yx$ for $y \in A$. Clearly $L_x, R_x \in E(A)$ and if A is commutative they belong to $M(A)$. However, if A is not commutative then the multiplication operators need not be multipliers of A. When A is commutative $L_x = R_x$ and the following theorem is easily established. A net $\{x_\alpha\}$ in a commutative Banach algebra A is an *approximate identity* if $\lim_\alpha \|x_\alpha x - x\| = 0$ for each $x \in A$. An approximate identity is *minimal* if $\lim_\alpha \|x_\alpha\| = 1$.

Theorem 1.1.4. *Let A be a commutative Banach algebra without order. Then the mapping $x \to L_x = R_x$ is a continuous isomorphism of A onto* *the ideal* $\{L_x|x\in A\}=\{R_x|x\in A\}$ *in* $M(A)$. *If* A *possesses a minimal approximate identity then the mapping is an isometry.*

The proof is straightforward and will be omitted.

The fact that A need not be embeddable in $M(A)$ in this canonical way for noncommutative A is contained in the next theorem. A commutative subalgebra B of an algebra A is called a *maximal commutative subalgebra* if B is not properly contained in any proper commutative subalgebra of A. If A has an identity e then e belongs to any maximal commutative subalgebra of A.

Theorem 1.1.5. *Let* A *be a Banach algebra without order. Then the following are equivalent:*

(i) A *is commutative.*
(ii) $M(A)$ *is a maximal commutative subalgebra of* $E(A)$.
(iii) $\{L_x|x\in A\}\subset M(A)$.
(iv) $\{R_x|x\in A\}\subset M(A)$.

Proof. Obviously (i) implies (iii) and (iv). Conversely, if (iii) holds and $x, y, z\in A$ then $(yx)z=y(xz)=y(L_x z)=(L_x y)z=(xy)z$. Since A is without order we conclude that $yx=xy$ for all $x, y\in A$, and A is commutative. Thus (iii) implies (i). Similarly (iv) implies (i). Hence (i), (iii), and (iv) are equivalent and we need only demonstrate the equivalence of (i) and (ii).

Suppose $M(A)$ is a maximal commutative subalgebra of $E(A)$. If $x, y\in A$ and $T\in M(A)$ then $(L_x T)y=x(Ty)=T(xy)=(TL_x)y$. Consequently $L_x T=TL_x$ for all $x\in A$, $T\in M(A)$. The maximality of $M(A)$ implies that $\{L_x|x\in A\}\subset M(A)$ and hence A is commutative.

Conversely, assume A is commutative. Then $\{L_x|x\in A\}\subset M(A)$. If $M(A)$ were not a maximal commutative subalgebra of $E(A)$ then, since $E(A)$ contains an identity, we may appeal to Zorn's lemma to guarantee the existence of a maximal commutative subalgebra $MC(A)$ of $E(A)$ which properly contains $M(A)$. But if $T\in MC(A)$ then for each $x, y\in A$, $x(Ty)=L_x(Ty)=(L_x T)y=(TL_x)y=T(xy)$. Since A is commutative, it follows that $x(Ty)=T(xy)=T(yx)=(Tx)y$, that is, $T\in M(A)$. This however contradicts the construction of $MC(A)$ and we conclude that $M(A)$ is maximal. $\square$

If the Banach algebra A has an identity e then every multiplier is both a left and a right multiplication operator. Indeed, if $x\in A$ and $T\in M(A)$ then $L_{Te}x=(Te)x=T(ex)=Tx=T(xe)=x(Te)=R_{Te}x$. Thus when A is also commutative we see that $M(A)=\{L_x|x\in A\}$. If A is without identity this of course need not be the case. One can, however, say more about the relationship between $\{L_x|x\in A\}$ and $M(A)$ for commutative A when A possesses an approximate identity.

Theorem 1.1.6. *Let A be a commutative Banach algebra without order. Then the following are equivalent:*

(i) $\{L_x | x \in A\}$ *is dense in $M(A)$ in the strong operator topology.*

(ii) *A has an approximate identity.*

Proof. If (i) holds then let $\{L_{x_\alpha}\}$ be a net converging to the identity operator $I \in M(A)$ in the strong operator topology. Then $\{x_\alpha\}$ is an approximate identity for A. Conversely, if $\{x_\alpha\}$ is an approximate identity then $\{L_{Tx_\alpha}\}$ converges in the strong operator sense to $T \in M(A)$ for each $T \in M(A)$. Hence $\{L_x | x \in A\}$ is strong operator dense in $M(A)$. ▯

In general then for commutative A the strong operator closure of $\{L_x | x \in A\}$ need not be all of $M(A)$. Some additional information is however provided by the next theorem.

Theorem 1.1.7. *Let A be a commutative Banach algebra without order, and denote by B the strong operator closure in $M(A)$ of $\{L_x | x \in A\}$. Then B is a commutative Banach algebra without order and $M(A) \subset M(B)$. Moreover if A is equal to the norm closure of A^2 then $M(A) = M(B)$.*

Proof. Clearly B is a commutative algebra without order, and, since the strong operator topology is weaker than the norm topology, it is closed in the norm topology and hence a Banach algebra. Let $T \in M(A)$, $S \in B$ and suppose $\{L_{x_\alpha}\}$ converges to S in the strong operator topology. Then for each $x \in A$, $\|(TS)x - (TL_{x_\alpha})x\| \leqq \|T\| \, \|Sx - L_{x_\alpha}x\|$ and we conclude that $\{TL_{x_\alpha}\} = \{L_{Tx_\alpha}\}$ converges in the strong operator sense to TS. Thus $TS \in B$. The commutativity of $M(A)$ implies that $T \in M(B)$. Hence $M(A) \subset M(B)$. Incidentally we have also shown that B is an ideal in $M(A)$.

Now assume A is the norm closure of $A^2 = \{xy | x, y \in A\}$. Then by Theorem 1.1.4 the set $\{L_{xy} | x, y \in A\}$ is norm dense in $\{L_z | z \in A\}$. Let $T \in M(B)$ and $x, y \in A$. Then $TL_x \in B \subset M(A)$ and so $TL_x L_y = TL_{xy} \in \{L_z | z \in A\}$. Since $\{L_{xy} | x, y \in A\}$ is norm dense in $\{L_z | z \in A\}$, it follows that $TL_z \in \{L_x | x \in A\}$ for each $z \in A$. Then, defining Tz by $(Tz)x = (TL_z)x$ for each $x \in A$, we conclude that $T \in M(A)$ and $M(B) \subset M(A)$. But $M(A) \subset M(B)$ and thus $M(A) = M(B)$. ▯

Whenever A is commutative and without order we have seen that A may be considered as a normed subalgebra of $M(A)$. However, A so considered need not be complete in the operator norm inherited from $M(A)$.

Indeed, let φ be a real valued continuous function on the real line such that $\varphi(x) \geqq 1$, $x \in R$, and $\lim_{x \to +\infty} \varphi(x) = \lim_{x \to -\infty} \varphi(x) = +\infty$. Define A to be the collection of all complex valued continuous functions f on R for which $\|f\| = \sup_x |f(x)\,\varphi(x)| < \infty$. It is apparent that A forms a commutative normed algebra under pointwise operations and the indicated norm. If $\{f_n\} \subset A$ is a Cauchy sequence then $\{f_n\}$ and $\{f_n \varphi\}$ are Cauchy sequences in the usual supremum norm. Let f and g be, respectively, the uniform limits of these sequences. Then an elementary argument shows that $\sup_x |g(x)| < \infty$ and $f\varphi = g$. Hence A is a Banach algebra. Moreover, it is obvious that A is without order. It is apparent that $A \subset C_0(R)$ but $A \neq C_0(R)$. Indeed any real valued function $f \in C_0(R)$ such that $f(x)\sqrt{\varphi(x)} \geqq 1$ belongs to $C_0(R) \sim A$. However, A with the operator norm is dense in $C_0(R)$, and so A is not complete in the operator norm. For instance, let $f \in C_0(R)$ and $\{f_n\} \subset A$ be a sequence of functions with compact support such that $\lim_n \|f_n - f\|_\infty = 0$. Then

$$\begin{aligned}\sup_{\|g\| \leqq 1} \|f_n g - f g\| &= \sup_{\|g\| \leqq 1} \sup_x \big|\big(f_n(x) - f(x)\big)\, g(x)\, \varphi(x)\big| \\ &\leqq \|f_n - f\|_\infty \sup_{\|g\| \leqq 1} \sup_x |g(x)\, \varphi(x)| \leqq \|f_n - f\|_\infty,\end{aligned}$$

and the indicated assertion is evident.

In 1.4 we shall obtain some additional information on the norm closure of A in $M(A)$.

1.2. Characterizations of Multipliers. Following the proof of Theorem 1.1.1 we remarked that if $T \in M(A)$ then $T(xy) = x(Ty) = (Tx)y$ for all $x, y \in A$. We now wish to examine these equations in more detail. We shall say that $T \in E(A)$ is a *left (right) multiplier* of A if $T(xy) = (Tx)y$ $\big(= x(Ty)\big)$, and denote by $L(A)$ $\big(R(A)\big)$ the collection of all left (right multipliers of A. If $T, S \in L(A)$ then $(TS)(xy) = T[S(xy)] = T[(Sx)y] = [(TS)x]y$ for all $x, y \in A$. Thus $L(A)$ is a subalgebra of $E(A)$. The same is, of course, true for $R(A)$.

These results immediately establish the next theorem.

Theorem 1.2.1. *Let A be a Banach algebra without order. Then $M(A) = L(A) \cap R(A)$. If A is commutative then $M(A) = L(A) = R(A)$.*

Thus we see that $T \in M(A)$ if and only if $T(xy) = x(Ty) = (Tx)y$ and for commutative A if and only if $T(xy) = x(Ty)$ or $T(xy) = (Tx)y$ for all $x, y \in A$. We shall use this result shortly to obtain some additional characterizations of $M(A)$.

If $A = L_1(G)$ then, as was indicated in the preceding chapter, the multiplier algebra of A can be considered as an algebra of continuous functions defined on the regular maximal ideal space of A. In particular, for $A = L_1(G)$ one can consider $M(A) = M(G)^\wedge$. The next theorem shows that such an identification can be made whenever A is a semi-simple commutative Banach algebra.

Given a commutative Banach algebra A we shall denote by $\Delta(A)$ the *regular maximal ideal space* of A. The *Gelfand transform* $\hat{x}$ of an element $x \in A$ is defined on $\Delta(A)$ by $\hat{x}(m) = \langle x, \mu \rangle$ where μ is the multiplicative linear functional on A such that $\mu^{-1}(0) = m$. As is well known $\|\hat{x}\|_\infty \leqq \|x\|$ (E.2). Set $\hat{A} = \{\hat{x} | x \in A\}$.

Theorem 1.2.2. *Let A be a commutative Banach algebra without order and let $T \in M(A)$. Then there exists a unique bounded continuous function φ on $\Delta(A)$ such that:*

(i) $(Tx)\hat{} = \varphi \hat{x}$ for all $x \in A$.

(ii) $\|\varphi\|_\infty \leqq \|T\|$.

Proof. Let $m \in \Delta(A)$ and $x, y \in A$ be such that $\hat{x}(m) \neq 0$, $\hat{y}(m) \neq 0$. Since $(Tx)y = x(Ty)$, it follows that

$$(Tx)\hat{}(m)/\hat{x}(m) = (Ty)\hat{}(m)/\hat{y}(m).$$

For each $m \in \Delta(A)$ choose $x \in A$ such that $\hat{x}(m) \neq 0$, and define

$$\varphi(m) = (Tx)\hat{}(m)/\hat{x}(m).$$

The preceding equation shows that the definition is independent of x and hence φ is a well-defined continuous function on $\Delta(A)$. Moreover, if $\hat{x}(m) = 0$ and $\hat{y}(m) \neq 0$ then $(Tx)\hat{}(m)\,\hat{y}(m) = \hat{x}(m)(Ty)\hat{}(m) = 0$ implies that $(Tx)\hat{}(m) = 0$. Hence the equation $(Tx)\hat{}(m) = \varphi(m)\,\hat{x}(m)$ is valid for all $m \in \Delta(A)$ and $x \in A$. That is $(Tx)\hat{} = \varphi \hat{x}$ for all $x \in A$.

If ψ is a second continuous function on $\Delta(A)$ for which $(Tx)\hat{} = \psi \hat{x}$ then $[\varphi(m) - \psi(m)]\,\hat{x}(m) = 0$ for all $x \in A$ implies that $\varphi(m) = \psi(m)$. Thus φ is unique.

To prove φ is bounded we set $K_m = \sup_{\|x\|=1} |\hat{x}(m)|$ for each $m \in \Delta(A)$. Clearly $0 < K_m \leqq 1$. Furthermore for each $x \in A$,

$$|\varphi(m)\,\hat{x}(m)| = |(Tx)\hat{}(m)| \leqq K_m \|Tx\| \leqq K_m \|T\|\,\|x\|.$$

In particular, restricting our attention to only those $x \in A$ such that $\|x\| = 1$ we obtain

$$|\varphi(m)| \leqq \inf_{\|x\|=1} K_m \|T\|/|\hat{x}(m)| = K_m \|T\|/ \sup_{\|x\|=1} |\hat{x}(m)| = \|T\|.$$

Therefore φ is bounded and $\|\varphi\|_\infty \leqq \|T\|$. □

If A is a commutative Banach algebra without order we shall denote by $\mathcal{M}(A)$ *the normed algebra of all bounded continuous functions φ on $\Delta(A)$ such that $\varphi \hat{A} \subset \hat{A}$*. The norm of course being the supremum norm. The content of the previous theorem is that there exists a continuous linear mapping from $M(A)$ into $\mathcal{M}(A)$. However, it is not clear that the mapping carries distinct elements of $M(A)$ onto distinct functions in

$\mathcal{M}(A)$. But if A is *semi-simple*, that is, $\hat{x}(m)=0$ for all $m\in\Delta(A)$ implies $x=0$, then it is clearly without order and a simple argument shows that the mapping is one-to-one.

These remarks establish the following corollary.

Corollary 1.2.1. *Let A be a semi-simple commutative Banach algebra. Then the equation $(Tx)\hat{}=\varphi\hat{x}$, $x\in A$, defines a continuous isomorphism of $M(A)$ onto $\mathcal{M}(A)$.*

In general this mapping between $M(A)$ and $\mathcal{M}(A)$ is not an isometry. However if A is a *supremum norm algebra*, that is, $\|x\|=\|\hat{x}\|_\infty$ for all $x\in A$, then the isomorphism is isometric.

Theorem 1.2.3. *Let A be a semi-simple commutative supremum norm algebra. If $T\in M(A)$ and $\varphi\in\mathcal{M}(A)$ is such that $(Tx)\hat{}=\varphi\hat{x}$ for all $x\in A$ then $\|T\|=\|\varphi\|_\infty$.*

Proof. From Theorem 1.2.2 we know that $\|\varphi\|_\infty\leqq\|T\|$. On the other hand

$$\|T\|=\sup_{\|x\|=1}\|Tx\|=\sup_{\|\hat{x}\|_\infty=1}\|\varphi\hat{x}\|_\infty\leqq\|\varphi\|_\infty.$$

Hence $\|T\|=\|\varphi\|_\infty$. □

Thus we see that for semi-simple commutative Banach algebras we can consider $M(A)$ as a space of bounded continuous functions defined on the regular maximal ideal space of A. This is an extremely useful result and we shall frequently employ it in the sequel. In particular, whenever $T\in M(A)$ we shall always denote by φ the function in $\mathcal{M}(A)$ for which $(Tx)\hat{}=\varphi\hat{x}$.

It should also be noted that if A is commutative semi-simple and has an identity then $T\in M(A)$ if and only if $\varphi\in\hat{A}$, that is, $\mathcal{M}(A)=\hat{A}$. Consequently the theory of multipliers for such algebras is trivial. We shall always tacitly assume in what follows that A *does not possess an identity*.

Now we wish to give some further characterizations of elements of $M(A)$.

Theorem 1.2.4. *Let A be a semi-simple commutative Banach algebra and T a linear mapping from A to A. Then the following are equivalent:*

(i) $T\in M(A)$.
(ii) $Tm\subset m$ *for all* $m\in\Delta(A)$.
(iii) $Tx^2=x(Tx)$ *for all* $x\in A$.

Proof. Suppose $T\in M(A)$ and $m\in\Delta(A)$. If $x\notin m$ then for any $y\in m$ we have $\hat{x}(m)(Ty)\hat{}(m)=\hat{x}(m)[\varphi(m)\hat{y}(m)]=0$. Since $\hat{x}(m)\neq 0$, it follows that $Ty\in m$ (E.2), and so $Tm\subset m$.

On the other hand, suppose $Tm\subset m$ for all $m\in\Delta(A)$. For each $m\in\Delta(A)$ if either x or y belongs to m then it is apparent that $[x(Ty)-(Tx)y]\hat{}(m)=0$. If $x, y\notin m$ then since m is maximal there exists $z\notin m$, scalars α, β and elements $u, v \in m$ for which $x = \alpha z + u$, $y = \beta z + v$. Elementary calculation reveals that $[x(Ty) - (Tx)y]\hat{}(m) = \alpha \hat{z}(m) \beta (Tz)\hat{}(m) - \alpha (Tz)\hat{}(m) \beta \hat{z}(m) = 0$. Since this argument is valid for each $m \in \Delta(A)$ we conclude that $[x(Ty) - (Tx)y]\hat{}(m) = 0$ for all $m \in \Delta(A)$ and $x, y \in A$. Hence from the semi-simplicity of A it follows that $x(Ty) = (Tx)y$ for all $x, y \in A$, that is, $T \in M(A)$.

From the definition of multiplier it is obvious that (i) implies (iii). Conversely, suppose $Tx^2 = x(Tx)$ for all $x \in A$. By Theorem 1.2.1 it is sufficient to show that $T(xy) = x(Ty)$ for all $x, y \in A$.

If $x, z \in A$ then

$$\begin{aligned} x(Tx) + 2T(xz) + z(Tz) &= T[(x+z)^2] \\ &= (x+z)\, T(x+z) \\ &= x(Tx) + x(Tz) + z(Tx) + z(Tz), \end{aligned}$$

from which we conclude that $(2TL_x - L_x T)z = (L_{Tx})z$ for all $z \in A$. Since A is commutative, $L_{Tx} \in M(A)$ and it follows that $2TL_x - L_x T \in M(A)$ for all $x \in A$. Consequently, $2TL_x L_z - L_x TL_z = 2L_z TL_x - L_z L_x T$ for all $z \in A$ as $M(A)$ is commutative. Applying both sides of this identity to z and appealing to the assumption $Tz^2 = z(Tz)$, a straightforward, but lengthy, computation reveals that $T(xz^2) = zT(xz)$ for all $x, z \in A$.

Using this last identity and expanding both sides of $T[(x+z)^2 y] = (x+z)\, T[(x+z)y]$ one concludes that $2TL_x L_z = L_z TL_x + L_x TL_z$ for all $x, z \in A$. Substituting this equation in the identity $2TL_x L_z - L_x TL_z = 2L_z TL_x - L_z L_x T$ we deduce that $L_z TL_x = L_z L_x T$ for all $x, z \in A$.

Therefore $zT(xy) = zx(Ty)$ and so $T(xy) = x(Ty)$ for all $x, y \in A$ as A is semi-simple. □

It should be noted that the proof actually establishes the equivalence of (i) and (iii) for commutative algebras without order.

1.3. An Application: Multiplications which Preserve the Regular Maximal Ideals. As an application of these characterization theorems we shall determine all multiplication operations on certain semi-simple commutative Banach algebras which preserve the regular maximal ideal space. If A is a commutative Banach algebra then A is said to *admit factorization* if for each $z \in A$ there exists $x, y \in A$ such that $z = xy$. If A has a minimal approximate identity then it admits factorization (E.5).

Theorem 1.3.1. *Let A be a commutative semi-simple Banach algebra. Let $T \in M(A)$ be such that $\varphi(m) \neq 0$ for all $m \in \Delta(A)$. Define $x \circ y = x(Ty)$ and $\|x\|_T = \|T\|\,\|x\|$ for all $x, y \in A$. Then $\|\cdot\|_T$ is a norm on A equivalent to $\|\cdot\|$, and A forms a commutative semi-simple Banach algebra A_T with the norm $\|\cdot\|_T$ and the multiplication $\circ$ such that $\Delta(A) = \Delta(A_T)$.*

Conversely, if A forms under distinct multiplications and equivalent norms two commutative semi-simple Banach algebras A and A_1 which have the same regular maximal ideal space and if A admits factorization then there exists a multiplier $T\in M(A)$ such that $\varphi(m)\neq 0$ for all $m\in\Delta(A)$ and $A_1=A_T$.

Proof. Let A and $T\in M(A)$ be given. Clearly the norm $\|\cdot\|_T$ is equivalent to $\|\cdot\|$. And if $x, y\in A$,

$$\|x\circ y\|_T=\|T\|\,\|x\circ y\|=\|T\|\,\|x(Ty)\|\leqq\|T\|\,\|x\|\,\|T\|\,\|y\|=\|x\|_T\,\|y\|_T.$$

Thus A forms a commutative Banach algebra A_T under $\|\cdot\|_T$ and the multiplication $\circ$.

Suppose $m\in\Delta(A)$ and let μ denote the nonzero multiplicative linear functional on A such that $\mu^{-1}(0)=m$. Define $\mu_T=\varphi(m)\,\mu$. Clearly μ_T is nonzero, linear and $\mu_T^{-1}(0)=m$. Moreover,

$$\begin{aligned}\langle x\circ y, \mu_T\rangle&=\varphi(m)\,\langle x(Ty), \mu\rangle=\varphi(m)\,\hat{x}(m)\,\varphi(m)\,\hat{y}(m)\\&=\langle x; \mu_T\rangle\langle y, \mu_T\rangle.\end{aligned}$$

Thus μ_T defines a multiplicative linear functional on A_T and so $m\in\Delta(A_T)$. Hence $\Delta(A)\subset\Delta(A_T)$. This also demonstrates the semi-simplicity of A_T.

To establish the reverse inclusion we first note that T is also a multiplier for A_T. Indeed, if $x\in A_T$ then $T(x\circ x)=T[x(Tx)]=(Tx)(Tx)=x\circ Tx$, and the assertion follows immediately from Theorem 1.2.4. Let $\varphi_T\in\mathcal{M}(A_T)$ denote the bounded continuous function on $\Delta(A_T)$ associated with $T\in M(A_T)$ by Theorem 1.2.2.

Now if $m_T\in\Delta(A_T)$ and μ_T is the multiplicative linear functional on A_T such that $\mu_T^{-1}(0)=m_T$ then $\langle Tx, \mu_T\rangle=\varphi_T(m_T)\langle x, \mu_T\rangle$ for all $x\in A_T$. Thus if $x, y\in A_T$,

$$\langle x, \mu_T\rangle\langle y, \mu_T\rangle=\langle x\circ y, \mu_T\rangle=\langle T(x\,y), \mu_T\rangle=\varphi_T(m_T)\langle x\,y, \mu_T\rangle.$$

We deduce at once from this identity that $\varphi_T(m_T)\neq 0$ for all $m_T\in\Delta(A_T)$.

Define $\mu=\varphi_T(m_T)^{-1}\,\mu_T$. Clearly μ is a linear functional on A such that $\mu^{-1}(0)=\mu_T^{-1}(0)=m_T$. Furthermore, the previous equations show that

$$\langle x\,y, \mu\rangle=\varphi_T(m_T)^{-1}\langle x\,y, \mu_T\rangle=\varphi_T(m_T)^{-2}\langle x, \mu_T\rangle\langle y, \mu_T\rangle=\langle x, \mu\rangle\langle y, \mu\rangle.$$

Therefore $m_T\in\Delta(A)$.

Consequently $\Delta(A)=\Delta(A_T)$, and the first portion of the proof is complete.

Conversely, suppose A and A_1 are as given in the statement of the theorem, and let μ_1 be a multiplicative linear functional on A_1. Since $\Delta(A)=\Delta(A_1)$, $\mu_1^{-1}(0)\in\Delta(A)$ and so there exists a multiplicative linear functional μ on A and a nonzero constant $a(\mu)$ such that $\mu_1=a(\mu)\,\mu$. Define the function φ on $\Delta(A)$ by $\varphi\big(\mu^{-1}(0)\big)=a(\mu)$.

Since A admits factorization for any $z \in A$, there exist $x, y \in A$ such that $z = xy$. Then for each multiplicative linear functional μ on A

$$\begin{aligned}\varphi(\mu^{-1}(0))\,\hat{z}(\mu^{-1}(0)) &= \varphi(\mu^{-1}(0))\,\hat{x}(\mu^{-1}(0))\,\hat{y}(\mu^{-1}(0))\\ &= a(\mu)\,\langle x, \mu\rangle\langle y, \mu\rangle\\ &= a(\mu)^{-1}\langle x, \mu_1\rangle\langle y, \mu_1\rangle\\ &= a(\mu)^{-1}\langle x\circ y, \mu_1\rangle\\ &= (x\circ y)\hat{}\,(\mu^{-1}(0)),\end{aligned}$$

where $\circ$ denotes the multiplication in A_1. Thus $\varphi\hat{z} = \varphi\hat{x}\hat{y} = (x\circ y)\hat{}\in\hat{A}$ for all $z\in A$, and the equation $(Tz)\hat{} = \varphi\hat{z}$ defines a multiplier for A. Clearly $x\circ y = x(Ty)$ for all $x, y\in A$ and the proof is complete. ∎

We remark that though the regular maximal ideals for A and A_T are identical, the multiplicative linear functionals μ and μ_T corresponding to a given ideal in $\Delta(A) = \Delta(A_T)$ are not the same. They do, however, only differ by a multiplicative constant.

If $A = L_1(G)$, G a locally compact Abelian group, and $M(G) = M(A)$ then the theorem shows that the only new multiplications which can be introduced into $L_1(G)$ which preserve the regular maximal ideal space are those of the form $f\circ g = f*(v*g)$ where $v\in M(G)$ and $\hat{v}(\gamma)\neq 0$ for all $\gamma\in\hat{G}$. However, such a multiplication need not preserve the self-adjointness of $L_1(G)$ (Birtel [1], Coddington [1]).

Two corollaries of the preceding theorem are given below. The first one asserts that the space of multipliers for A is unchanged under a change of multiplication as effected in Theorem 1.3.1.

Corollary 1.3.1. *Let A be a commutative semi-simple Banach algebra. Then $M(A) = M(A_T)$.*

Proof. If $S\in M(A)$ then for each $x\in A_T$, $S(x\circ x) = S[x(Tx)] = x(TSx) = x\circ Sx$ and by Theorem 1.2.4 we see that $S\in M(A_T)$. Thus $M(A)\subset M(A_T)$.

Now suppose $S\in M(A_T)$ and let $\psi\in\mathcal{M}(A_T)$ be such that $(Sx)\hat{} = \psi\hat{x}$ for all $x\in A_T$. If μ_T is a multiplicative linear functional on A_T then the corresponding functional on A is $\mu = a(\mu)^{-1}\mu_T$, and we have

$$\langle Sx, \mu_T\rangle/\langle x, \mu_T\rangle = a(\mu)\langle Sx, \mu\rangle/a(\mu)\langle x, \mu\rangle = \langle Sx, \mu\rangle/\langle x, \mu\rangle.$$

But $\langle Sx, \mu_T\rangle = \psi(\mu_T^{-1}(0))\langle x, \mu_T\rangle$ and hence the equation $(Sx)\hat{} = \psi\hat{x}$ for all $x\in A$ defines a multiplier for A, that is, $S\in M(A)$.

Therefore $M(A) = M(A_T)$. ∎

Corollary 1.3.2. *Let A be a commutative semi-simple Banach algebra which admits factorization. Then the following are equivalent:*

(i) *A_T admits factorization.*

(ii) *T is invertible.*

Proof. Suppose A_T admits factorization. By Theorem 1.3.1 there there exists an $S \in M(A_T) = M(A)$ such that $xy = x \circ Sy$ for all $x, y \in A$. Let $z \in A$. Then, since A and A_T admit factorization, there exist $x, y \in A$ such that $z = xy$ and $u, v \in A$ such that $z = u \circ v$. On the one hand, we have $z = xy = x \circ Sy = S(x \circ y) = S[x(Ty)] = ST(xy) = STz$. And on the other hand, $z = u \circ v = u(Tv) = T(uv) = T(u \circ Sv) = TS(u \circ v) = TSz$. Consequently T is invertible.

Conversely, if T^{-1} exists then for any $z = xy \in A$, $z = T^{-1}T(xy) = T^{-1}[x(Ty)] = (T^{-1}x) \circ y$ since $T^{-1} \in M(A)$ by Theorem 1.1.3. Therefore A_T admits factorization. □

1.4. Maximal Ideal Spaces. We saw in Section 1.1 that the multiplier algebra $M(A)$ of a commutative Banach algebra without order was itself a commutative Banach algebra with identity under the operator norm. Our first concern in this section will be to study the relation between the maximal ideal spaces of these algebras.

Since we shall consider only commutative A, it is always the case that A is isomorphic to the set $\{L_x | x \in A\} \subset M(A)$.

Theorem 1.4.1. *Let A be a commutative Banach algebra without order. If μ is a nonzero multiplicative linear functional on A then there exists a unique multiplicative linear functional μ' on $M(A)$ such that $\langle L_x, \mu' \rangle = \langle x, \mu \rangle$ for all $x \in A$. And if v is a multiplicative linear functional on $M(A)$ then either $\langle L_x, v \rangle = 0$ for all $x \in A$ or there is a unique μ' such that $v = \mu'$.*

Proof. Suppose μ is a nonzero multiplicative linear functional on A and let $x \in A$ be such that $\langle x, \mu \rangle \neq 0$. Then define $\langle T, \mu' \rangle = \langle Tx, \mu \rangle / \langle x, \mu \rangle$ for $T \in M(A)$. The definition is independent of the choice of x since if $y \in A$, $\langle y, \mu \rangle \neq 0$, then

$$\begin{aligned} \langle T, \mu' \rangle \langle y, \mu \rangle &= \langle Tx, \mu \rangle \langle y, \mu \rangle / \langle x, \mu \rangle \\ &= \langle (Tx)y, \mu \rangle / \langle x, \mu \rangle \\ &= \langle x(Ty), \mu \rangle / \langle x, \mu \rangle \\ &= \langle Ty, \mu \rangle . \end{aligned}$$

Clearly then μ' is a linear functional on $M(A)$ and

$$\langle L_y, \mu' \rangle = \langle L_y x, \mu \rangle / \langle x, \mu \rangle = \langle yx, \mu \rangle / \langle x, \mu \rangle = \langle y, \mu \rangle$$

for all $y \in A$. Moreover, if $T, S \in M(A)$ then

$$\begin{aligned} \langle TS, \mu' \rangle &= \langle TSx, \mu \rangle / \langle x, \mu \rangle = \langle T, \mu' \rangle \langle Sx, \mu \rangle / \langle x, \mu \rangle \\ &= \langle T, \mu' \rangle \langle S, \mu' \rangle . \end{aligned}$$

Thus μ' is multiplicative.

Finally, μ' is unique. As if μ^* is any other multiplicative linear functional on $M(A)$ such that $\langle L_y, \mu^*\rangle = \langle y, \mu\rangle$ for all $y \in A$ then choosing any $x \in A$ for which $\langle x, \mu\rangle \neq 0$ we see that for each $T \in M(A)$,

$$\begin{aligned}\langle T, \mu^*\rangle \langle x, \mu\rangle &= \langle T L_x, \mu^*\rangle = \langle L_{Tx}, \mu^*\rangle \\ &= \langle Tx, \mu\rangle = \langle T, \mu'\rangle \langle x, \mu\rangle.\end{aligned}$$

Hence $\mu^* = \mu'$.

On the other hand, if v is a multiplicative linear functional on $M(A)$ and $\langle L_y, v\rangle \neq 0$ for some $y \in A$ then clearly the equation $\langle x, \mu\rangle = \langle L_x, v\rangle$ defines a multiplicative linear functional μ on A. From the first portion of the proof it then follows that $v = \mu'$. ▯

As is well known (E.2), the collection of multiplicative linear functionals on a commutative Banach algebra B with the weak* topology forms a locally compact Hausdorff topological space. The space is compact if B has an identity. This space is generally identified in a natural manner with the regular maximal ideal space $\Delta(B)$ of B, namely, μ corresponds to $\mu^{-1}(0)$. In the statement of the next theorem we shall assume such an identification has been made.

The previous result asserts that the correspondence $\mu \to \mu'$ defines a bijective mapping of $\Delta(A)$ onto those points of $\Delta(M(A))$ which do not contain the ideal $\{L_x \mid x \in A\}$, that is, with those multiplicative functionals which do not vanish identically on $\{L_x \mid x \in A\}$. We shall denote this subset of $\Delta(M(A))$ by $\Delta'(A)$.

If I is an ideal in a commutative Banach algebra B then the *hull of I*, denoted by $h(I)$, is the collection of all the regular maximal ideals which contain I. $h(I)$ is always a closed subset of $\Delta(B)$ (E.4).

Theorem 1.4.2. *Let A be a commutative Banach algebra without order and let $H(A)$ be the hull of $\{L_x \mid x \in A\}$. Then $\Delta(M(A)) = H(A) \cup \Delta'(A)$. Moreover $\Delta(A)$ is homeomorphic to the open set $\Delta'(A)$ and $H(A)$ is compact.*

Proof. From Theorem 1.4.1 and the preceding remarks it is apparent that $\Delta(M(A)) = H(A) \cup \Delta'(A)$ and that $\mu \to \mu'$ defines a bijective mapping of $\Delta(A)$ onto $\Delta'(A)$. The mapping is continuous in the respective weak* topologies since $\langle T, \mu'\rangle = \langle Tx, \mu\rangle / \langle x, \mu\rangle$. The inverse mapping is also continuous because $\langle L_x, \mu'\rangle = \langle x, \mu\rangle$ and $\{L_x \mid x \in A\} \subset M(A)$.

Therefore the mapping $\mu \to \mu'$ is a homeomorphism.

Since $\Delta(M(A))$ is compact and $\mu \to \mu'$ is a homeomorphism it follows immediately that $H(A)$ is compact. ▯

It is clear that the correspondence between $\Delta(A)$ and $\Delta'(A)$ can also be described in the following way. For each $m' \in \Delta(M(A))$ such that m' does not contain $\{L_x \mid x \in A\}$, we associate $m \in \Delta(A)$ by setting $m = \{y \mid y \in A,$

$L_y \in m' \cap \{L_x | x \in A\}\}$, and conversely. Then the correspondence $m \leftrightarrow m'$ defines the same homeomorphism as in the theorem.

Some easy corollaries of Theorem 1.4.2 are given below.

Corollary 1.4.1. *Let A be a commutative Banach algebra without order. Then for each $x \in A$ the Gelfand transform of L_x vanishes identically on $\Delta(M(A)) \sim \Delta'(A) = H(A)$.*

Corollary 1.4.2. *Let A be a semi-simple commutative Banach algebra. Then $M(A)$ is semi-simple.*

If $T \in E(A)$ then the *adjoint* T^* of T is the linear operator on the space of continuous linear functionals v on A defined by $\langle x, T^* v\rangle = \langle Tx, v\rangle$ for all $x \in A$ (D.3).

Corollary 1.4.3. *Let A be a semi-simple commutative Banach algebra and $T \in E(A)$. Then the following are equivalent:*

(i) *$T \in M(A)$.*

(ii) *For each multiplicative linear functional μ on A, there exists a constant $c(\mu)$ such that $T^* \mu = c(\mu)\, \mu$.*

Proof. Suppose $T \in M(A)$ and μ is a multiplicative linear functional. Then for each $x \in A$,

$$\langle x, T^* \mu\rangle = \langle Tx, \mu\rangle = \langle T, \mu'\rangle \langle x, \mu\rangle,$$

and (ii) follows upon setting $c(\mu) = \langle T, \mu'\rangle$.

Conversely, suppose $T^* \mu = c(\mu)\, \mu$ for each multiplicative linear functional μ. Define $\varphi[\mu^{-1}(0)] = c(\mu)$. Then for each $x \in A$,

$$\begin{aligned}(Tx)\hat{}\,[\mu^{-1}(0)] &= \langle Tx, \mu\rangle = \langle x, T^* \mu\rangle = c(\mu) \langle x, \mu\rangle \\ &= \varphi[\mu^{-1}(0)]\, \hat{x}[\mu^{-1}(0)].\end{aligned}$$

Hence $T \in M(A)$ by Corollary 1.2.1. □

Corollary 1.4.4. *Let A be a semi-simple commutative supremum norm algebra. Then $M(A)$ is a supremum norm algebra.*

Proof. From Theorem 1.2.3 we see that for each $T \in M(A)$,

$$\begin{aligned}\|\hat{T}\|_\infty \leqq \|T\| &= \sup_{\|x\|=1} \|Tx\| = \sup_{\|\hat{x}\|_\infty=1} \|(Tx)\hat{}\,\|_\infty = \sup_{\|\hat{x}\|_\infty=1} \|\hat{T}\hat{L}_x\|_\infty \\ &\leqq \sup_{\|\hat{x}\|_\infty=1} \|\hat{L}_x\|_\infty \|\hat{T}\|_\infty = \|\hat{T}\|_\infty.\end{aligned}$$

Therefore $\|\hat{T}\|_\infty = \|T\|$. □

For a commutative Banach algebra B the regular maximal ideal space is often given another topology known as the hull-kernel topology. If

$E \subset \Delta(B)$ then the *kernel* of E, denoted by $k(E)$. is the intersection of all the regular maximal ideals in E. The *hull-kernel topology* is defined on $\Delta(B)$ by the closure operation which assigns to a set $E \subset \Delta(B)$ the closure $\bar{E} = h[k(E)]$ (E.4).

Theorem 1.4.3. *Let A be a semi-simple commutative Banach algebra. If $\Delta(A)$ and $\Delta(M(A))$ are given the hull-kernel topologies then $\Delta(M(A)) = H(A) \cup \Delta'(A)$ and $\Delta(A)$ is homeomorphic to the open subset $\Delta'(A)$ and $H(A)$ is compact. Moreover $\Delta'(A)$ is dense in $\Delta(M(A))$.*

Proof. In view of Theorem 1.4.2 it is evident that $\Delta(M(A)) = H(A) \cup \Delta'(A)$ and that $H(A)$ is compact and $\Delta'(A)$ open. Let $K \subset \Delta(A)$ and set $K' = \{m' | m \in K\}$. It is easy to show that $k(K) = \{x | x \in A, L_x \in k(K')\}$, and that $m \supset k(K)$ if and only if $m' \supset k(K')$. From these facts one concludes that

$$\{h[k(K)]\}' = \Delta'(A) \cap h[k(K')].$$

It is then obvious that the mapping $m \to m'$ is a homeomorphism.

Let $L_x \in k[\Delta'(A)]$. Then $\langle L_x, \mu' \rangle = \langle x, \mu \rangle = 0$ for all multiplicative linear functionals μ on A. Hence from the semi-simplicity of A it follows that $x = 0$. Thus $k[\Delta'(A)] = \{0\}$ and $\Delta(M(A)) = h\{k[\Delta'(A)]\}$. That is, $\Delta'(A)$ is dense in $\Delta(M(A))$ in the hull-kernel topology. □

It should be noted that the first portion of the theorem is also valid with only the assumption that A is without order. On the other hand, the second portion of the result is not valid in general for semi-simple A with the weak* topologies on the regular maximal ideal spaces. Furthermore, Banach algebra properties of A are not always possessed by $M(A)$.

If B is a commutative Banach algebra then B is said to be *regular* if the weak* and the hull-kernel topologies agree on the regular maximal ideal space (E.4). B is *self-adjoint* if for each $x \in B$ there exists an $x^+ \in B$ such that $\langle x^+, \mu \rangle = \overline{\langle x, \mu \rangle}$ for every multiplicative linear functional μ (E.3). The bar, of course, denotes complex conjugation. Now A may be both regular and self-adjoint without $M(A)$ being either.

For example, suppose $A = L_1(G)$ where G is a nondiscrete locally compact Abelian group. Then $M(A) = M(G)$, both A and $M(A)$ are commutative semi-simple Banach algebras, and A is regular and self-adjoint. On the other hand, it is well known that $\Delta(A) = \hat{G}$ is not dense in the maximal ideal space of $M(G)$ under the weak* topologies (F.6). In view of Theorem 1.4.3 this implies that $M(A) = M(G)$ is not regular. Moreover, $M(G)$ is self-adjoint if and only if G is discrete (F.7b) and hence $M(A)$ is not self-adjoint.

Some results relating the properties of A and $M(A)$ are collected in the next theorem.

Theorem 1.4.4. *Let A be a semi-simple commutative Banach algebra. If $M(A)$ is regular then A is regular. If $M(A)$ is regular and A is self-adjoint then $M(A)$ is self-adjoint.*

Proof. The first assertion is immediate from Theorem 1.4.2 and the regularity of $M(A)$.

Let $T\in M(A)$ and define φ^+ on $\Delta(A)$ by $\varphi^+(m)=\overline{\hat{T}(m')}$ for all $m\in\Delta(A)$ where $m'\in\Delta(M(A))$ is the regular maximal ideal associated with m by the mapping in Theorem 1.4.2. Then for each $x\in A$ using the self-adjointness of A we see that

$$\varphi^+(m)\,\hat{x}(m)=\overline{\hat{T}(m')}\,\hat{x}(m)=\overline{\hat{T}(m')\,\overline{\hat{x}(m)}}$$
$$=\overline{(Tx^+)^\wedge(m)}=[(Tx^+)^+]^\wedge(m).$$

Hence $\varphi^+\hat{A}\subset\hat{A}$. Clearly φ^+ is bounded and continuous and hence, by Corollary 1.2.1, φ^+ defines a multiplier $T^+\in M(A)$ such that $\hat{T}^+(m')=\overline{\hat{T}(m')}$.

However, since $M(A)$ is regular, the set $\Delta'(A)$ is dense in $\Delta(M(A))$ and so $\hat{T}^+=\overline{\hat{T}}$. Therefore $M(A)$ is self-adjoint. □

The preceding example with $A=L_1(G)$ shows that the converse to the first portion of the theorem is false, and that in the second part the regularity condition on $M(A)$ cannot be eliminated.

Generally, as indicated in 1.1, the algebra A considered as a normed subalgebra of $M(A)$ is not complete. The following result provides some information on the location of the closure of A in $M(A)$.

Theorem 1.4.5. *Let A be a commutative Banach algebra without order. Then the norm closure of $\{L_x \mid x\in A\}$ in $M(A)$ is contained in $k[H(A)]$.*

Proof. Let T belong to the norm closure of $\{L_x \mid x\in A\}$. Then for each $\varepsilon>0$ there exists $y\in A$ such that $\|T-L_y\|<\varepsilon$. From Corollary 1.4.1 we know that $\hat{L}_y$ vanishes identically on $H(A)$ and hence for each $n\in H(A)$ we have

$$|\hat{T}(n)|=|\hat{T}(n)-\hat{L}_y(n)|\leqq\|T-L_y\|<\varepsilon.$$

Therefore $\hat{T}$ vanishes identically on $H(A)$ and $T\in k[H(A)]$. □

1.5. Integral Representations of Multipliers. Now we wish to return to a more detailed consideration of supremum norm algebras. We have seen that when A is a supremum norm algebra then the correspondence between $M(A)$ and $\mathscr{M}(A)$ is an isometry and $M(A)$ is itself a supremum norm algebra. First we shall give two elementary results concerning supremum norm algebras and then examine the possibility of integral representations for elements of $M(A)$.

Theorem 1.5.1. *Let A be a semi-simple commutative supremum norm algebra and suppose $\{L_x \mid x \in A\}$ is strong operator dense in $M(A)$. If φ is a bounded function on $\Delta(A)$ such that $\varphi \hat{A} \subset M(A)^\wedge$ then $\varphi \in M(A)^\wedge$, that is φ defines a multiplier for $M(A)$.*

Proof. By Theorem 1.1.6 the algebra A contains an approximate identity $\{x_\alpha\}$. Since $\varphi \hat{A} \subset M(A)^\wedge$, there exists $\{T_\alpha\} \subset M(A)$ such that $\varphi (L_{x_\alpha} x)^\wedge = (T_\alpha x)^\wedge$ for all $x \in A$. Moreover, for each $x \in A$,

$$\begin{aligned} \|T_\alpha x - T_\beta x\| &= \|(T_\alpha x)^\wedge - (T_\beta x)^\wedge\|_\infty = \|\varphi \hat{x}_\alpha \hat{x} - \varphi \hat{x}_\beta \hat{x}\|_\infty \\ &\leqq \|\varphi\|_\infty \|x_\alpha x - x_\beta x\|. \end{aligned}$$

Hence $\{T_\alpha\}$ is a Cauchy net in the strong operator topology. Since $M(A)$ is complete in the strong operator topology, there is a $T \in M(A)$ such that $\{T_\alpha\}$ converges to T in the strong operator sense.

It is elementary to show that $\varphi = \hat{T} \in M(A)^\wedge$, and the proof is complete. □

Recall that for each semi-simple commutative Banach algebra A there exists a unique minimal closed subset $\partial\Delta(A)$ of $\Delta(A)$ on which every function in $\hat{A}$ assumes its maximum modulus (E.6). That is,

$$\|\hat{x}\|_\infty = \sup_{m \in \Delta(A)} |\hat{x}(m)| = \sup_{m \in \partial\Delta(A)} |\hat{x}(m)|.$$

The set $\partial\Delta(A)$ is called the *Šilov boundary* of A. As is well known (Theorem E.6.3), for each $m_0 \in \Delta(A)$ there exists a regular nonnegative Baire measure μ_0 defined on $\partial\Delta(A)$ such that $\mu_0(\partial\Delta(A)) \leqq 1$ and

$$\hat{x}(m_0) = \int_{\partial\Delta(A)} \hat{x}(m)\, d\mu_0(m) \qquad (x \in A).$$

Ultimately we wish to establish such an integral representation for elements of $M(A)$ where the integral is taken over the Šilov boundary of A rather than that of $M(A)$. For an arbitrary supremum norm algebra such a representation does not seem to be possible. However, under certain conditions the type of theorem indicated can be established.

The first step towards this end is the next result. $\partial\Delta'(A)$ denotes the image in $\Delta(M(A))$ under the homeomorphism studied in Theorem 1.4.2 of $\partial\Delta(A)$. Similarly m' is the maximal ideal in $\Delta(M(A))$ corresponding to the regular maximal ideal $m \in \Delta(A)$.

Theorem 1.5.2. *Let A be a semi-simple commutative supremum norm algebra. Then the closure of $\partial\Delta'(A)$ in $\Delta(M(A))$ is equal to $\partial\Delta(M(A))$.*

Proof. By Corollary 1.4.1 for each $x \in A$ the function $\hat{L}_x$ vanishes identically off $\Delta'(A)$. Since $\hat{L}_x(m') = \hat{x}(m)$, it follows that $\partial\Delta'(A) \subset \partial\Delta(M(A))$, and hence the closure of $\partial\Delta'(A)$ is also contained in $\partial\Delta(M(A))$.

Conversely, since $M(A)$ is a supremum norm algebra, if $T\in M(A)$ then

$$\begin{aligned}\|\hat{T}\|_\infty = \|T\| &= \sup_{\|x\|=1}\|Tx\| = \sup_{\|x\|=1}\|(Tx)\hat{}\|_\infty = \sup_{\|x\|=1}\ \sup_{m\in\partial\Delta(A)}|(Tx)\hat{}(m)|\\ &= \sup_{\|x\|=1}\ \sup_{m\in\partial\Delta(A)}|\hat{T}(m')\,\hat{L}_x(m')|\\ &\leqq \sup_{\|x\|=1}\ \sup_{m\in\partial\Delta(A)}|\hat{L}_x(m')|\ \sup_{m\in\partial\Delta(A)}|\hat{T}(m')|\\ &= \sup_{\|x\|=1}\ \sup_{m\in\partial\Delta(A)}|\hat{x}(m)|\ \sup_{m\in\partial\Delta(A)}|\hat{T}(m')|\\ &= \sup_{\|x\|=1}\|\hat{x}\|_\infty\ \sup_{m'\in\partial\Delta'(A)}|\hat{T}(m')| = \sup_{m'\in\partial\Delta'(A)}|\hat{T}(m')|.\end{aligned}$$

Consequently $\partial\Delta(M(A))$ is contained in the closure of $\partial\Delta'(A)$ as $\partial\Delta(M(A))$ is the minimal closed boundary in $\Delta(M(A))$.

Therefore $\partial\Delta(M(A))$ is the closure of $\partial\Delta'(A)$. ▯

Corollary 1.5.1. *Let A be a semi-simple commutative supremum norm algebra. If $T\in M(A)$ and $\hat{T}$ vanishes identically on $\partial\Delta'(A)$ then T is the zero multiplier.*

The next theorem gives one condition under which an integral representation does exist.

Theorem 1.5.3. *Let A be a semi-simple commutative supremum norm algebra. If $\partial\Delta(A)$ is compact then for each $n\in\Delta(M(A))$ there exists a regular nonnegative Baire measure μ_n on $\partial\Delta'(A)$ such that $\mu_n(\partial\Delta'(A))\leqq 1$ and*

$$\hat{T}(n) = \int_{\partial\Delta'(A)} \hat{T}(m')\,d\mu_n(m') \qquad (T\in M(A)).$$

Proof. If $\partial\Delta(A)$ is compact then $\partial\Delta'(A)$ is closed and so $\partial\Delta'(A) = \partial\Delta(M(A))$. We may then choose μ_n to be the measure on $\partial\Delta(M(A))$ indicated in the remarks preceding Theorem 1.5.2. ▯

If we consider $\mathscr{M}(A)$ in place of $M(A)$ then a slightly different representation formula can be obtained. Since $M(A)$ is a supremum norm algebra whenever A is a supremum norm algebra, the correspondence between T and φ determined by $(Tx)\hat{}(m)=\varphi(m)\,\hat{x}(m)$ is an isometric isomorphism. Indeed by Theorem 1.2.2 we have $\|\varphi\|_\infty\leqq\|T\|$. On the other hand, the proof of Theorem 1.5.2 shows that

$$\|T\| = \|\hat{T}\|_\infty = \sup_{m'\in\partial\Delta'(A)}|\hat{T}(m')| = \sup_{m\in\partial\Delta(A)}|\varphi(m)| \leqq \|\varphi\|_\infty.$$

Thus $\mathscr{M}(A)$ is a semi-simple Banach algebra of bounded continuous functions on $\Delta(A)$.

These observations enable us to state and prove the following result.

Theorem 1.5.4. *Let A be a semi-simple commutative supremum norm algebra. Suppose there exists a sequence $\{x_k\} \subset A$ such that $\sup_k \|\hat{x}_k\|_\infty < \infty$ and $\lim_k \hat{x}_k(m) = 1$ for each $m \in \Delta(A)$. Then for each $n \in \Delta(A)$ there exists a regular nonnegative Baire measure μ_n on $\partial\Delta(A)$ such that $\mu_n(\partial\Delta(A)) \leqq 1$ and*

$$\varphi(n) = \int_{\partial\Delta(A)} \varphi(m)\, d\mu_n(m) \qquad (\varphi \in \mathcal{M}(A)).$$

Proof. Let μ_n be the measure on $\partial\Delta(A)$ such that

$$\hat{x}(n) = \int_{\partial\Delta(A)} \hat{x}(m)\, d\mu_n(m) \qquad (x \in A).$$

If $\{x_k\}$ is the bounded pointwise approximate identity of the hypotheses, and $\varphi \in \mathcal{M}(A)$ then $\varphi\, \hat{x}_k \in \hat{A}$. Hence

$$\int_{\partial\Delta(A)} \varphi(m)\, \hat{x}_k(m)\, d\mu_n(m) = \varphi(n)\, \hat{x}_k(n).$$

An application of the Lebesgue Dominated Convergence Theorem (Theorem C.5.2) immediately yields the desired result. □

If an integral representation as discussed in Theorem 1.5.3 is available then it can be used to further illuminate the structure of $\Delta(M(A))$ for certain algebras A even when A is not a supremum norm algebra. In what follows it will be convenient to assume that the measure μ_n is defined on $\partial\Delta(A)$ rather than $\partial\Delta'(A)$. There is a slight loss of precision in such a convention but this is offset by the greater simplicity of notation.

By $\mathrm{Re}\,\hat{A}|_{\partial\Delta(A)}$ we shall mean the algebra over the real numbers consisting of the real parts of functions in $\hat{A}$ restricted to $\partial\Delta(A)$.

It is evident for supremum norm algebras that if $\Delta(M(A)) = \Delta'(A)$ then $\partial\Delta'(A)$ is closed in $\Delta(M(A))$ and hence $\partial\Delta'(A) = \partial\Delta(M(A))$. In this case an integral representation of the desired form clearly exists. The next theorem is a partial converse to this fact.

Theorem 1.5.5. *Let A be a semi-simple commutative Banach algebra such that for each $n \in \Delta(M(A))$ there exists a regular nonnegative Baire measure μ_n on $\partial\Delta(A)$ for which $\mu_n(\partial\Delta(A)) \leqq 1$ and*

$$\hat{T}(n) = \int_{\partial\Delta(A)} \hat{T}(m)\, d\mu_n(m) \qquad (T \in M(A)).$$

If $\mathrm{Re}\,\hat{A}|_{\partial\Delta(A)}$ is uniformly dense in $C_0^R(\partial\Delta(A))$ then $\Delta(M(A)) = \Delta'(A)$. In particular, if $\partial\Delta'(A) = \partial\Delta(M(A))$ then $\Delta(M(A)) = \Delta'(A)$.

Proof. Suppose $n \in \Delta(M(A)) \sim \Delta'(A)$ and μ_n is the regular nonnegative Baire measure provided by the hypotheses. From Corollary 1.4.1 we see that for each $x \in A$,

$$0 = \hat{L}_x(n) = \int_{\partial\Delta(A)} \hat{x}(m)\, d\mu_n(m).$$

If $\mu_n \neq 0$ then by the regularity of μ_n there exists a compact set $K \subset \partial\Delta(A)$ such that $\mu_n(K) = \delta > 0$. Let f be a real valued nonnegative continuous function on $\partial\Delta(A)$ such that $f(m) = 1$ for each $m \in K$. Since $\operatorname{Re} \hat{A}|_{\partial\Delta(A)}$ is uniformly dense in $C_0^R(\partial\Delta(A))$, for each ε, $0 < \varepsilon < \delta$, there is an element $y \in A$ for which $\|\operatorname{Re}(\hat{y}) - f\|_\infty < \varepsilon$.

But then

$$\begin{aligned}\operatorname{Re}\Big[\int_{\partial\Delta(A)} \hat{y}(m)\, d\mu_n(m)\Big] &= \int_{\partial\Delta(A)} \operatorname{Re}[\hat{y}(m)]\, d\mu_n(m) \\ &= \int_{\partial\Delta(A)\sim K} \operatorname{Re}[\hat{y}(m)]\, d\mu_n(m) + \int_K \operatorname{Re}[\hat{y}(m)]\, d\mu_n(m) \\ &> -\varepsilon(1-\delta) + (1-\varepsilon)\,\delta = \delta - \varepsilon > 0,\end{aligned}$$

which contradicts $\int_{\partial\Delta(A)} \hat{x}(m)\, d\mu_n(m) = 0$ for all $x \in A$.

Consequently, $\mu_n = 0$ and thus $\hat{T}(n) = 0$ for all $T \in M(A)$, which is impossible.

Therefore $\Delta(M(A)) \sim \Delta'(A) = \emptyset$, that is, $\Delta(M(A)) = \Delta'(A)$.

The last assertion of the theorem is now apparent in view of the proof just given. □

It is also evident from the preceding theorem that integral representations of the multipliers do not generally exist. For example, if $A = L_1(R)$ then $\Delta(A) = \partial\Delta(A) = R$ and $\operatorname{Re}[L_1(R)^\wedge]$ is uniformly dense in $C_0^R(R)$. But $\Delta(M(R)) \neq R$ (F.6).

The main difficulty in an attempt to obtain integral representations for the multipliers lies in the fact that $\partial\Delta'(A)$ is generally a proper subset of $\partial\Delta(M(A))$. This leads us to investigate other boundaries for A and $M(A)$ in the hope that this obstacle can be circumvented.

Let B be a semi-simple commutative Banach algebra. A set $E \subset \Delta(B)$ is called a *boundary for* B if $\|\hat{x}\|_\infty = \sup_{m \in E} |\hat{x}(m)|$ for each $x \in B$. If B has a minimal boundary then we call it the *Bishop boundary* and denote it by $\rho\Delta(B)$. It should be noted that a boundary need not be a closed set. If it exists the Bishop boundary is clearly a subset of the Šilov boundary. However, the Bishop boundary may fail to exist, that is, there may be no minimal boundary (E.6).

For supremum norm algebras the relationship between the Bishop boundaries of A and $M(A)$ is described by the next theorem. If $\rho\Delta(A)$ exists we denote by $\rho\Delta'(A)$ its image under the homeomorphism which maps $\Delta(A)$ into $\Delta(M(A))$.

Theorem 1.5.6. *Let A be a semi-simple commutative supremum norm algebra. If $\rho\Delta(A)$ exists then $\rho\Delta'(A)=\rho\Delta(M(A))$.*

Proof. As in the proof of Theorem 1.5.2 it is evident that $\rho\Delta'(A)\subset \rho\Delta(M(A))$.

Similarly, since $M(A)$ is a supremum norm algebra, if $T\in M(A)$ then

$$\|\hat{T}\|_\infty=\|T\|=\sup_{\|x\|=1}\|Tx\|=\sup_{\|x\|=1}\|(Tx)\hat{}\|_\infty=\sup_{\|x\|=1}\sup_{m\in\rho\Delta(A)}|(Tx)\hat{}(m)|$$
$$\leq \sup_{\|x\|=1}\sup_{m\in\rho\Delta(A)}|\hat{x}(m)|\sup_{m\in\rho\Delta(A)}|\hat{T}(m')|=\sup_{m'\in\rho\Delta'(A)}|\hat{T}(m')|.$$

Thus $\rho\Delta(M(A))\subset\rho\Delta'(A)$.

Therefore $\rho\Delta'(A)=\rho\Delta(M(A))$. □

Now by Corollary 1.4.4 we see that $M(A)$ is a supremum norm algebra with identity whenever A is a supremum norm algebra. Thus the following theorem is an easy application of general results on the existence of the Bishop boundary and the type of integral representation we seek (Theorems E.6.2 and E.6.4).

Theorem 1.5.7. *Let A be a semi-simple commutative supremum norm algebra. If $\rho\Delta(A)$ exists and $\Delta(M(A))$ is metrizable then for each $n\in\Delta(M(A))$ there exists a regular nonnegative Baire measure μ_n on $\Delta(M(A))$ with total mass one such that*

$$\hat{T}(n)=\int_{\Delta(M(A))}\hat{T}(m)\,d\mu_n(m) \qquad (T\in M(A))$$

and $\mu_n(E)=0$ for each Baire set E disjoint from $\rho\Delta'(A)$.

In closing we note that a theorem such as Theorem 1.5.5 does not seem to be generally valid with the Bishop boundary in place of the Šilov boundary. The difficulty resides in the fact that though $\rho\Delta(M(A))$ is a G_δ set (E.6) it need not be locally compact in the relative topology.

1.6. Isometric Multipliers. When $A=L_1(G)$ the multiplier algebra $M(A)$ is isometrically isomorphic to $M(G)$, the space of bounded regular Borel measures on G. Our concern in this section is to examine some algebras A for which $M(A)$ is isometrically isomorphic to a homomorphic image of $M(G)$ for a certain compact group G. The algebras A will be essentially homomorphic images of $L_1(G)$ where G is the group of isometric multipliers. Moreover, the conditions we shall impose on A characterize all algebras for which the indicated results are valid.

We shall always assume that A is a semi-simple commutative Banach algebra, and denote by $I(A)$ the family of all *isometric multipliers* of A onto A, that is, multipliers which are onto mappings and for which $\|Tx\|=\|x\|$ for all $x\in A$.

Lemma 1.6.1. *Let A be a semi-simple commutative Banach algebra. Then $I(A)$ with operator composition as multiplication is an Abelian topological group in the strong operator topology.*

Proof. Appealing to Theorem 1.1.3 it is apparent that $I(A)$ forms an Abelian group. Suppose $\{T_\alpha\} \subset I(A)$ and $\{S_\beta\} \subset I(A)$ are nets which converge to the identity multiplier I in the strong operator topology. Then for each $x \in A$ we have

$$\|T_\alpha(S_\beta x) - x\| \leqq \|T_\alpha(S_\beta x) - T_\alpha x\| + \|T_\alpha x - x\|$$
$$= \|S_\beta x - x\| + \|T_\alpha x - x\|,$$

and

$$\|T_\alpha^{-1} x - x\| = \|T_\alpha(T_\alpha^{-1} x) - T_\alpha x\| = \|x - T_\alpha x\|.$$

From these observations one concludes immediately that multiplication is jointly continuous and inversion is continuous in the strong operator topology. One shows as easily that $I(A)$ is Hausdorff. Thus $I(A)$ is a topological group. □

We wish to establish under various assumptions that $I(A)$ is a compact topological group. For each $x \in A$ we define $O_x = \{Tx \mid T \in I(A)\}$.

Lemma 1.6.2. *Let A be a semi-simple commutative Banach algebra. If O_x has compact norm closure in A for each $x \in A$ then $I(A)$ is a compact Abelian topological group in the strong operator topology.*

Proof. Denote by $\bar{O}_x$ the norm closure of O_x in A, and consider the compact product space $Q = \Pi \bar{O}_x$. Let π_x be the projection of Q onto $\bar{O}_x$. Define τ: $I(A) \to Q$ by $\pi_x[\tau(T)] = Tx$ and ω: $I(A) \to Q$ by $\pi_x[\omega(T)] = T^{-1}x$ for $x \in A$ and $T \in I(A)$.

Now let $\{T_\alpha\}$ be a net in $I(A)$. Then $\{\tau(T_\alpha)\}$ and $\{\omega(T_\alpha)\}$ are nets in the compact space Q and hence have convergent subnets. Without loss of generality, we may assume that $\{T_\beta\}$ is a subnet of $\{T_\alpha\}$ such that both $\{\tau(T_\beta)\}$ and $\{\omega(T_\beta)\}$ converge in Q, say to t and t', respectively. Define T and T' by $Tx = \pi_x(t) = \lim_\beta T_\beta x$ and $T'x = \pi_x(t') = \lim_\beta T_\beta^{-1} x$ for each $x \in A$.

Clearly T and T' are linear mappings of A into A. Moreover, if $x, y \in A$ then $T(xy) = \lim_\beta T_\beta(xy) = \lim_\beta x(T_\beta y) = x(Ty)$ and $\|Tx\| = \lim_\beta \|T_\beta x\| = \lim_\beta \|x\| = \|x\|$. Thus T is an isometric multiplier for A. The same conclusion is easily seen to hold for T'.

Furthermore, if $x \in A$ then

$$\|T(T'x) - x\| \leqq \|T(T'x) - T_\beta(T'x)\| + \|T_\beta(T'x) - T_\beta(T_\beta^{-1} x)\|$$
$$= \|T(T'x) - T_\beta(T'x)\| + \|T'x - T_\beta^{-1} x\|.$$

It follows at once from the definition of T and T' that $T(T'x)=x$ for all $x \in A$.

Therefore T is an isometric onto multiplier for A and $\{T_\beta\}$ converges to T in the strong operator topology, that is, $I(A)$ is compact. □

Corollary 1.6.1. *Let A be a semi-simple commutative Banach algebra with a minimal approximate identity. If L_x is a compact operator for each $x \in A$ then $I(A)$ is compact in the strong operator topology.*

Proof. Without loss of generality we may assume $\{x_\alpha\}$ is an approximate identity such that $\|x_\alpha\|=1$. Then if $T \in I(A)$ and $x \in A$ we have $\lim_\alpha (T x_\alpha) x = T x$. Noting that $\|T x_\alpha\| = \|x_\alpha\| = 1$ we conclude that $Tx \in \{yx \mid y \in A, \ \|y\|=1\}^-$ for all $T \in I(A)$, that is, $O_x \subset \{yx \mid y \in A, \ \|y\|=1\}^-$, where the bar denotes the norm closure in A. However, since left multiplications are assumed to be compact operators this latter set is compact.

Therefore $\bar{O}_x$ is compact for all $x \in A$ and $I(A)$ is compact. □

As will be seen shortly, the next criterion for the compactness of $I(A)$ will be the most useful one for our investigation. However, before giving it, we wish to make another definition. An element $x \in A$ is said to be *almost invariant* if O_x spans a finite dimensional subspace of A. The *set of all almost invariant elements in A* will be denoted by $D(A)$. Clearly $D(A)$ is an ideal in A since, in general, $O_{yx} = y O_x$.

Corollary 1.6.2. *Let A be a semi-simple commutative Banach algebra. If $D(A)$ is norm dense in A then $I(A)$ is compact in the strong operator topology.*

Proof. Let $x \in A$ and $\varepsilon > 0$. Choose $y \in D(A)$ such that $\|x-y\| < \varepsilon/3$. Since O_y spans a finite dimensional space, it follows that $\bar{O}_y$ is norm compact, that is, O_y is totally bounded. Consequently there exists $T_1, T_2, \ldots, T_n \in I(A)$ for which $O_y \subset \bigcup_{i=1}^{n} \{z \mid z \in A, \ \|z - T_i y\| < \varepsilon/3\}$. But then for each $T \in I(A)$, there exists a j, $1 \leqq j \leqq n$, such that $\|Ty - T_j y\| < \varepsilon/3$, and hence

$$\begin{aligned} \|Tx - T_j x\| &\leqq \|Tx - Ty\| + \|Ty - T_j y\| + \|T_j y - T_j x\| \\ &= \|x-y\| + \|Ty - T_j y\| + \|y - x\| \\ &< \varepsilon, \end{aligned}$$

since T and T_j are isometries. Thus $O_x \subset \bigcup_{i=1}^{n} \{z \mid z \in A, \ \|z - T_i x\| < \varepsilon\}$, that is, O_x is totally bounded.

Therefore $\bar{O}_x$ is norm compact for each $x \in A$ and so $I(A)$ is compact in the strong operator topology by Lemma 1.6.2. □

We are now in a position to prove the theorem indicated at the beginning of this section.

Theorem 1.6.1. *Let A be a semi-simple commutative Banach algebra such that:*

(i) *A has a minimal approximate identity.*

(ii) *$D(A)$ is norm dense in A.*

(iii) *If $x\in A$, $\|x\|\leqq 1$, then L_x is the strong operator limit of convex combinations of elements of $I(A)$.*

Then there exists a weak closed ideal $N\subset M[I(A)]$, the convolution algebra of bounded regular Borel measures on the compact Abelian group $I(A)$, such that $M(A)$ is isometrically isomorphic to $M[I(A)]/N$. Furthermore, if γ: $M[I(A)]\to M[I(A)]/N$ is the canonical homomorphism defined by N and $M[I(A)]/N$ is given the quotient norm then A is isometrically isomorphic to the norm closure in $M[I(A)]/N$ of $\gamma(L_1[I(A)])$.*

Proof. By assumption (ii) and Corollary 1.6.2 we see that $I(A)$ is a compact Abelian group in the strong operator topology. For each $x\in A$ it is evident that the function $\tilde{x}$: $I(A)\to A$ defined by $\tilde{x}(T)=Tx$ is continuous. Consequently, if $\mu\in M[I(A)]$ we define for each $x\in A$ the vector valued integral

$$(F\mu)x=\int_{I(A)}\tilde{x}(T)\,d\mu(T),$$

and note that $(F\mu)x\in A$ (Theorem D.12.1). Clearly this defines a linear mapping $F\mu$ from A to A such that $\|F\mu\|\leqq\|\mu\|=|\mu|(I(A))$.

Moreover, for each $x^*\in A^*$ we have (Theorem D.12.2) that

$$\langle (F\mu)x, x^*\rangle=\int_{I(A)}\langle Tx, x^*\rangle\,d\mu(T) \qquad (x\in A).$$

Thus let $x, y\in A$. Then for any $x^*\in A^*$ we have

$$\begin{aligned}\langle F\mu(yx), x^*\rangle &= \int_{I(A)}\langle T(yx), x^*\rangle\,d\mu(T)=\int_{I(A)}\langle y(Tx), x^*\rangle\,d\mu(T)\\ &= \int_{I(A)}\langle Tx, L_y^* x^*\rangle\,d\mu(T)=\langle (F\mu)x, L_y^* x^*\rangle\\ &= \langle L_y(F\mu)x, x^*\rangle=\langle y(F\mu)x, x^*\rangle\end{aligned}$$

where, of course, L_y^* denotes the adjoint operator to L_y. Consequently $(F\mu)(yx)=y(F\mu)x$ for all $x, y\in A$ and $F\mu\in M(A)$.

Thus we have defined a mapping F: $M[I(A)]\to M(A)$. Clearly F is linear. Moreover, if $\mu, \nu\in M[I(A)]$ then for each $x\in A$, $x^*\in A^*$,

$$\begin{aligned}\langle F\mu(F\nu(x)), x^*\rangle &= \langle (F\nu)x, (F\mu)^* x^*\rangle=\int_{I(A)}\langle Tx, (F\mu)^* x^*\rangle\,d\nu(T)\\ &= \int_{I(A)}\langle (F\mu)(Tx), x^*\rangle\,d\nu(T)\\ &= \int_{I(A)}\Big[\int_{I(A)}\langle STx, x^*\rangle\,d\mu(S)\Big]\,d\nu(T)\\ &= \int_{I(A)}\langle Ux, x^*\rangle\,d\mu*\nu(U)=\langle F(\mu*\nu)x, x^*\rangle.\end{aligned}$$

Thus $F: M[I(A)] \to M(A)$ is a homomorphism.

Let N be the kernel of this homomorphism. Obviously N is an ideal in $M[I(A)]$. We claim that N is also weak* closed in $M[I(A)]$. In order to establish this we first show that on norm bounded sets F is continuous from $M[I(A)]$ with the weak* topology to $M(A)$ with the strong operator topology. Clearly we may restrict our attention to the unit ball in $M[I(A)]$. First we need a few preliminary results.

Denote by $\overline{\text{co}}[I(A)]$ the strong operator convex closure of $I(A)$. $\overline{\text{co}}[\mathrm{I}(A)]$ is compact in $M(A)$ with the strong operator topology since it is the strong operator convex closure of a strong operator compact set (Theorem D.7.4). We assert that if $\|\mu\| \leqq 1$ then $F\mu \in \overline{\text{co}}[I(A)]$. Indeed, since the unit ball of $M[I(A)]$ is compact in the weak* topology there is, by virtue of the Krein-Milman Theorem (Theorem D.7.2), a net $\{v_\alpha\}$ of convex combinations of extreme points of the unit ball which converges to μ in the weak* topology. Since the extreme points are unit point masses multiplied by complex numbers of modulus one, it follows easily that $F v_\alpha \in \overline{\text{co}}[I(A)]$ because $F\delta_T = T$ where δ_T is the unit mass concentrated at T. The compactness of $\overline{\text{co}}[I(A)]$ implies the existence of a subnet $\{F v_\beta\}$ and an $S \in \overline{\text{co}}[I(A)]$ such that $\{F v_\beta\}$ converges to S in the strong operator sense.

For each $x \in A$, $x^* \in A^*$ we have on the one hand

$$\langle Sx, x^*\rangle = \lim_\beta \langle (F v_\beta)\, x, x^*\rangle = \lim_\beta \int_{I(A)} \langle Tx, x^*\rangle \, d v_\beta(T).$$

While on the other hand, since $\{v_\beta\}$ converges to μ in the weak* sense,

$$\lim_\beta \int_{I(A)} \langle Tx, x^*\rangle \, d v_\beta(T) = \int_{I(A)} \langle Tx, x^*\rangle \, d\mu(T) = \langle (F\mu)\, x, x^*\rangle.$$

Therefore $F\mu = S \in \overline{\text{co}}[I(A)]$.

Moreover, the preceding argument shows that $F\mu$ is the only limit point of $\{F v_\alpha\}$ and hence $\{F v_\alpha\}$ converges to $F\mu$ in the strong operator topology. Thus, if $\{\mu_\alpha\} \subset M[I(A)]$, $\|\mu_\alpha\| \leqq 1$ and $\{\mu_\alpha\}$ converges to μ in the weak* topology, then since $\{F\mu_\alpha\} \subset \overline{\text{co}}[I(A)]$ we can repeat the previous argument to deduce that $\{F\mu_\alpha\}$ converges to $F\mu$ in the strong operator sense, thereby establishing the asserted continuity property for the mapping F.

In particular, if $\{\mu_\alpha\} \subset N$ is such that $\|\mu_\alpha\| \leqq 1$ and $\{\mu_\alpha\}$ converges to μ in the weak* topology then $F\mu = \lim_\alpha F\mu_\alpha = 0$ implies $\mu \in N$. Appealing to the Krein-Šmulian Theorem (Theorem D.7.1) we conclude that N is weak* closed.

To complete the proof of the first portion of the theorem we need only show that F maps onto $M(A)$ and for each $S \in M(A)$ there exists some $\mu \in M[I(A)]$ for which $\|\mu\| = \|S\|$. Since A has a minimal approximate identity, the left multiplications $\{L_x | x \in A\}$ are strong operator dense in $M(A)$ and $\|L_x\| = \|x\|$. Hence from assumption (iii) it follows that $\overline{\text{co}}[I(A)]$ is all of the unit ball in $M(A)$. In particular, if $S \in M(A)$ and $\|S\| \leqq 1$ then there exists a net of convex combinations of elements of $I(A)$, say

$$\{S_\alpha\} = \left\{ \sum_{i=1}^{n(\alpha)} c(\alpha, T_i) \, T_i \right\},$$

which converges to S in the strong operator topology. Let

$$\mu_\alpha = \sum_{i=1}^{n(\alpha)} c(\alpha, T_i) \, \delta_{T_i}.$$

Then clearly $\{\mu_\alpha\}$ is contained in the unit ball of $M[I(A)]$ and $F\mu_\alpha = S_\alpha$. Using the weak* compactness of the unit ball of $M[I(A)]$ we select a subnet $\{\mu_\beta\}$ and a μ such that $\{\mu_\beta\}$ converges to μ in the weak* topology. Because F is continuous on norm bounded sets from $M[I(A)]$ in the weak* topology to $M(A)$ in the strong operator topology, we conclude that $\{F\mu_\beta\}$ converges to $F\mu$ in the strong operator sense. But $\{F\mu_\beta\} = \{S_\beta\}$ converges to S in the strong operator topology and hence $F\mu = S$. It is now apparent that F maps $M[I(A)]$ onto $M(A)$. To show that $\|S\| = \|\mu\|$ we may without loss of generality assume that $\|S\| = 1$. Evidently $1 = \|S\| = \|F\mu\| \leqq \|\mu\|$. While on the other hand, since $\|\mu_\beta\| \leqq 1$ for each β and the unit ball in $M[I(A)]$ is weak* closed, we conclude that $\|\mu\| \leqq 1$. Thus $\|\mu\| = \|S\|$.

Combining all of the previous arguments we see that F defines an isometric isomorphism of $M[I(A)]/N$ onto $M(A)$.

Next we wish to show that A is isometrically isomorphic to the norm closure of $\gamma(L_1[I(A)])$. We begin by showing that the norm closure in $M(A)$ of $F(L_1[I(A)])$ is equal to $\{L_x | x \in A\}$.

Suppose $f \in D(L_1[I(A)])$. Then, since δ_T for each $T \in I(A)$ defines an isometric multiplier for $L_1[I(A)]$, we see that the set $\{(Ff)T = F(f * \delta_T) | T \in I(A)\}$ spans a finite dimensional space $X(f)$. If $\|x\| \leqq 1$ then there exists a net $\{S_\alpha\}$ of convex combinations of elements in $I(A)$ which converges in the strong operator sense to L_x. Hence $\{F(f) S_\alpha\}$ converges to $F(f) L_x$, and $F(f) L_x \in X(f)$ because $X(f)$ is finite dimensional. More generally it is then clear that $F(f) L_x \in X(f)$ for all $x \in A$. Furthermore, the set $\{F(f) L_x | x \in A\}$ is clearly a subspace of $X(f)$ which belongs to $\{L_x | x \in A\}$ since the latter is an ideal in $M(A)$. Consequently, if $\{x_\alpha\} \subset A$ is an approximate identity we see that $\{F(f) L_{x_\alpha}\}$ converges to $F(f)$ in the strong operator topology and hence $F(f) \in \{F(f) L_x | x \in A\} \subset \{L_x | x \in A\}$ as $\{F(f) L_x | x \in A\}$ is finite dimensional. Now let $f \in L_1[I(A)]$.

Since the trigonometric polynomials are norm dense in $L_1[I(A)]$ (F.7d), it follows that there exists a sequence $\{f_n\} \subset D(L_1[I(A)])$ such that

$$\lim_n \|f_n - f\|_1 = 0.$$

Clearly $\|F(f_n) - F(f)\| \leqq \|f_n - f\|_1$ implies that

$$\lim_n \|F(f_n) - F(f)\| = 0.$$

However, by Theorem 1.1.4, the space $\{L_x | x \in A\}$ is isometric to A and is thus complete in $M(A)$. Hence $\{L_x | x \in A\}$ is a closed subspace of $M(A)$ and so $F(f) \in \{L_x | x \in A\}$ since $\{F(f_n)\} \subset \{L_x | x \in A\}$.

Finally we claim that $\{L_x | x \in A\}$ belongs to the norm closure of $F(L_1[I(A)])$. Indeed, let $\{f_\alpha\} \subset L_1[I(A)]$ be an approximate identity. Then $\{f_\alpha\}$ converges to δ_I in the weak* topology on $M[I(A)]$ and thus $\{F(f_\alpha)\}$ converges to I in the strong operator topology on $M(A)$. For each $x \in A$ we claim that $L_{F(f_\alpha)x} \in F(L_1[I(A)])$. Because if $\mu \in M[I(A)]$ is such that $F\mu = L_x$, then $L_{F(f_\alpha)x} = F(f_\alpha) F\mu = F(f_\alpha * \mu)$ and $f_\alpha * \mu \in L_1[I(A)]$. But then $\lim_\alpha \|L_{F(f_\alpha)x} - L_x\| = \lim_\alpha \|F(f_\alpha)x - x\| = 0$. Hence $\{L_x | x \in A\}$ is equal to the norm closure in $M(A)$ of $F(L_1[I(A)])$.

Now consider $M[I(A)]/N$ with the quotient norm and define $\beta\colon M[I(A)]/N \to M(A)$ by the relation $F = \beta \circ \gamma$. This is meaningful since the kernels of F and γ are both N. Clearly β is an onto isomorphism. Moreover, if $S \in M(A)$ and $\mu \in M[I(A)]$ is such that $\|\mu\| = \|S\|$ then on the one hand,

$$\|\gamma(\mu)\| = \inf\{\|\mu + \nu\| \,|\, \nu \in N\} \leqq \|\mu\| = \|F\mu\|;$$

while on the other hand,

$$\|F\mu\| = \inf\{\|F(\mu + \nu)\| \,|\, \nu \in N\} \leqq \inf\{\|\mu + \nu\| \,|\, \nu \in N\} = \|\gamma(\mu)\|,$$

since N is the kernel of F. Hence $\|\gamma(\mu)\| = \|\beta[\gamma(\mu)]\|$, that is, β is an isometry. It follows at once that A is isometrically isomorphic to the norm closure in $M[I(A)]/N$ of $\gamma(L_1[I(A)])$.

This completes the proof of the theorem. □

The algebras which are the subject of Theorem 1.6.1 are discussed extensively in Greenleaf [3] where the concept of a QCG algebra is introduced. A Banach algebra A is called a *QCG algebra* if there exists a compact group G and a weak* closed two-sided ideal N in $M(G)$ such that A is isometrically isomorphic with the norm closure of $\gamma[L_1(G)]$ where γ is the canonical homomorphism defined by N and $M(G)/N$ is given the quotient norm. It is shown in Greenleaf [3], pp. 249–259, that the properties (i)–(iii) of Theorem 1.6.1 completely characterize QCG algebras, and hence the conclusion of the previous theorem is valid if and only if A is a QCG algebra. The group algebras of compact groups form a subclass of QCG algebras. If $A=L_1(G)$ where G is a compact Abelian group then, of course, $I(A)=\{T|Tf=\zeta\,\delta_s*f, \zeta\in C, |\zeta|=1, s\in G\}$. In the strong operator topology $I(A)$ is topologically isomorphic to $\Gamma\times G$ where $\Gamma=\{\zeta|\zeta\in C, |\zeta|=1\}$. Thus for $\mu\in M[I(A)]$ we can write $\mu=\mu_1\times\mu_2$ where $\mu_1\in M(\Gamma)$ and $\mu_2\in M(G)$. Then for each $f\in L_1(G)$ we have

$$(F\mu)f(t)=\int_{I(A)}(\zeta\,\delta_s)f(t)\,d\mu(\zeta\,\delta_s)=\int_\Gamma\int_G \zeta f(t\,s^{-1})\,d\mu_1(\zeta)\,d\mu_2(s)$$
$$=\int_\Gamma \zeta\,d\mu_1(\zeta)\int_G f(t\,s^{-1})\,d\mu_2(s)=f*\big(c(\mu_1)\,\mu_2\big)(t)$$

where $c(\mu_1)=\int_\Gamma \zeta\,d\mu_1(\zeta)$. Hence $(F\mu)f=f*\big(c(u_1)\,\mu_2\big)$ as one would expect from the results of the introductory chapter.

An examination of the proof of Theorem 1.6.1 reveals that in establishing the isometric isomorphism between $M(A)$ and $M[I(A)]/N$ the denseness assumption of the almost invariant elements was used only to infer the compactness of $I(A)$. However, it was employed in identifying A with the norm closure of $\gamma\big(L_1[I(A)]\big)$. Nevertheless the characterization of $M(A)$ only depends on the first and third hypotheses and the compactness of $I(A)$. Consequently the following theorem is also valid.

Theorem 1.6.2. *Let A be a semi-simple commutative Banach algebra such that*

(i) *A has a minimal approximate identity.*

(ii) *$I(A)$ is a compact topological group in the strong operator topology.*

(iii) *If $x\in A$, $\|x\|\leqq 1$, then L_x is the strong operator limit of convex combinations of elements of $I(A)$.*

Then there exists a weak closed ideal $N\subset M[I(A)]$ such that $M(A)$ is isometrically isomorphic to $M[I(A)]/N$.*

Some conditions which are weaker than the assumption of the denseness of the almost invariant elements and which insure that $I(A)$ is compact were given in Lemma 1.6.2 and Corollary 1.6.1.

We shall not pursue these matters further but refer interested readers to Greenleaf [3] where the general noncommutative theory is developed.

Instead we wish to examine several results which follow from the assumption of compactness for $I(A)$. In view of the Corollary 1.6.2, all of these results are valid for QCG algebras and the algebras in Theorem 1.6.2.

Theorem 1.6.3. *Let A be a semi-simple commutative Banach algebra and suppose $I(A)$ is compact in the strong operator topology. Then the linear space spanned by the common eigenvectors of $I(A)$ is norm dense in A.*

Proof. Let λ denote normalized Haar measure on $I(A)$. If $f: I(A) \to A$ is continuous then as in the proof of Theorem 1.6.1 we denote by $\int_{I(A)} f(T)\, d\lambda(T)$ that unique element $z \in A$ such that

$$\langle z, x^* \rangle = \int_{I(A)} \langle f(T), x^* \rangle \, d\lambda(T) \qquad (x^* \in A^*).$$

Let $\{f_\alpha\}$ be a net of trigonometric polynomials such that for each continuous f the net $\int_{I(A)} f_\alpha(T) f(T)\, d\lambda(T)$ converges in norm to $f(I)$ (F.7d). Thus if $x \in A$ then the function $\tilde{x}(T) = Tx$ is continuous and the net $\{\int_{I(A)} f_\alpha(T)\, \tilde{x}(T)\, d\lambda(T)\}$ converges in norm to x.

Denote the dual group of $I(A)$ by $\hat{I}(A)$ and by $(\cdot,\cdot)$ the usual pairing between elements $T \in I(A)$ and $T' \in \hat{I}(A)$. Then if $T_0 \in I(A)$ and $S' \in \hat{I}(A)$, we have

$$\begin{aligned} T_0 \int_{I(A)} (T, S')\, \tilde{x}(T)\, d\lambda(T) &= \int_{I(A)} (T, S')(T_0 T)\, x\, d\lambda(T) \\ &= \int_{I(A)} (T, S')\, \tilde{x}(T_0 T)\, d\lambda(T) \\ &= \int_{I(A)} (T_0^{-1} T, S')\, \tilde{x}(T)\, d\lambda(T) \\ &= (T_0^{-1}, S') \int_{I(A)} (T, S')\, \tilde{x}(T)\, d\lambda(T). \end{aligned}$$

Thus if we set $x_s = \int_{I(A)} (T, S')\, \tilde{x}(T)\, d\lambda(T)$ the equation

$$T_0\, x_s = (T_0^{-1}, S')\, x_s \qquad (T_0 \in I(A))$$

shows that x_s is an eigenvector for all of $I(A)$.

Since each f_α is a linear combination of continuous characters, the conclusion of the theorem is now obvious. □

We shall say that $I(A)$ separates the points of $\Delta(A)$ provided $\{\hat{T} \mid T \in I(A)\}$ does so. An element $x \in A$ is *idempotent* if $x^2 = x$.

Theorem 1.6.4. *Let A be a semi-simple commutative Banach algebra and suppose $I(A)$ is compact in the strong operator topology. If $I(A)$ separates the points of $\Delta(A)$ then $\Delta(A)$ is discrete and A is spanned by its idempotent elements.*

Proof. It is evident that any $x \in A$ such that $\hat{x}$ has one-point support is an eigenvector for $M(A)$, and in particular for $I(A)$. Conversely, suppose $x \in A$ is a common eigenvector for $I(A)$. Then for each $T \in I(A)$ there exists a complex number $c(T)$ such that $Tx = c(T)\, x$. Clearly $\hat{T}(m) = c(T)$ for all $m \in \Delta(A)$ such that $\hat{x}(m) \neq 0$. Now suppose $m_1, m_2 \in \Delta(A)$, $m_1 \neq m_2$, $\hat{x}(m_1) \neq 0$, $\hat{x}(m_2) \neq 0$. Then for all $T \in I(A)$ we would have

$\hat{T}(m_1)=\hat{T}(m_2)=c(T)$, contradicting the assumption that $I(A)$ separates the points of $\Delta(A)$.

Consequently, if x is a common eigenvector of $I(A)$ then $\hat{x}$ has a one-point support.

Let E be the set of all common eigenvectors of $I(A)$ such that for some $m\in\Delta(A)$ we have $\hat{x}(m)=1$ and $\hat{x}$ vanishes identically on $\Delta(A)\sim\{m\}$. Every element of E is an idempotent and from Theorem 1.6.3 it is apparent that the linear space spanned by E is norm dense in A. Therefore A is spanned by its idempotents.

Moreover, since $\hat{A}$ separates the points of $\Delta(A)$, it follows that there is no point in $\Delta(A)$ at which $\hat{x}$ vanishes for all $x\in E$. Thus if $m_0\in\Delta(A)$ there exists an $x\in E$ for which $\hat{x}(m_0)=1$. But then $\{m\,|\,|\hat{x}(m)-1|<\frac{1}{2}\}=\{m_0\}$ is an open neighborhood of m_0. Hence $\Delta(A)$ is discrete. □

Of course the results are valid when $A=L_1(G)$ and G is a compact Abelian group. In this case the idempotents are the continuous characters on G.

1.7. Multipliers and Dual Spaces. In the previous section we investigated the multipliers of certain algebras which are similar to group algebras. In particular, if A is a QCG algebra we saw that $M(A)$ is isometrically isomorphic to a certain algebra of bounded regular Borel measures. When $A=L_1(G)$ and $\mu\in M(G)$ corresponds to $T\in M[L_1(G)]$ then we also know that $Tf=\mu*f$ for all $f\in L_1(G)$. Our concern in this short section will be to establish this sort of a characterization of the multipliers for a certain class of Banach algebras which are dual spaces.

We shall say that an ideal I in a Banach algebra B is *without order in* B if $x\in B$ and $xI=\{0\}$ imply that $x=0$.

Theorem 1.7.1. *Let B be a semi-simple commutative Banach algebra such that:*

(i) *$B=X^*$ for some Banach space X.*
(ii) *Every regular maximal ideal in B is weak* closed.*

If A is a closed ideal in B which contains a minimal approximate identity and is without order in B then $M(A)$ is algebraically and topologically isomorphic to B. Moreover if the correspondence between $M(A)$ and B is denoted by $T_x\leftrightarrow x$ then $T_x y=xy$ for all $y\in A$.

Proof. Clearly A is a commutative Banach algebra without order. Hence $M(A)$ is well defined and every $T\in M(A)$ is a bounded linear operator. If $x\in B$ and we define $T_x y=xy$ for all $y\in A$ then, since A is an ideal, it is apparent that T_x defines a multiplier of A. The mapping $x\to T_x$ is obviously an injective isomorphism of B into $M(A)$ because A is without order in B.

On the other hand, let $T \in M(A)$ and $\{y_\alpha\} \subset A$ be a minimal approximate identity. Without loss of generality we may assume that $\|y_\alpha\| = 1$. Then $\{Ty_\alpha\}$ is a bounded net in $B = X^*$ and hence by Alaoglu's Theorem (Theorem D.4.3) it has a weak* convergent subnet. Let this subnet be denoted by $\{Ty_\beta\}$ and let its weak* limit be x.

If μ is a multiplicative linear functional on B then $\mu^{-1}(0)$ is a regular maximal ideal in B and so weak* closed. Hence μ is a weak* continuous linear functional (D.4). In particular, $\lim_\beta \langle Ty_\beta, \mu\rangle = \langle x, \mu\rangle$ for each multiplicative linear functional μ on B.

Consequently for each such μ,

$$\lim_\beta \langle (Ty_\beta)\, z, \mu\rangle = \lim_\beta \langle Ty_\beta, \mu\rangle \langle z, \mu\rangle = \langle x, \mu\rangle \langle z, \mu\rangle = \langle x\, z, \mu\rangle \quad (z \in A),$$

and

$$\lim_\beta \langle (Ty_\beta)\, z, \mu\rangle = \lim_\beta \langle y_\beta(Tz), \mu\rangle = \langle Tz, \mu\rangle \quad (z \in A).$$

That is, $(Tz)\hat{}\,(m) = \hat{x}(m)\,\hat{z}(m)$ for all $m \in \Delta(B)$ and $z \in A$. Therefore, since B is semi-simple, we see that $Tz = x\,z$ for all $z \in A$ and the mapping $x \to T_x$ is surjective.

Obviously $\|T_x\| \leqq \|x\|$. Consequently by the Open Mapping Theorem (Theorem D.6.2) the inverse mapping is also continuous.

Therefore $M(A)$ is algebraically and topologically isomorphic to B. □

Corollary 1.7.1. *Let A be a semi-simple commutative Banach algebra with a minimal approximate identity such that:*

(i) *$A = X^*$ for some Banach space X.*

(ii) *Every regular maximal ideal in A is weak* closed.*

Then $M(A)$ is isometrically isomorphic to A.

It should be noted that this theorem does not apply to $A = L_1(G)$ and $B = M(G)$ since the maximal ideals in $M(G)$ are not all weak* closed. Nor does it apply to $A = B = L_p(G)$, $1 < p < \infty$, where G is a compact Abelian group, because these algebras do not possess minimal approximate identities (F.7d).

A second corollary is the following result.

Corollary 1.7.2. *Let A be a semi-simple commutative Banach algebra with minimal approximate identity $\{y_\alpha\}$ and such that*

(i) *$A = X^*$ for some Banach space X.*

(ii) *Every regular maximal ideal in A is weak* closed.*

Suppose $\{T_\alpha\} \subset M(A)$ and denote by $\varphi_\alpha \in C[\Delta(A)]$ those functions for which $(T_\alpha y)\hat{} = \varphi_\alpha \hat{y}$ for all $y \in A$. If the family $\{T_\alpha\}$ is uniformly bounded and $\{\varphi_\alpha\}$ converges pointwise to φ then there exists a $T \in M(A)$ such that $(Ty)\hat{} = \varphi\,\hat{y}$ for all $y \in A$.

Proof. As before we may assume that $\|y_\alpha\|=1$. Then $\{T_\alpha y_\alpha\}$ is a bounded subset of $A=X^*$ and so has a weak* convergent subnet. Call this subnet $\{T_\beta y_\beta\}$ and its limit x. Let $T\in M(A)$ be the multiplier defined by x. Then for each multiplicative linear functional μ on A,

$$\lim_\beta \langle (T_\beta y_\beta)\, y, \mu\rangle = \lim_\beta \langle T_\beta y_\beta, \mu\rangle \langle y, \mu\rangle = \langle x\, y, \mu\rangle = \langle T y, \mu\rangle \quad (y\in A),$$

and if $m=\mu^{-1}(0)$ then

$$\begin{aligned}\lim_\beta \langle (T_\beta y_\beta)\, y, \mu\rangle &= \lim_\beta (T_\beta y_\beta)\hat{}\,(m)\, \hat{y}(m) = \lim_\beta \varphi_\beta\, \hat{y}_\beta(m)\, \hat{y}(m)\\ &= \varphi(m)\, \hat{y}(m) \qquad (y\in A).\end{aligned}$$

Thus $(T y)\hat{}=\varphi\, \hat{y}$ for all $y\in A$ and the proof is complete. □

1.8. The Derived Algebra. In this section instead of focusing our attention on the algebra A and its multiplier algebra we shall investigate a certain subalgebra of A, called the derived algebra, and its multipliers. We shall show that under certain conditions the multipliers of the derived algebra consist of all bounded continuous functions on the maximal ideal space of the algebra, and that a continuous function which vanishes at infinity always defines a multiplier for the derived algebra. In the following section we shall examine the derived algebra of certain concrete Banach algebras.

Recall that if A is a semi-simple commutative Banach algebra and $T\in M(A)$ then there exists a unique bounded continuous function φ on $\Delta(A)$ such that $(T x)\hat{}=\varphi\, \hat{x}$ for all $x\in A$. In general the set of such functions φ, which we denoted by $\mathcal{M}(A)$, is not all of $C[\Delta(A)]$, the algebra of bounded continuous functions on $\Delta(A)$. Nor does $\mathcal{M}(A)$ generally contain $C_0[\Delta(A)]$. Given a semi-simple commutative Banach algebra A we shall denote by A_0 the set of all $x\in A$ such that $\psi\, \hat{x}\in \hat{A}$ for all $\psi\in C_0[\Delta(A)]$. Clearly A_0 is a subalgebra of A. We call A_0 the *derived algebra* of A.

When A possesses a minimal approximate identity then A_0 can be made into a Banach algebra. To do this we need the following lemma. We note first that when A is semi-simple then $\|x\|_\omega=\|\hat{x}\|_\infty$ defines a norm on A which induces a topology on A which is weaker than the norm topology. A_ω shall denote A with this topology.

Lemma 1.8.1. *Let A be a semi-simple self-adjoint commutative Banach algebra. Then the following are equivalent:*

(i) $x\in A_0$.

(ii) $\|x\|_0=\sup\limits_y \|x\, y\|/\|\hat{y}\|_\infty<\infty$.

(iii) *The mapping from $A_\omega\to A$ defined by $y\to x\, y$ is continuous.*

Proof. The equivalence of (ii) and (iii) is immediate. If x satisfies (iii) then the mapping $\hat{y} \to x\,y$ from $\hat{A}$ to A is continuous and hence can be extended to a continuous mapping from the completion of $\hat{A}$ to A. Since A is self-adjoint and semi-simple the completion of $\hat{A}$ is all of $C_0[\Delta(A)]$ (Theorems E.2.2 and E.7.1), and hence $x \in A_0$.

Finally, suppose $x \in A_0$. Denote by $T_\psi x$ that element in A such that $(T_\psi x)\hat{} = \psi\,\hat{x}$ for each $\psi \in C_0[\Delta(A)]$. Clearly $\psi \to T_\psi x$ is a linear mapping from $C_0[\Delta(A)]$ to A. Let $\lim_n \|\psi_n - \psi\|_\infty = 0$ and $\lim_n \|T_\psi x - z\|_\infty = 0$. Then

$$\begin{aligned}\|(T_\psi x)\hat{} - \hat{z}\|_\infty &\leqq \|(T_\psi x)\hat{} - (T_{\psi_n} x)\hat{}\|_\infty + \|(T_{\psi_n} x)\hat{} - \hat{z}\|_\infty \\ &\leqq \|\hat{x}\|_\infty \|\psi_n - \psi\|_\infty + \|T_{\psi_n} x - z\|.\end{aligned}$$

Hence $(T_\psi x)\hat{} = \hat{z}$. From the semi-simplicity of A we conclude that $T_\psi x = z$, and an application of the Closed Graph Theorem (Theorem D.6.1) shows that the mapping $\psi \to T_\psi x$ is continuous.

In particular, there exists a constant K such that $\|x\,y\| = \|T_{\hat{y}} x\| \leqq K\|\hat{y}\|_\infty$ for all $y \in A$. Therefore $\|x\|_0 < \infty$. ▯

Theorem 1.8.1. *Let A be a semi-simple self-adjoint commutative Banach algebra with a minimal approximate identity. Then $\|\cdot\|_0$ defines a complete norm for A_0 under which A_0 is a Banach algebra.*

Proof. It is obvious that A_0 is a normed linear space under $\|\cdot\|_0$. We also note that by Theorem 1.1.4

$$\|x\|_0 = \sup_y \|x\,y\|/\|\hat{y}\|_\infty \geqq \sup_y \|x\,y\|/\|y\| = \|L_x\| = \|x\|$$

since A has a minimal approximate identity. Thus for $x, z \in A_0$,

$$\|x\,z\|_0 = \sup_y \|x\,z\,y\|/\|\hat{y}\|_\infty \leqq \|x\|\,\|z\|_0 \leqq \|x\|_0\,\|z\|_0,$$

and A_0 is a normed algebra.

Suppose $\{x_n\} \subset A_0$ is a Cauchy sequence. Since $\|x\| \leqq \|x\|_0$ it follows that there exists an $x \in A$ such that $\lim_n \|x_n - x\| = 0$. Given $\varepsilon > 0$ there exists a positive integer N such that $\|x_n - x_m\|_0 \leqq \varepsilon$ for $n, m \geqq N$, that is,

$$\|x_n\,y - x_m\,y\|/\|\hat{y}\|_\infty \leqq \varepsilon \qquad (n, m \geqq N,\ y \in A).$$

Thus for each $n \geqq N$ and each $y \in A$ we see, utilizing the fact that $\lim_m \|x_m - x\| = 0$, that

$$\|x_n\,y - x\,y\|/\|\hat{y}\|_\infty \leqq \varepsilon \qquad (n \geqq N,\ y \in A).$$

Hence $x_n - x \in A_0$ and $\|x_n - x\|_0 \leqq \varepsilon$ for $n \geqq N$. Moreover, $x = x_N - (x_N - x) \in A_0$ as A_0 is a vector space.

Therefore A_0 is complete. ▯

The assumption that A possesses a minimal approximate identity is made primarily so that the statement of the results on derived algebras is consistent with the previous development. It is important to note that this assumption is used only to establish the inequality $\|x\|_0 \geqq \|x\|$, and that this result is valid under a slightly weaker hypothesis. Indeed, suppose A possesses a net $\{x_\alpha\}$ such that $\lim_\alpha \|x\,x_\alpha - x\| = 0$ for each $x \in A$ and $\lim_\alpha \|\hat{x}_\alpha\|_\infty = 1$. One can then also assume without loss of generality that $\|\hat{x}_\alpha\|_\infty = 1$. If $x \in A_0$ and $\varepsilon > 0$ we choose α_0 such that $\|x\,x_{\alpha_0}\| > \|x\| - \varepsilon$. Then we have

$$\|x\|_0 = \sup_y \|x\,y\| / \|\hat{y}\|_\infty \geqq \sup_\alpha \|x\,x_\alpha\| \geqq \|x\,x_{\alpha_0}\| > \|x\| - \varepsilon.$$

Since ε is arbitrary it follows that $\|x\|_0 \geqq \|x\|$.

We bring these facts to the reader's attention since in the next section we shall study, for compact G, the derived algebra of $L_p(G)$, $1 < p < \infty$, and these algebras do not have minimal approximate identities. However they do possess approximate identities as described in the previous paragraph (F.7d). Thus the derived algebra for these algebras is well-defined.

Let $m_0 \in \Delta(A_0)$ and denote by μ_0 the multiplicative linear functional defined by m_0. For $x \in A_0$ such that $\langle x, \mu_0 \rangle \neq 0$ set $\langle y, \mu \rangle = \langle y\,x, \mu_0 \rangle / \langle x, \mu_0 \rangle$. The right-hand side makes sense since it is evident that A_0 is an ideal in A. Moreover by essentially the same argument as given in the proofs of Theorems 1.4.1, 1.4.2 and 1.4.3, we see that the definition is independent of the choice of x and the correspondence $\mu_0 \to \mu$ defines a bijective mapping from $\Delta(A_0)$ to $\Delta(A) \sim H(A_0)$, where

$$H(A_0) = \bigcap_{x \in A_0} \{m \mid m \in \Delta(A),\ \hat{x}(m) = 0\}$$

is the hull of A_0 in A. If we denote the Gelfand transform of an element $x \in A_0$ determined by $\Delta(A_0)$ as $\mathring{x}$ then, as before, $\mathring{x}(m_0) = \langle x, \mu_0 \rangle = \langle x, \mu \rangle = \hat{x}(m)$, and the mapping $m_0 \to m$ defines a homeomorphism from $\Delta(A_0)$ to the open subset $\Delta^0(A_0) = \Delta(A) \sim H(A_0)$ of $\Delta(A)$ in either the weak* or hull-kernel topologies. The previous construction also shows that when $A_0 \neq \{0\}$ then $m \in \Delta^0(A_0)$ if and only if there exists an $x \in A_0$ such that $\hat{x}(m) \neq 0$. Moreover, as in Theorem 1.4.3, if one uses the hull-kernel topology then $\Delta^0(A_0)$ is dense in $\Delta(A)$.

We summarize these results in the next theorem.

Theorem 1.8.2. *Let A be a semi-simple self-adjoint commutative Banach algebra with a minimal approximate identity. Then $\Delta(A) = H(A_0) \cup \Delta^0(A_0)$ where $H(A_0)$ is the hull of A_0 in A and $\Delta^0(A_0)$ is an open subset of $\Delta(A)$ which is homeomorphic to $\Delta(A_0)$ when either the weak* or hull-kernel* *topologies are given to* $\Delta(A)$ *and* $\Delta(A_0)$. *Moreover in the hull-kernel topologies* $\Delta^0(A_0)$ *is dense in* $\Delta(A)$.

The next theorem collects some elementary results about the derived algebra, one of which says that every function in $C_0[\Delta(A_0)]$ defines a multiplier for A_0. $\mathring{A}_0 = \{\mathring{x} \mid x \in A_0\}$.

Theorem 1.8.3. *Let A be a semi-simple self-adjoint commutative Banach algebra with a minimal approximate identity. Then:*

(i) A_0 *is semi-simple and self-adjoint.*
(ii) $\mathring{A}_0$ *is an ideal in* $C_0[\Delta(A_0)]$.
(iii) A_0 *is regular.*
(iv) $C_c[\Delta(A_0)] \subset \mathring{A}_0$.
(v) *The mapping* $\mathring{x} \to \hat{x}$ *is an isometric isomorphism of* $\mathring{A}_0$ *into* $\hat{A}$ *with the usual supremum norms on* $\mathring{A}_0$ *and* $\hat{A}$.

Proof. The self-adjointness of A_0 is obvious and the semi-simplicity follows from the preceding theorem.

Let $x \in A_0$ and $\psi \in C_0[\Delta(A_0)]$. Considering $\Delta(A_0)$ as an open subset of $\Delta(A)$ we can extend ψ to a function $\tilde{\psi}$ in $C_0[\Delta(A)]$ by defining $\tilde{\psi} = \psi$ on $\Delta(A_0)$ and $\tilde{\psi} \equiv 0$ on $\Delta(A) \sim \Delta(A_0)$. Then $\hat{x}\tilde{\psi} \in \hat{A}$ and for each $\chi \in C_0[\Delta(A)]$ we also have $\hat{x}\tilde{\psi}\chi \in \hat{A}$. From the definition of A_0 it follows that there exists a $y \in A_0$ for which $\hat{y} = \hat{x}\tilde{\psi}$, and hence $\mathring{y} = \mathring{x}\psi \in \mathring{A}_0$. Therefore $\mathring{A}_0$ is an ideal in $C_0[\Delta(A_0)]$.

Since $\mathring{A}_0$ is an ideal in $C_0[\Delta(A_0)]$, which is a regular algebra with regular maximal ideal space $\Delta(A_0)$, we conclude at once that A_0 is regular.

Let $\psi \in C_c[\Delta(A_0)]$ and suppose ψ has compact support K. Since A_0 is self-adjoint there exists an $x \in A_0$ such that $\mathring{x} > 0$ on K. Let $U \supset K$ be an open set on which $\mathring{x} > 0$. Choose $\chi \in C_c[\Delta(A_0)]$ such that $\chi \equiv 0$ on $\Delta(A_0) \sim U$, $\chi \equiv 1$ on K and $|\chi| \leqq 1$. Clearly $\varphi = \chi/\mathring{x} \in C_0[\Delta(A_0)]$, $\varphi = 1/\mathring{x}$ on K and $\psi = \mathring{x}\varphi\psi$. Thus since $\mathring{A}_0$ is an ideal in $C_0[\Delta(A_0)]$ it follows immediately that $\psi \in \mathring{A}_0$, that is, $C_c[\Delta(A_0)] \subset \mathring{A}_0$.

Finally, let $x \in A_0$. By Theorem 1.8.2,

$$\|\mathring{x}\|_\infty = \sup_{m_0 \in \Delta(A_0)} |\mathring{x}(m_0)| = \sup_{m \in \Delta^0(A_0)} |\hat{x}(m)| = \sup_{m \in \Delta(A)} |\hat{x}(m)| = \|\hat{x}\|_\infty . \quad \square$$

Next we wish to investigate several situations in which $\mathring{A}_0$ is not only an ideal in $C_0[\Delta(A_0)]$ but is also an ideal in $C[\Delta(A_0)]$, that is, instances where $M(A_0)$ can be identified with all of $C[\Delta(A_0)]$.

We begin by establishing an important lemma.

Lemma 1.8.2. *Let A be a semi-simple self-adjoint commutative Banach algebra with a minimal approximate identity. If* $x^* \in A_0^*$ *is a continuous linear functional then there exists a unique complex valued regular Borel* *measure* μ_{x^*} *on* $\Delta(A_0)$ *such that*

$$\langle x\,y, x^*\rangle = \int_{\Delta(A_0)} \mathring{x}\,\mathring{y}(m_0)\, d\mu_{x^*}(m_0) \qquad (x, y\in A_0).$$

Moreover, if $x\in A_0$ *then* $\mathring{x}$ *is integrable with respect to* μ_{x^*}, *and if* $\mathring{x}$ *has compact support then*

$$\langle x, x^*\rangle = \int_{\Delta(A_0)} \mathring{x}(m_0)\, d\mu_{x^*}(m_0).$$

Proof. For $x, y\in A_0$ we see, using part (v) of the previous theorem, that

$$\|x\,y\|_0 = \sup_z \|x\,y\,z\|/\|\hat{z}\|_\infty \leqq \sup_z \|x\,y\,z\|\;\|\hat{y}\|_\infty/\|\hat{y}\,\hat{z}\|_\infty \leqq \|x\|_0\,\|\hat{y}\|_\infty$$
$$= \|x\|_0\,\|\mathring{y}\|_\infty .$$

Thus for $x^*\in A_0^*$ we have

$$|\langle x\,y, x^*\rangle| \leqq \|x^*\|\;\|x\,y\|_0 \leqq \|x^*\|\;\|x\|_0\,\|\mathring{y}\|_\infty .$$

Hence if $x^*\in A_0^*$ is fixed then each $x\in A_0$ defines a continuous linear functional on $\mathring{A}_0$ whose value at $\mathring{y}$ is $\langle x\,y, x^*\rangle$. However since A_0 is self-adjoint, $\mathring{A}_0$ is norm dense in $C_0[\Delta(A_0)]$ (Theorems E.2.2 and E.7.1) and hence the functional can be extended uniquely to all of $C_0[\Delta(A_0)]$.

Thus there exists a unique $\mu_x\in M[\Delta(A_0)]$ (Theorem D.9.4) such that

$$\text{(1)} \qquad \langle x\,y, x^*\rangle = \int_{\Delta(A_0)} \mathring{y}(m_0)\, d\mu_x(m_0) \qquad (y\in A_0),$$

and $\|\mu_x\| \leqq \|x^*\|\;\|x\|_0$.

Moreover for each $x, z\in A_0$ and all $y\in A_0$, it is evident that

$$\int_{\Delta(A_0)} \mathring{y}\,\mathring{z}(m_0)\, d\mu_x(m_0) = \langle x\,y\,z, x^*\rangle = \langle z\,y\,x, x^*\rangle$$
$$= \int_{\Delta(A_0)} \mathring{y}\,\mathring{x}(m_0)\, d\mu_z(m_0).$$

Consequently by the uniqueness of the measures we conclude that $\mathring{x}\,\mu_z = \mathring{z}\,\mu_x$ for each $x, z\in A_0$.

For each $x\in A_0$ define $S_x = \{m_0 \mid \mathring{x}(m_0) \neq 0\}$. Clearly S_x is an open subset of $\Delta(A_0)$ and thus is a locally compact subspace of $\Delta(A_0)$. For every $\psi\in C_c(S_x)$ the integral

$$\text{(2)} \qquad \int_{S_x} \psi(m_0)/\mathring{x}(m_0)\, d\mu_x(m_0)$$

exists. If K is a compact subset of S_x denote by $C_c^K(S_x)$ all those functions in $C_c(S_x)$ whose support lies in K. It is evident that for each compact K the integral (2) defines a continuous linear functional on $C_c^K(S_x)$ in the topology of uniform convergence, and hence defines a continuous linear functional in the inductive limit topology on $C_c(S_x)$ (D.1 and D.9).

Therefore there exists a unique complex valued regular Borel measure $\mu_{x^*}^x$ on S_x (Theorem D.9.5) such that

$$\int_{S_x} \psi(m_0)/\mathring{x}(m_0)\, d\mu_x(m_0) = \int_{S_x} \psi(m_0)\, d\mu_{x^*}^x(m_0) \qquad (\psi \in C_c(S_x)).$$

Obviously $\{S_x \mid x \in A_0\}$ forms an open covering of $\Delta(A_0)$, and for each $x, z \in A_0$ the equation $\mathring{x}\,\mu_z = \mathring{z}\,\mu_x$ shows that the measures $\mu_{x^*}^x$ and $\mu_{x^*}^z$ have identical restrictions to the set $S_x \cap S_z$. Consequently (C.3) there exists a unique complex valued regular Borel measure μ_{x^*} on $\Delta(A_0)$ whose restriction to each S_x is $\mu_{x^*}^x$.

Now from (1),

$$\langle x\,x, x^* \rangle = \int_{\Delta(A_0)} \mathring{x}(m_0)\, d\mu_x(m_0) = \int_{S_x} \mathring{x}(m_0)\, d\mu_x(m_0) \qquad (x \in A_0),$$

and by (2),

$$(3) \qquad \int_{S_x} \psi(m_0)\, \mathring{x}(m_0)\, d\mu_{x^*}(m_0) = \int_{S_x} \psi(m_0)\, d\mu_x(m_0) \qquad (\psi \in C_c(S_x)).$$

But clearly $\mathring{x}$ is the uniform limit of a sequence $\{\psi_n\} \subset C_c(S_x)$ and so

$$\begin{aligned}\langle x\,x, x^* \rangle &= \int_{S_x} \mathring{x}(m_0)\, d\mu_x(m_0) = \lim_n \int_{S_x} \psi_n(m_0)\, d\mu_x(m_0)\\ &= \lim_n \int_{S_x} \psi_n(m_0)\, \mathring{x}(m_0)\, d\mu_{x^*}(m_0) = \int_{S_x} \mathring{x}\,\mathring{x}(m_0)\, d\mu_{x^*}(m_0)\\ &= \int_{\Delta(A_0)} \mathring{x}\,\mathring{x}(m_0)\, d\mu_{x^*}(m_0).\end{aligned}$$

From the identity $4xy = (x+y)^2 - (x-y)^2$ and the preceding equation we conclude immediately that

$$\langle x\,y, x^* \rangle = \int_{\Delta(A_0)} \mathring{x}\,\mathring{y}(m_0)\, d\mu_{x^*}(m_0) \qquad (x, y \in A_0).$$

If $\mathring{x}$ has compact support then using a similar construction as that employed in the proof of part (iv) of Theorem 1.8.3 we can find a $y \in A_0$ such that $\mathring{x} = \mathring{x}\,\mathring{y}$. Since A_0 is semi-simple $x = x\,y$ and

$$\langle x, x^* \rangle = \langle x\,y, x^* \rangle = \int_{\Delta(A_0)} \mathring{x}\,\mathring{y}(m_0)\, d\mu_{x^*}(m_0) = \int_{\Delta(A_0)} \mathring{x}(m_0)\, d\mu_{x^*}(m_0).$$

Furthermore, the preceding equation, which is valid for all $x \in A_0$ such that $\mathring{x}$ has compact support, and the fact that $C_c[\Delta(A_0)] \subset \mathring{A}_0$ together imply that the measure μ_{x^*} constructed for each $x^* \in A_0^*$ is unique.

Finally we must show that each $\mathring{x} \in \mathring{A}_0$ is integrable with respect to μ_{x^*}, that is, with respect to each component of the Jordan decomposition (Theorem C.1.2) of μ_{x^*}. We note first that if $x^+ \in A_0$ is such that $\mathring{x}^+ = \overline{\mathring{x}}$ then the mapping $x \to x^+$ from A_0 to A_0 is continuous. Consequently the linear functional on A_0 defined by $\langle x, (x^*)^+ \rangle = \overline{\langle x^+, x^* \rangle}$

is continuous. From the uniqueness of the μ_{x^*} we then conclude that $\mu_{(x^*)^+}=\overline{\mu_{x^*}}$ and $\mu_{x^*+y^*}=\mu_{x^*}+\mu_{y^*}$. In particular we can write

$$x=\tfrac{1}{2}(x+x^+-i[i(x-x^+)]) \quad\text{and}\quad \mu_{x^*}=\tfrac{1}{2}(\mu_{x^*+(x^*)^+}-i\,\mu_{i[x^*-(x^*)^+]}).$$

Hence to establish the integrability of $\mathring{x}$ we may assume without loss of generality that $\mathring{x}$ and μ_{x^*} are real valued.

Denote the Jordan decomposition of μ_{x^*} by $\mu_{x^*}=\mu_{x^*}^+-\mu_{x^*}^-$. For each set $E\subset\Delta(A_0)$ which is a countable union of Borel sets with finite μ_{x^*}-measure, we set $\mu_{x^*}^+(E)=\sup\mu_{x^*}^+(K)$ and $\mu_{x^*}^-(E)=\sup\mu_{x^*}^-(K)$ where the suprema are taken over all compact subsets K of E. This is possible as $\mu_{x^*}^+$ and $\mu_{x^*}^-$ are regular. Setting $K_0=\emptyset$ and $K_n=\{m_0\,|\,|\mathring{x}(m_0)|\geqq 1/n\}$ for each positive integer $n\geqq 1$, we see that $K_{n-1}\subset K_n$ and $S_x=\bigcup_{n=1}^{\infty}K_n$. Moreover each K_n is compact since $\mathring{x}\in C_0[\Delta(A_0)]$ by Theorem 1.8.3 (ii). Clearly $\mu_{x^*}^+\left(S_x\sim\bigcup_{n=1}^{\infty}K_n\right)=\mu_{x^*}^-\left(S_x\sim\bigcup_{n=1}^{\infty}K_n\right)=0$. Each $K_n\sim K_{n-1}$ has finite $\mu_{x^*}^+$ and $\mu_{x^*}^-$ measure and can be written according to the Hahn decomposition (Theorem C.1.1) as $K_n\sim K_{n-1}=E_n^+\cup E_n^-$, where E_n^+ and E_n^- are disjoint Borel sets for which $\mu_{x^*}^+(E_n^-)=\mu_{x^*}^-(E_n^+)=0$. Set $E^+=\bigcup_{n=1}^{\infty}E_n^+$ and $E^-=\bigcup_{n=1}^{\infty}E_n^-$.

Suppose $\mathring{x}$ is not integrable with respect to μ_{x^*}. Then either $\int_{\Delta(A_0)}|\mathring{x}(m_0)|\,d\mu_{x^*}^+(m_0)$ or $\int_{\Delta(A_0)}|\mathring{x}(m_0)|\,d\mu_{x^*}^-(m_0)$ is infinite. Assume the former is the case and let $K\subset E^+$ be a compact set such that

(4) $$\int_K|\mathring{x}(m_0)|\,d\mu_{x^*}(m_0)=\int_K|\mathring{x}(m_0)|\,d\mu_{x^*}^+(m_0)\geqq 2\,\|x\|_0\,\|x^*\|.$$

Then choose a continuous function ψ with compact support in E^+ and $\|\psi\|_\infty<2$ for which

(5) $$\left|\int_{\Delta(A_0)}\psi\,\mathring{x}(m_0)\,d\mu_{x^*}(m_0)\right|\geqq\int_K|\mathring{x}(m_0)|\,d\mu_{x^*}(m_0).$$

Since $E^+\subset S_x$ we conclude from (3) that

$$\left|\int_{\Delta(A_0)}\psi\,\mathring{x}(m_0)\,d\mu_{x^*}(m_0)\right|=\left|\int_{\Delta(A_0)}\psi(m_0)\,d\mu_x(m_0)\right|$$
$$\leqq\|\psi\|_\infty\,\|\mu_x\|\leqq\|\psi\|_\infty\,\|x\|_0\,\|x^*\|<2\,\|x\|_0\,\|x^*\|.$$

Whereas from (4) and (5) we have

$$\left|\int_{\Delta(A_0)}\psi\,\mathring{x}(m_0)\,d\mu_{x^*}(m_0)\right|\geqq 2\,\|x\|_0\,\|x^*\|,$$

which provides a contradiction.

Therefore $\mathring{x}$ is integrable with respect to μ_{x^*} for each $x \in A_0$ and each $x^* \in A_0^*$. □

As our first application of this lemma we prove the following theorem.

Theorem 1.8.4. *Let A be a semi-simple self-adjoint commutative Banach algebra with a minimal approximate identity. If the linear span of $A_0^2 = \{x \mid x = yz,\ y, z \in A_0\}$ is norm dense in A_0 then $\mathscr{M}(A_0) = C[\Delta(A_0)]$.*

Proof. We first note that the set of elements $z \in A_0$ for which $\mathring{z}$ has compact support is dense in A_0. Indeed, since $C_c[\Delta(A_0)] \subset \mathring{A}_0$, if $x \in A_0$ and $\varepsilon > 0$ there exists a $y \in A_0$ such that $\mathring{y}$ has compact support and $\|\mathring{x} - \mathring{y}\|_\infty < \varepsilon / \|x\|_0$. But then as indicated in the beginning of the proof of the preceding lemma, $\|xy - x^2\|_0 \leqq \|x\|_0 \|\mathring{y} - \mathring{x}\|_\infty < \varepsilon$. The assertion then follows from the denseness of the linear span of A_0^2 in A_0, the equation $4uv = (u+v)^2 - (u-v)^2$ and the fact that $\mathring{x}\mathring{y}$ has compact support.

Let $\psi \in C[\Delta(A_0)]$. If $x^* \in A_0^*$ and $y \in A_0$ is such that $\mathring{y} \in C_c[\Delta(A_0)]$ then set $\langle y, \beta(x^*)\rangle = \int_{\Delta(A_0)} \mathring{y}\psi(m_0)\, d\mu_{x^*}(m_0)$, where μ_{x^*} is the regular Borel measure constructed in Lemma 1.8.2. For each $y \in A_0$ such that $\mathring{y} \in C_c[\Delta(A_0)]$ let $K(y)$ denote the compact support of $\mathring{y}$. Since $C_c[\Delta(A_0)] \subset \mathring{A}_0$ and A_0 is self-adjoint, if $U(y) \supset K(y)$ is an open set with compact closure then there exists a $z \in A_0$ for which $\mathring{z} \in C_c[\Delta(A_0)]$, $\mathring{z} \equiv 1$ on $U(y)$ and $\|\mathring{z}\|_\infty = 1$. Clearly $\mathring{y} = \mathring{z}\mathring{y}$, and since $\mathring{z}\psi \in C_c[\Delta(A_0)]$ there is a $z_\psi \in A_0$ such that $\mathring{z}_\psi = \mathring{z}\psi$. But then using Lemma 1.8.2,

$$\begin{aligned}\langle y, \beta(x^*)\rangle &= \int_{\Delta(A_0)} \mathring{y}\psi(m_0)\, d\mu_{x^*}(m_0) = \int_{\Delta(A_0)} \mathring{y}\mathring{z}_\psi(m_0)\, d\mu_{x^*}(m_0)\\ &= \langle y z_\psi, x^*\rangle = \int_{\Delta(A_0)} \mathring{z}_\psi(m_0)\, d\mu_y(m_0).\end{aligned}$$

The last equality is valid because of formula (1) in the proof of Lemma 1.8.2. Hence,

$$|\langle y, \beta(x^*)\rangle| \leqq \|\mathring{z}\psi\|_\infty \|\mu_y\| \leqq \|\psi\|_\infty \|x^*\| \, \|y\|_0$$

for each $y \in A_0$ such that $\mathring{y} \in C_c[\Delta(A_0)]$. Clearly then $\beta(x^*)$ defines a bounded linear functional on $\{y \mid y \in A_0,\ \mathring{y} \in C_c[\Delta(A_0)]\}$. In view of the remark at the beginning of the proof it follows that $\beta(x^*)$ can be uniquely extended to all of A_0 without increasing the norm.

The mapping $\beta\colon A_0^* \to A_0^*$ is clearly linear. Moreover, if $\lim_n \|x_n^* - x^*\| = 0$ and $\lim_n \|\beta(x_n^*) - z^*\| = 0$ then for any $y \in A_0$ such that $\mathring{y} \in C_c[\Delta(A_0)]$ we have by Lemma 1.8.2 that

$$|\langle y, \beta(x_n^*)\rangle - \langle y, \beta(x^*)\rangle| = |\langle y_\psi, x_n^* - x^*\rangle| \leqq \|y_\psi\|_0 \, \|x_n^* - x^*\|.$$

Thus $\lim_n \langle y, \beta(x_n^*)\rangle = \langle y, \beta(x^*)\rangle$ and $\lim_n \langle y, \beta(x_n^*)\rangle = \langle y, z^*\rangle$ for all y with $\mathring{y} \in C_c[\Delta(A_0)]$. Consequently $\langle y, \beta(x^*)\rangle = \langle y, z^*\rangle$ for all such y and hence $\beta(x^*) = z^*$. An application of the Closed Graph Theorem (Theorem D.6.1) then reveals that β is a continuous mapping. Denote by β^* the continuous adjoint mapping of the second dual space A_0^{**} to A_0^{**}, and consider A_0 as isometrically embedded in A_0^{**} in the canonical manner (D.6.e).

Now let $x \in A_0$ and choose a sequence $\{x_n\} \subset A_0$ such that $\lim_n \|x_n - x\|_0 = 0$ and $\{\mathring{x}_n\} \subset C_c[\Delta(A_0)]$. Let $z_n \in A_0$ be such that $\mathring{x}_n \psi = \mathring{z}_n$. Then we have for each $x^* \in A_0^*$ that

$$\langle \beta^*(x_n), x^*\rangle = \langle x_n, \beta(x^*)\rangle = \langle z_n, x^*\rangle.$$

Thus $\beta^*(x_n) = z_n$. Since $\lim_n \|x_n - x\|_0 = 0$ and β^* is continuous there exists a $z \in A_0$ such that $\lim_n z_n = \lim_n \beta^*(x_n) = \beta^*(x) = z$.

It is apparent then that $\mathring{z} = \mathring{x}\psi$ and hence $\mathring{x}\psi \in \mathring{A}_0$, that is, $\mathring{A}_0$ is an ideal in $C[\Delta(A_0)]$. □

Our second application of the lemma will be to obtain an explicit representation of $\mathring{A}_0$ for certain algebras A. For these algebras it will again be the case that $\mathring{A}_0$ is an ideal in $C[\Delta(A_0)]$.

Recall that a Banach space B is *weakly complete* if for every sequence $\{x_n\} \subset B$ for which $\{\langle x_n, x^*\rangle\}$ converges for each $x^* \in B^*$ there exists an $x \in B$ such that $\lim_n \langle x_n, x^*\rangle = \langle x, x^*\rangle$ for each $x^* \in B^*$ (D.10).

Theorem 1.8.5. *Let A be a semi-simple self-adjoint commutative Banach algebra with a minimal approximate identity. If A is weakly complete then there exists a family $\{\mu_\alpha\}$ of nonnegative regular Borel measures on $\Delta(A_0)$ such that*

$$\mathring{A}_0 = \bigcap_\alpha \{L_1(\mu_\alpha) \cap C_0[\Delta(A_0)]\}.$$

In particular $\mathscr{M}(A_0) = C[\Delta(A_0)]$.

Before giving the proof we remark that, in this instance, $L_1(\mu_\alpha)$ denotes the space of functions absolutely integrable with respect to μ_α and not equivalence classes of such functions.

Proof. For each $x^* \in A_0^*$ let μ_{x^*} be the measure constructed in Lemma 1.8.2. If $\mu_{x^*} = \mu_{x^*}^1 - \mu_{x^*}^2 + i\mu_{x^*}^3 - i\mu_{x^*}^4$ is the Jordan decomposition (Theorem C.1.2) of μ_{x^*} then each $\mu_{x^*}^i$, $i = 1, 2, 3, 4$, is a nonnegative regular Borel measure on $\Delta(A_0)$. The family of measures $\{\mu_{x^*}^1 + \mu_{x^*}^2 + \mu_{x^*}^3 + \mu_{x^*}^4 | x^* \in A_0^*\}$ will be the desired family of measures $\{\mu_\alpha\}$. It is immediate from the lemma that $\mathring{A}_0 \subset \bigcap_\alpha \{L_1(\mu_\alpha) \cap C_0[\Delta(A_0)]\}$.

To prove the reverse inclusion let $\psi \in C_0[\Delta(A_0)]$ be absolutely integrable with respect to all of the μ_α. Since $C_c[\Delta(A_0)] \subset \mathring{A}_0$ is norm dense in $C_0[\Delta(A_0)]$ there exists a sequence $\{x_n\} \subset A_0$ such that $\{\mathring{x}_n\} \subset C_c[\Delta(A_0)]$, $|\mathring{x}_n(m_0)| \leqq |\psi(m_0)|$ for all $m_0 \in \Delta(A_0)$ and $\lim_n \|\mathring{x}_n - \psi\|_\infty = 0$. Combining the Lebesgue Dominated Convergence Theorem (Theorem C.5.2) and the last part of Lemma 1.8.2, we see that

$$\int_{\Delta(A_0)} \psi(m_0)\, d\mu_{x^*}(m_0) = \lim_n \int_{\Delta(A_0)} \mathring{x}_n(m_0)\, d\mu_{x^*}(m_0) = \lim_n \langle x_n, x^* \rangle$$

for each $x^* \in A_0^*$. But every continuous linear functional on A obviously defines a continuous linear functional on A_0 and hence $\{x_n\}$ is a weakly convergent sequence in A. Consequently there exists a $y \in A$ such that $\lim_n \langle x_n, x^* \rangle = \langle y, x^* \rangle$ for each $x^* \in A^*$. From Theorem 1.8.2 it is evident that $\hat{y}$ restricted to $\Delta(A_0)$ is equal to ψ.

Now for each $\varphi \in C_0[\Delta(A)]$, if $x \in A_0$ we denote by x^φ the element in A_0 such that $(x^\varphi)^\wedge = \hat{x}\varphi$. As in previous proofs an argument using the Closed Graph Theorem shows that the mapping $x \to x^\varphi$ is continuous. Hence the mapping $x \to \langle x^\varphi, x^* \rangle$ defines a continuous linear functional on A_0 for each $x^* \in A_0^*$. In particular, since $\{\mathring{x}_n \varphi\} \subset C_c[\Delta(A_0)]$, an application of the Lebesgue Dominated Convergence Theorem shows that for each $x^* \in A_0^*$,

$$\lim_n \langle x_n^\varphi, x^* \rangle = \lim_n \int_{\Delta(A_0)} \mathring{x}_n \varphi(m_0)\, d\mu_{x^*}(m_0) = \int_{\Delta(A_0)} \psi \varphi(m_0)\, d\mu_{x^*}(m_0).$$

Thus $\{x_n^\varphi\}$ is weakly convergent in A, and so there exists $z \in A$ such that $\lim_n \langle x_n^\varphi, x^* \rangle = \langle z, x^* \rangle$ for each $x^* \in A^*$. Furthermore, since $\Delta(A) \subset A^*$, it is clear for each $m \in \Delta(A)$ that we have

$$\hat{y}\varphi(m) = \lim_n \hat{x}_n(m)\, \varphi(m) = \lim_n (x_n^\varphi)^\wedge(m) = \hat{z}(m),$$

and hence $\hat{y}\varphi \in \hat{A}$.

Therefore $y \in A_0$ and $\mathring{y} = \psi$, which completes the proof that $\mathring{A}_0 = \bigcap_\alpha \{L_1(\mu_\alpha) \cap C_0[\Delta(A_0)]\}$. The assertion about $\mathscr{M}(A_0)$ is now obvious. $\square$

Our final application of Lemma 1.8.2 is also concerned with weakly complete algebras.

Theorem 1.8.6. *Let A be a semi-simple self-adjoint commutative Banach algebra with a minimal approximate identity. If A is separable and weakly complete than $\Delta(A_0)$ is discrete.*

Proof. Because the topology on $\Delta(A)$ is the weak* topology it is apparent that $\Delta(A)$, and hence $\Delta(A_0)$, is first countable. Thus if $n_0 \in \Delta(A_0)$

there exists a sequence of compact sets $\{K_n\}\subset\Delta(A_0)$ such that K_{n+1} is contained in the interior of K_n and $\bigcap_{n=1}^{\infty} K_n=\{n_0\}$. Since A_0 is regular we can find a sequence $\{x_n\}\subset A_0$ for which $\mathring{x}_n\equiv 1$ on K_{n+1}, $\mathring{x}_n\equiv 0$ on $\Delta(A_0)\sim K_n$ and $\|\mathring{x}_n\|_\infty\leqq 2$. Clearly $\{\mathring{x}_n\}$ converges pointwise to the characteristic function $\chi_{\{n_0\}}$ of the set $\{n_0\}$. Thus from the Lebesgue Dominated Convergence Theorem (Theorem C.5.2) and the last portion of Lemma 1.8.2, we conclude that for each $x^*\in A_0^*$,

$$\lim_n \langle x_n, x^*\rangle=\lim_n \int_{\Delta(A_0)} \mathring{x}_n(m_0)\, d\mu_{x^*}(m_0)=\mu_{x^*}(\{n_0\}).$$

Hence $\{x_n\}$ converges weakly in A_0 and *a fortiori* in A. Let $x\in A$ be the weak limit of $\{x_n\}$. Then $\hat{x}=\chi_{\{n_0\}}$ and so $\{n_0\}$ is an open set in $\Delta(A)$.

Therefore by Theorem 1.8.2 the set $\{n_0\}$ is open in $\Delta(A_0)$ and $\Delta(A_0)$ is discrete. ▯

We note that this theorem may fail if A is not weakly complete. Indeed, let μ be a positive regular Borel measure with total mass one on an infinite locally compact space S and consider $A=L_1(\mu)\cap C_0(S)$ under pointwise operations and the norm $\|f\|=\frac{1}{2}[\|f\|_1+\|f\|_\infty]$, where $\|\cdot\|_1$ denotes the usual norm in $L_1(\mu)$. It is easy to see that A satisfies all the hypotheses of the previous theorem except A is not weakly complete as $C_0(S)$ is not (D.10). Moreover it is evident that $A_0=A$ and so $\Delta(A_0)=S$.

1.9. The Derived Algebra for $L_p(G)$, $1\leqq p<\infty$. We wish to study the derived algebras of the classic Banach algebras $L_p(G)$, $1\leqq p<\infty$, where G is a compact Abelian group. It is well known when G is compact that $L_p(G)$, $1\leqq p<\infty$, with the usual convolution product is a semi-simple, self-adjoint commutative Banach algebra with regular maximal ideal space $\hat{G}$ (F.2 and F.5). In view of the comments following Theorem 1.8.1 it makes sense to discuss the derived algebra of $L_p(G)$ as developed in the preceding section even though for $1<p<\infty$ the algebras do not possess minimal approximate identities.

As usual we shall denote the group of continuous characters on G by $\hat{G}$, and the pairing between the elements of G and $\hat{G}$ by (t,γ). Haar measure λ on a compact Abelian group G will always be assumed normalized such that $\lambda(G)=1$.

Theorem 1.9.1. *Let G be a compact Abelian group and $A=L_p(G)$, $1\leqq p<\infty$.*

(i) *If $1\leqq p\leqq 2$ then A_0 is algebraically and topologically isomorphic to $L_2(G)$ and*

$$2^{-\frac{1}{2}}\|f\|_2\leqq\|f\|_0\leqq\|f\|_2 \qquad (f\in A_0).$$

(ii) *For any* p, $1 < p < \infty$, *if* $\varphi \in \hat{A}_0$ *and* ψ *is a function on* $\hat{G}$ *such that* $|\psi(\gamma)| \leqq |\varphi(\gamma)|$, $\gamma \in \hat{G}$, *then* $\psi \in \hat{A}_0$.

Proof. Let $p=1$ and suppose $f \in L_2(G)$, $g \in L_1(G)$. Then applying the Plancherel Theorem (Theorem F.8.2) we see that

$$\|f*g\|_1 \leqq \|f*g\|_2 = \|\hat{f}\hat{g}\|_2 \leqq \|\hat{f}\|_2 \, \|\hat{g}\|_\infty = \|f\|_2 \, \|\hat{g}\|_\infty .$$

Thus $\|f\|_0 \leqq \|f\|_2$, and by Lemma 1.8.1 we conclude that $f \in A_0$, that is, $L_2(G) \subset A_0$.

Conversely, let Γ^n denote the n-torus and μ normalized Haar measure on Γ^n. Elements of Γ^n will be written as $\zeta = (\zeta_1, \zeta_2, \ldots, \zeta_n)$. Suppose $f(t) = \sum_{k=1}^{n} a_k (t, \gamma_k)$ is an arbitrary trigonometrical polynomial on G. Clearly $f \in A_0$ since $f \in L_\infty(G) \subset L_2(G) \subset A_0$ as G is compact. Moreover

$$\begin{aligned}
\int_{\Gamma^n} \left| \sum_{k=1}^{n} a_k \zeta_k \right| d\mu(\zeta) &= \int_G \left[\int_{\Gamma^n} \left| \sum_{k=1}^{n} a_k \zeta_k (t, \gamma_k) \right| d\mu(\zeta) \right] d\lambda(t) \\
&= \int_{\Gamma^n} \left[\int_G \left| \sum_{k=1}^{n} a_k \zeta_k (t, \gamma_k) \right| d\lambda(t) \right] d\mu(\zeta) \\
&\leqq \sup_\zeta \int_G \left| \sum_{k=1}^{n} a_k \zeta_k (t, \gamma_k) \right| d\lambda(t) \\
&= \sup_\zeta \left\| \sum_{k=1}^{n} a_k (\cdot, \gamma_k) * \sum_{j=1}^{n} \zeta_j (\cdot, \gamma_j) \right\|_1 \\
&= \sup_\zeta \left\| f * \sum_{j=1}^{n} \zeta_j (\cdot, \gamma_j) \right\|_1 \Big/ \left\| \left[\sum_{j=1}^{n} \zeta_j (\cdot, \gamma_j) \right]^\wedge \right\|_\infty \\
&\leqq \sup_g \|f*g\|_1 / \|\hat{g}\|_\infty = \|f\|_0 .
\end{aligned}$$

The validity of these equations of course depends on the translation invariance of μ and the orthogonality relations of the characters on G (F.13).

Now utilizing the orthogonality relations of the characters on Γ^n we see, via a straighforward but tedious computation, that

$$\int_{\Gamma^n} \left| \sum_{k=1}^{n} a_k \zeta_k \right|^4 d\mu(\zeta) = 2 \left(\sum_{k=1}^{n} |a_k|^2 \right)^2 - \sum_{k=1}^{n} |a_k|^4 .$$

Moreover, if $h(\zeta) = \sum_{k=1}^{n} a_k \zeta_k$ then Hölders inequality shows that

$$\int_{\Gamma^n} |h(\zeta)|^2 d\mu(\zeta) \leqq \Big[\int_{\Gamma^n} |h(\zeta)| \, d\mu(\zeta) \Big]^{\frac{2}{3}} \Big[\int_{\Gamma^n} |h(\zeta)|^4 d\mu(\zeta) \Big]^{\frac{1}{3}} .$$

Combining these two results we immediately conclude that

$$\int_{\Gamma^n}\left|\sum_{k=1}^{n} a_k\zeta_k\right| d\mu(\zeta) \geqq \left[\int_{\Gamma^n}\left|\sum_{k=1}^{n} a_k\zeta_k\right|^2 d\mu(\zeta)\right]^{\frac{3}{2}} \Big/ \left[\int_{\Gamma^n}\left|\sum_{k=1}^{n} a_k\zeta_k\right|^4 d\mu(\zeta)\right]^{\frac{1}{2}}$$

$$\geqq \left(\sum_{k=1}^{n}|a_k|^2\right)^{\frac{3}{2}} \Big/ \left[2\left(\sum_{k=1}^{n}|a_k|^2\right)^2\right]^{\frac{1}{2}}$$

$$= 2^{-\frac{1}{2}}\left(\sum_{k=1}^{n}|a_k|^2\right)^{\frac{1}{2}}.$$

Consequently, for the trigonometric polynomial f, we have $\|f\|_0 \geqq 2^{-\frac{1}{2}}\|f\|_2$ since $\|f\|_2 = \left(\sum_{k=1}^{n}|a_k|^2\right)^{\frac{1}{2}}$. More generally it then follows that for any $f\in L_2(G)$ *we have*

$$2^{-\frac{1}{2}}\|f\|_2 \leqq \|f\|_0 \leqq \|f\|_2,$$

since the trigonometric polynomials are dense in $L_2(G)$. Hence the norms on $L_2(G)$ and A_0 are equivalent when restricted to $L_2(G)\cap A_0 = L_2(G)$, and $L_2(G)$ is a closed subspace of A_0. Suppose $x^*\in A_0^*$ is such that $\langle f, x^*\rangle = 0$ for all $f\in L_2(G)$. Let μ_{x^*} be the measure given by Lemma 1.8.2 corresponding to x^*. Then for each trigonometric polynomial $f\in L_2(G)$ we would have

$$0 = \langle f, x^*\rangle = \int_{\hat{G}} \hat{f}(\gamma)\, d\mu_{x^*}(\gamma).$$

Since μ_{x^*} is regular and $\hat{G}$ is discrete we conclude that $\mu_{x^*}=0$, that is, $x^*=0$.

Therefore $L_2(G)=A_0$, and the estimates $2^{-\frac{1}{2}}\|f\|_2 \leqq \|f\|_0 \leqq \|f\|_2$, $f\in A_0$, show that the identity map is an algebraic and topological isomorphism.

Suppose $1<p\leqq 2$. Since G is compact $L_2(G)\subset L_p(G)\subset L_1(G)$. If $f\in L_2(G)$ then by Hölder's inequality, the Plancherel Theorem and the fact that $L_2(G)$ is an ideal in $L_1(G)$ we see that for each $g\in L_p(G)$,

$$\|f*g\|_p \leqq \|f*g\|_2 = \|(f*g)\hat{}\,\|_2 \leqq \|f\|_2\,\|\hat{g}\|_\infty.$$

Thus $\|f\|_0 \leqq \|f\|_2$ and so $L_2(G)\subset A_0$.

Conversely, if $f\in A_0$ then applying Hölder's inequality we have

$$\sup_{g\in L_p(G)} \|f*g\|_p/\|\hat{g}\|_\infty \geqq \sup_{g\in L_p(G)} \|f*g\|_1/\|\hat{g}\|_\infty = \sup_{g\in L_1(G)} \|f*g\|_1/\|\hat{g}\|_\infty.$$

Hence f belongs to $[L_1(G)]_0$ and so $f\in L_2(G)$. Thus $A_0 = L_2(G)$. Moreover from the results for $L_1(G)$ we see that

$$2^{-\frac{1}{2}}\|f\|_2 \leqq \|f\|_0 \leqq \|f\|_2 \qquad (f\in A_0).$$

Consequently the identity mapping is an algebraic and topological isomorphism between A_0 and $L_2(G)$.

Now let $1 < p < \infty$ and let p' be such that $1/p + 1/p' = 1$. We first note that if $f \in A_0$ and $h \in L_{p'}(G)$ then $f * h$ has an absolutely convergent Fourier series. To see this define

$$F(g) = \int_G f * g(t)\, h(t^{-1})\, d\lambda(t) \qquad \big(g \in L_p(G)\big).$$

Clearly F is linear, and since $f \in A_0$ by Hölder's inequality we have

$$|F(g)| \leqq \|f * g\|_p \, \|h\|_{p'} \leqq \|f\|_0 \, \|h\|_{p'} \, \|\hat{g}\|_\infty .$$

Thus F defines a bounded linear functional on $\hat{A}$ and hence on all of $C_0(\hat{G})$. Consequently (Theorem D.9.4) there exists a bounded regular Borel measure μ on the discrete group $\hat{G}$ such that

$$\int_G f * g(t)\, h(t^{-1})\, d\lambda(t) = \sum_{\gamma \in G} \mu(\{\gamma\})\, \hat{g}(\gamma) \qquad \big(g \in L_p(G)\big).$$

In particular, for each $\gamma \in \hat{G}$, $(\cdot, \gamma) \in L_p(G)$ and

$$\begin{aligned} \mu(\{\gamma\}) &= \int_G f * (\cdot, \gamma)(t)\, h(t^{-1})\, d\lambda(t) \\ &= \int_G \Big[\int_G f(t s^{-1})(s, \gamma)\, d\lambda(s)\Big]\, h(t^{-1})\, d\lambda(t) = (f * h)\hat{}\,(\gamma), \end{aligned}$$

upon interchanging the order of integration. Since μ is a bounded measure it is obvious that $f * h$ has an absolutely convergent Fourier series.

Suppose $f \in A_0$, $\hat{f} = \varphi$ and ψ is such that $|\psi(\gamma)| \leqq |\varphi(\gamma)|$, $\gamma \in \hat{G}$. In light of the previous remarks it is apparent that for each $h \in L_{p'}(G)$, $\sum_{\gamma \in \hat{G}} |\varphi(\gamma)\, \hat{h}(\gamma)| < \infty$, and hence $\sum_{\gamma \in \hat{G}} |\psi(\gamma)\, \hat{h}(\gamma)| < \infty$. Clearly then $\psi(\gamma)$ vanishes except for countably many $\gamma \in \hat{G}$, say for $\gamma_1, \gamma_2, \gamma_3, \ldots$. For each positive integer n define $g_n(t) = \sum_{i=1}^{n} \psi(\gamma_i)(t, \gamma_i)$. Then for each $h \in L_{p'}(G)$,

$$\begin{aligned} \int_G g_n(t)\, h(t^{-1})\, d\lambda(t) &= \sum_{i=1}^{n} \psi(\gamma_i) \int_G (t^{-1}, \gamma_i)\, h(t)\, d\lambda(t) \\ &= \sum_{i=1}^{n} \psi(\gamma_i)\, \hat{h}(\gamma_i). \end{aligned}$$

It follows at once that $\{g_n\}$ is a weakly convergent sequence in $L_p(G)$. Since $L_p(G)$ is weakly complete (D.10) there is a $g \in L_p(G)$ for which

$$\lim_n \int_G g_n(t)\, h(t^{-1})\, d\lambda(t) = \int_G g(t)\, h(t^{-1})\, d\lambda(t) \qquad \big(h \in L_{p'}(G)\big).$$

It is evident from this that $\hat{g} = \psi$.

Finally, since $f \in A_0$, for any $k \in L_p(G)$ we have $\|f*k\|_p \leqq \|f\|_0 \|\hat{k}\|_\infty$. Thus if $k(t)=\sum \hat{k}(\gamma)(t,\gamma)$ is an arbitrary trigonometric polynomial in $L_p(G)$, a straightforward computation shows that

$$\begin{aligned}\|g*k\|_p &= \|\sum \hat{k}(\gamma)\,\psi(\gamma)(\cdot,\gamma)\|_p = \|f*\sum \hat{k}(\gamma)\,\psi(\gamma)\,[\varphi(\gamma)]^{-1}(\cdot,\gamma)\|_p \\ &\leqq \|f\|_0 \,\|\{\sum \hat{k}(\gamma)\,\psi(\gamma)\,[\varphi(\gamma)]^{-1}(\cdot,\gamma)\}^\wedge\|_\infty \\ &\leqq \|f\|_0 \,\|\psi\,\varphi^{-1}\,\hat{k}\|_\infty \leqq \|f\|_0 \,\|\hat{k}\|_\infty\end{aligned}$$

as $|\psi(\gamma)| \leqq |\varphi(\gamma)|$. But since the trigonometric polynomials are norm dense in $L_p(G)$ (F.7d) we can also conclude that $\|g*k\|_p \leqq \|f\|_0 \|\hat{k}\|_\infty$ for all $k \in L_p(G)$.

Therefore $\|g\|_0 < \infty$, $g \in A_0$ and $\hat{g}=\psi$. □

If G is a noncompact locally compact Abelian group then $L_p(G)$, $1<p<\infty$, is not a Banach algebra so that the theory of derived algebras can not be developed in quite the same way as in this chapter. We shall return to a discussion of this sort of problem in 4.6. Among other results, we shall show there that if G is noncompact then $(L_1(G))_0=\{0\}$.

1.10. Notes. Some of the earliest papers dealing with the theory of multipliers for abstract Banach algebras seem to be those of Wermer [1], Helgason [1] and Foias [1]. Subsequently many people have studied multipliers in such a context, including Birtel [1–3], Ching and Wong [1], Greenleaf [3], Johnson [1, 4], Kellogg [1], Máté [1–5], Reid [1], Wang [1] and Wells [1]. [1]

One of the most general approaches to the notation of multipliers is to be found in Johnson [1] where the starting point of the theory is certain mappings on associative semi-groups. Applications of this development to specific topological algebras are available in Johnson [1, 2] and Reid [1]. Máté [3, 4] has also studied multipliers from a purely algebraic viewpoint.

Theorem 1.1.1 is based on Wang [1]. Other proofs of the continuity of multipliers can be found in Johnson [1, 2, 4], Reid [1], Rieffel [3] and Varopoulos [1]. The main characterization theorem (Theorem 1.2.2) of multipliers as bounded continuous functions is due to Wang [1], although the validity of such a characterization appears to have been realized before his proof was given. The result apparently was first observed by H. Mirkil (Helgason [1], p. 248). The characterization given in Theorem 1.2.4 is due to Birtel [1].

The maximal ideal space of the multiplier algebra has been investigated by Birtel [1], Foias [1], Helgason [1], Máté [4] and Wang [1]. Theorems 1.4.2 and 1.4.3 are based on the work of Wang, and the characterization of the multipliers which the first of these theorems provides

[1] See also Tomiuk and Wong [1].

is due to Máté [4]. The study of the maximal ideal space has been carried furthest by Wang, especially for supremum norm algebras. In particular Theorems 1.5.2, 1.5.3 and 1.5.4 are found in Wang [1]. The material on the Bishop boundary appears to be new.

If A is a supremum norm algebra than $M(A)$ can be identified with a function algebra. Wang [1] has also investigated, in this case, the relationship between the supremum norm topology, the strong operator topology and the topology given by uniform convergence on compact subsets of the regular maximal ideal space $\Delta(A)$ of A. We have not discussed any of these results and the interested reader is referred to Wang [1]. [1]

At about the same time that Wang's work was done, Birtel [1] examined a certain subalgebra of multipliers, namely, the algebra $M_0(A)$ of those multipliers which correspond to continuous functions which vanish at infinity. For these algebras of multipliers he established theorems analogous to the main results of 1.4 and 1.5. In particular he obtained an integral representation for the multipliers which vanish at infinity in terms of integrals involving measures on $\partial\Delta(A)$. The development in 1.3 is also based on a portion of Birtel [1].

Isometric multipliers have been considered by Birtel [2], Greenleaf [3, 4], Parrot [1], Strichartz [2], Wendel [2] and Wermer [1]. Of the latter five papers all except that of Wermer are concerned with isometric multipliers for the spaces $L_p(G)$, $1 \leqq p \leqq \infty$, and their connection with the problem of describing the isomorphisms of such spaces. The characterization of the multipliers for QCG algebras (Theorem 1.6.1) is based on the development in Greenleaf [3]. The results are actually valid, with essentially the same proof, for noncommutative QCG algebras. Greenleaf [3] also uses the group of isometric onto multipliers to give a characterization of all group algebras on any compact group. A different approach to this problem can be found in Rieffel [1]. Theorems 1.6.3 and 1.6.4 are modeled after a similar development in Birtel [2]. The theorems are established in this paper, however, only under the assumption that $I(A)$ is a compact topological group in the weak operator topology. This is the case, for example, whenever the unit ball in $M(A)$ is compact in the weak operator topology.

Birtel [2] has also proved an extension of Theorem 0.3.4 cited in the notes to the preceding chapter regarding the relationship between isomorphisms of group algebras and their multiplier algebras. To be precise, he showed that if there is an algebra isomorphism between two semi-simple commutative Banach algebras A and B then the isomorphism can be extended to one between $M(A)$ and $M(B)$. Conversely, if A and B are regular semi-simple Tauberian commutative Banach algebras then an algebra isomorphism of $M_0(A)$ onto $M_0(B)$ induces an algebra

[1] See also Taylor [2].

isomorphism of A onto B provided that A and B are norm closed in $M(A)$ and $M(B)$, respectively. The latter assertion is not generally valid if $M_0(A)$ and $M_0(B)$ are replaced by $M(A)$ and $M(B)$. An algebra A is *Tauberian* if $\{x \mid x \in A, \hat{x} \in C_c[\Delta(A)]\}$ is norm dense in A. The reader is referred to Birtel [2] for the details.

The description of the multipliers of certain Banach algebras which are dual spaces is based on Mátё [1], while the discussion of the derived algebra is taken from Helgason [1]. An interesting alternative interpretation of a portion of Theorem 1.9.1 has been discussed by Helgason [1, 2] and Sakai [1]. Let G be a locally compact Abelian group and let $T: L_1(G) \to L_1(G)$ be a linear mapping which commutes with translations and such that there exists a constant $B > 0$ for which $\|Tf\|_1 \leqq B \|\hat{f}\|_\infty$ for all $f \in L_1(G)$. That is, T is a spectrally continuous multiplier of $L_1(G)$. It is then of course also a multiplier of the algebra $L_1(G)$ as discussed previously. The first part of Theorem 1.9.1 says that the spectrally continuous multipliers of $L_1(G)$, G a compact Abelian group, are precisely those operators given by convolution with an element of $L_2(G)$. This was proved for arbitrary compact groups in Helgason [2]. If G is a noncompact group then the only spectrally continuous multiplier on $L_1(G)$ is the zero multiplier. For noncompact Abelian groups this is proven in Figà-Talamanca and Gaudry [2], and for noncompact non-Abelian groups in Sakai [1]. The same results but with various restrictions on the group G are available in Gaudry [2], [6], VI.7, Grothendieck [1] and Helgason [1, 2]. We shall return to a further consideration of the derived algebra of $L_1(G)$ and its extension to $L_p(G)$ in 4.6 and 5.5. Another description of the derived algebra for $L_p(G)$, $1 \leqq p < \infty$, G a compact Abelian group, is given in Bachelis [1]. He shows that $(L_p(G))_0$ coincides with the ideal of elements in $L_p(G)$ whose Fourier series are unconditionally convergent in the L_p-norm.

Several authors have studied multipliers by considering the multiplier algebras as a commutative extension of A. For example, this is the case in Birtel [3], Foias [1] and Máté [2].

Chapter 2

The Multipliers for Commutative H^*-Algebras

2.0. Introduction. An *H^*-algebra* is a Banach algebra A with involution* which is a Hilbert space under a scalar product $\langle \cdot, \cdot \rangle$ such that a) $\|x\| = \sqrt{\langle x, x \rangle}$, that is, the Hilbert space norm agrees with the Banach algebra norm, b) $\|x^*\| = \|x\|$, c) $x^* x \neq 0$ if $x \neq 0$ and d) $\langle x y, z \rangle = \langle y, x^* z \rangle$ for all $x, y, z \in A$. The standard example of an H^*-algebra is the algebra $L_2(G)$ for a compact group G with the usual convolution multiplication and scalar product. A general discussion of H^*-algebras can be found in Loomis [1] and Naimark [1].

We wish to apply some of the general theory of the previous chapter to give a description of the multipliers for commutative H^*-algebras. For such algebras it will be shown that $M(A)$ is isometrically isomorphic to $C[\Delta(A)]$. The proof is an application of the general characterization theorem for multipliers established in 1.2 and the structure theory for H^*-algebras. A characterization of the compact multipliers for H^*-algebras will also be given.

2.1. Multipliers for Commutative H^*-Algebras. First we shall call to mind some basic facts about H^*-algebras. If A is an H^*-algebra then $x \in A$ is *self-adjoint* if $x^* = x$, and it is an idempotent if $x^2 = x$. If $e \in A$ is an idempotent then e is said to be *reducible* if there exists nonzero idempotents $e_1, e_2 \in A$ such that $e = e_1 + e_2$ and $e_1 e_2 = e_2 e_1 = 0$. If e is an idempotent which is not reducible then it is called *irreducible*. Two immediate consequences of the defining conditions of an H^*-algebra are that $\langle x^*, y^* \rangle = \langle y, x \rangle$ and $\langle x y, z \rangle = \langle x, z y^* \rangle$ for all $x, y, z \in A$.

The structure theory of *commutative* H^*-algebras tells us that if A is such an algebra then every closed minimal ideal n_α in A is one-dimensional and consists of the scalar multiples of some irreducible self-adjoint idempotent e_α. Moreover, the collection of all such irreducible self-adjoint idempotents $\{e_\alpha\}$ forms a complete orthogonal system in the Hilbert space of A. If n_α is a minimal closed ideal in A then its orthogonal complement m_α is a regular maximal ideal in A, and all regular maximal ideals are obtained in this fashion. Clearly there exists a one-to-one correspondence between points of $\Delta(A)$ and the family $\{e_\alpha\}$. Since $\hat{e}_\alpha=\chi_{\{m_\alpha\}}$, it follows that $\Delta(A)$ is discrete. Thus one may also consider $\{e_\alpha\}$ to be the regular maximal ideal space for the H^*-algebra A. The correspondence between regular maximal and minimal closed ideals also reveals that $\hat{x}(e_\alpha)=\hat{x}(m_\alpha)=\langle x, e_\alpha\rangle/\|e_\alpha\|^2$ for each $x\in A$. Since $\langle x^*, y^*\rangle=\langle y, x\rangle$ for each x, $y\in A$ the preceding equation shows that $\hat{x}^*(e_\alpha)=\langle x^*, e_\alpha\rangle/\|e_\alpha\|^2=\langle x^*, e_\alpha^*\rangle/\|e_\alpha\|^2=\overline{\langle x, e_\alpha\rangle}/\|e_\alpha\|^2=\overline{\hat{x}(e_\alpha)}$, where the bar denotes complex conjugation. Hence $\hat{x}^*=\bar{\hat{x}}$ for each $x\in A$.

The proofs of these results can be found in Loomis [1], 27, and Naimark [1], V. 25.5.

A $*$-*isomorphism* between two algebras with involutions is an isomorphism which preserves the involution, that is, if $x\to y$ then $x^*\to y^*$. In an algebra of complex valued functions under pointwise operations the natural involution is the one provided by complex conjugation, and in an algebra of bounded linear operators on a Hilbert space the natural involution is the adjoint operation, that is, if T is a bounded linear operator then T^* is the bounded linear operator such that $\langle Tx, y\rangle=\langle x, T^* y\rangle$ for each x, y in the Hilbert space.

If A is an H^*-algebra then A is semi-simple (Naimark [1], V.25.5), and so $M(A)$ is an algebra of bounded linear operators on A. For a commutative H^*-algebra A the semi-simplicity of the algebra follows immediately from Parseval's formula, the completeness of the orthonormal system $\{e_\alpha/\|e_\alpha\|\}$ and the equation $\hat{x}(e_\alpha)=\langle x, e_\alpha\rangle/\|e_\alpha\|^2$. Moreover, the adjoint operation defines an involution on $M(A)$. To see this it suffices to show that if $T\in M(A)$ then $T^*\in M(A)$. Since A is an H^*-algebra for each x, y, $z\in A$ we see that

$$\begin{aligned}\langle (T^* x)\, y, z\rangle &=\langle T^* x, z\, y^*\rangle=\langle x, T(z\, y^*)\rangle=\langle x, (Tz)\, y^*\rangle=\langle x\, y, Tz\rangle\\ &=\langle y, x^*(Tz)\rangle=\langle y, T(x^* z)\rangle=\langle T^* y, x^* z\rangle=\langle x(T^* y), z\rangle\end{aligned}$$

because T is a multiplier. Thus $(T^* x)\, y=x(T^* y)$ for all x, $y\in A$, and T^* is a multiplier. Consequently when A is an H^*-algebra $M(A)$ is a Banach algebra with involution.

We are now in a position to state and prove the result announced in the introduction.

Theorem 2.1.1. *Let A be a commutative H^*-algebra. Then $M(A)$ is isometrically $*$-isomorphic to $C[\Delta(A)]$.*

Proof. Let $\{e_\alpha\}$ be the complete orthogonal family of irreducible self-adjoint idempotents in A. Set $E=\{e'_\alpha \mid e'_\alpha=e_\alpha/\|e_\alpha\|\}$. Then E forms a complete orthonormal system for A and we may, without loss of generality, identify $\Delta(A)$ with the discrete space E. Having done this it follows immediately that $\hat{x}(e'_\alpha)=\langle x, e'_\alpha\rangle/\|e_\alpha\|$ for each $x\in A$.

Corollary 1.2.1 shows that there exists an algebra isomorphism from $M(A)$ onto $\mathscr{M}(A) \subset C(E)$ where the image of $T \in M(A)$ is the function $\varphi \in C(E)$ such that $(Tx)^\wedge = \varphi\,\hat{x}$ for each $x \in A$. Moreover $\|\varphi\|_\infty \leqq \|T\|$. On the other hand, if $\varphi \in C(E)$ then for each $x \in A$ define $Tx = \sum_\alpha \langle x, e'_\alpha \rangle \varphi(e'_\alpha)\, e'_\alpha$. Tx is an element of A since $\{e'_\alpha\}$ is a complete orthonormal system and φ is bounded (D.5). Furthermore $(Tx)^\wedge(e'_\beta) = \langle Tx, e'_\beta \rangle / \|e_\beta\| = \langle x, e'_\beta \rangle \varphi(e'_\beta)/\|e_\beta\| = \varphi(e'_\beta)\,\hat{x}(e'_\beta)$. Thus φ defines a multiplier for A and $\mathscr{M}(A) = C(E)$.

If $T \in M(A)$ and $(T^* x)^\wedge = \varphi^*\,\hat{x}$ then the equations

$$\begin{aligned} \varphi^*(e'_\alpha) &= \varphi^*(e'_\alpha)\,\|e_\alpha\|\,\hat{e}'_\alpha(e'_\alpha) = \|e_\alpha\|\,(T^* e'_\alpha)^\wedge(e'_\alpha) = \langle T^* e'_\alpha, e'_\alpha \rangle \\ &= \langle e'_\alpha, Te'_\alpha \rangle = \overline{\langle Te'_\alpha, e'_\alpha \rangle} = \|e_\alpha\|\,\overline{(Te'_\alpha)^\wedge(e'_\alpha)} = \overline{\varphi(e'_\alpha)} \end{aligned}$$

show that the isomorphism from $M(A)$ onto $C(E)$ is a $*$-isomorphism.

Given $T \in M(A)$, we established above that $\langle Tx, e'_\alpha \rangle = \langle x, e'_\alpha \rangle \varphi(e'_\alpha)$ for each $x \in A$ and all α. Using this fact and the complete orthonormality of $\{e'_\alpha\}$ (D.5), we have for each $x \in A$ that

$$\begin{aligned} \|Tx\|^2 &= \sum_\alpha |\langle Tx, e'_\alpha \rangle|^2 = \sum_\alpha |\langle x, e'_\alpha \rangle \varphi(e'_\alpha)|^2 \leqq (\|\varphi\|_\infty)^2 \sum_\alpha |\langle x, e'_\alpha \rangle|^2 \\ &= (\|\varphi\|_\infty)^2\,\|x\|^2. \end{aligned}$$

Hence $\|Tx\| \leqq \|\varphi\|_\infty\,\|x\|$ for each $x \in A$, that is, $\|T\| \leqq \|\varphi\|_\infty$.

Therefore the $*$-isomorphism between $M(A)$ and $C(E) = C[\Delta(A)]$ is an isometry. ▯

As a special case of this theorem we note that if G is a compact Abelian group then $M[L_2(G)]$ is isometrically $*$-isomorphic to $C(\hat{G})$. Of course this can be proved more directly by means of the Plancherel Theorem.

2.2. Compact Multipliers for Commutative H^*-Algebras. In this section we shall give a characterization of those multipliers for a commutative H^*-algebra A which are compact operators, that is, multipliers which map bounded subsets of A onto subsets with compact closure. It will be seen that the compact multipliers correspond precisely to the elements of $C_0[\Delta(A)]$.

For a commutative H^*-algebra A we shall denote by $K(A)$ the *set of all linear compact operators* from A to A. $K(A)$ is a closed two-sided ideal in $E(A)$, the Banach algebra of all bounded linear operators from A to A (E.8). Thus $K(A) \cap M(A)$ is a closed subalgebra of $M(A)$. Clearly every linear operator on A with a finite dimensional range is compact.

Theorem 2.2.1. *Let A be a commutative H^*-algebra. Then $K(A)\cap M(A)$ is isometrically $*$-isomorphic to $C_0[\Delta(A)]$.*

Proof. From Theorem 2.1.1 it is apparent that $K(A)\cap M(A)$ is isometrically $*$-isomorphic to some subalgebra of $C[\Delta(A)]$. We need only show that this subalgebra is $C_0[\Delta(A)]$. As in the proof of the preceding theorem we shall identify $\Delta(A)$ with the discrete space $E=\{e'_\alpha \mid e'_\alpha = e_\alpha/\|e_\alpha\|\}$ where $\{e_\alpha\}$ forms a complete orthogonal family of irreducible self-adjoint idempotents in A.

Let $T\in M(A)$ and suppose $\varphi\in C_c(E)$ where $(Tx)^\wedge=\varphi\hat{x}$. Since E is discrete φ has finite support, say $e'_{\alpha_1}, e'_{\alpha_2}, \ldots, e'_{\alpha_n}$. Recalling from Theorem 2.1.1 that $\langle Tx, e'_\alpha\rangle = \langle x, e'_\alpha\rangle\, \varphi(e'_\alpha)$, we see for each $x\in A$ that

$$Tx=\sum_\alpha \langle Tx, e'_\alpha\rangle\, e'_\alpha = \sum_{i=1}^{n} \langle x, e'_{\alpha_i}\rangle\, \varphi(e'_{\alpha_i})\, e'_{\alpha_i}.$$

Thus for each $x\in A$ the image Tx is contained in the direct sum of the n one-dimensional minimal closed ideals generated by $e'_{\alpha_1}, e'_{\alpha_2}, \ldots, e'_{\alpha_n}$. Consequently the range of T is finite dimensional, and hence T is compact.

Therefore the image of $K(A)\cap M(A)$ under the isometric $*$-isomorphism of $M(A)$ onto $C(E)$ contains the subalgebra $C_c(E)$. Since the image is closed in the uniform norm it must also contain $C_0(E)$.

Now suppose $T\in K(A)\cap M(A)$ and $\varphi\in C(E)$ is such that $(Tx)^\wedge=\varphi\hat{x}$ for each $x\in A$. If $\varphi\notin C_0(E)$ then since E is discrete there exists a $\delta>0$ and a sequence $\{\alpha_j\}$ for which $\alpha_i\neq\alpha_j$, $i\neq j$, and $|\varphi(e'_{\alpha_j})|\geqq\delta$, $j=1,2,3,\ldots$. Recalling that for all $x\in A$ one has $Tx=\sum_\alpha \varphi(e'_\alpha)\langle x, e'_\alpha\rangle\, e'_\alpha$ we see at once that for $k\neq j$,

$$\|Te'_{\alpha_j}-Te'_{\alpha_k}\| = \|\varphi(e'_{\alpha_j})\, e'_{\alpha_j} - \varphi(e'_{\alpha_k})\, e'_{\alpha_k}\| = \left(|\varphi(e'_{\alpha_j})|^2+|\varphi(e'_{\alpha_k})|^2\right)^{\frac{1}{2}} \geqq 2^{\frac{1}{2}}\delta.$$

Thus the sequence $\{Te'_{\alpha_j}\}$ has no norm convergent subsequence, which contradicts the compactness of T as $\|e'_{\alpha_j}\|=1$, $j=1,2,3,\ldots$. Hence $\varphi\in C_0(E)$.

Therefore, since the image of $K(A)\cap M(A)$ contains $C_0(E)$, we conclude that $K(A)\cap M(A)$ is isometrically $*$-isomorphic to $C_0(E)=C_0[\Delta(A)]$. □

2.3. Notes. Theorems 2.1.1 and 2.2.1 are due to Kellogg [1] but the relatively simple proof of the latter result is due to R. B. Burckel. Both the theorems have been extended to the case of noncommutative H^*-algebras in Ching and Wong [1]. In particular they prove the following two results.

Theorem 2.3.1. *Let A be an H^*-algebra and E be the topological space of all minimal closed two-sided ideals in A with the discrete topology. Then $M(A)$ is isometrically $*$-isomorphic to $C(E)$.*

Theorem 2.3.2. *Let A be an H^*-algebra whose minimal closed two-sided ideals are finite dimensional. Then $K(A) \cap M(A)$ is isometrically $*$-isomorphic to $C_0(E)$.*

The proofs of these theorems rely on the structure theory for noncommutative H^*-algebras due to Ambrose [1], rather than the Gelfand representation theory available in the commutative case. A discussion of the structure theory for arbitrary H^*-algebras can also be found in Loomis [1] and Naimark [1].[1]

The multipliers for the Banach algebra of *all* compact linear operators on a Banach space has been studied in Johnson [1]. As remarked in the notes to Chapter 0 the compact multipliers for the group algebra $L_1(G)$ have been described in Akemann [1] and Gaudry [3] when G is a compact group.

That compact multipliers of Banach algebras do not always correspond to continuous functions which vanish at infinity has been shown in Figà-Talamanca and Gaudry [3]. They show that whenever $A = L_p(\Gamma^n)$, $n \geqq 1$, $p \neq 1, 2$ and Γ^n is the n-fold direct product of the circle group $\Gamma = \{z \mid z \in C, |z| = 1\}$ with itself, that is, the n-torus, then there exist compact $T \in M(A)$ such that $\varphi \notin C_0(\Delta(A))$. φ of course denotes the bounded continuous function on $\Delta(A)$ given by Theorem 1.2.2 for which $(Tx)^\wedge = \varphi \hat{x}$ for all $x \in A$.

[1] See also Saworotnow [1].

Chapter 3

Multipliers for Topological Linear Spaces of Functions and Measures

3.0. Introduction. We now wish to shift our attention away from the study of multipliers for Banach algebras and begin a discussion of multipliers for topological linear spaces. In this chapter we shall study a variety of topological linear spaces of functions and measures for which a characterization of the multipliers is relatively accessible. In addition to its intrinsic interest we hope that this material will illustrate some of the differences between the study of multipliers for commutative Banach algebras and for topological linear spaces. A general treatment of the important topic of multipliers for the L_p-spaces will be delayed until a subsequent chapter.

First we must decide what we shall mean by a multiplier for a topological linear space since the various formulations of this notion which were used in the context of commutative Banach algebras are now as a rule neither equivalent nor meaningful. Of the possible definitions the one which is most generally meaningful, and thus seems most natural to adopt, defines a multiplier as a continuous linear operator which commutes with translation. It will however also be advantageous to extend the notion of a multiplier to include transformations between two distinct spaces.

Consequently, let G be a locally compact Abelian group, X, Y topological linear spaces of functions or measures defined on G, then a continuous linear transformation T from X to Y is called a multiplier for the pair (X, Y) whenever $T\tau_s = \tau_s T$ for each $s \in G$. The collection of all multipliers for the pair (X, Y) will be denoted by $M(X, Y)$. If $X = Y$ we shall write $M(X, X) = M(X)$. Clearly $M(X, Y)$ is a linear subspace of the space of all continuous linear transformations from X to Y, and if $X = Y$ it is also a subalgebra with identity. Moreover, it is evident that if τ_s is strong operator continuous as an operator on Y for each $s \in G$ then $M(X, Y)$ is a closed subspace under the strong operator topology. If X and Y are Banach spaces and translation in Y is norm continuous then $M(X, Y)$ is a normed closed subspace of the Banach space of all continuous linear transformations from X to Y.

For many of the spaces to be considered it is possible to define a Fourier transform. In these cases it is also plausible to define a multiplier T for the pair (X, Y) as a linear transformation such that $(Tf)\hat{} = \varphi \hat{f}$ for each $f \in X$ and some function φ on $\hat{G}$. From the relationship between a Fourier transform and translations it is evident that such a T would commute with translations, and in many cases an application of the Closed Graph Theorem shows that T is continuous. Consequently such a T would often define an element of $M(X, Y)$. However even when a Fourier transform is available it is not generally known whether to each $T \in M(X, Y)$ there corresponds a φ for which $(Tf)\hat{} = \varphi \hat{f}$. This gives some indication why the definition adopted above was chosen.

Though the defining relations $T(xy) = x(Ty) = (Tx)y$ of a multiplier for a commutative Banach algebra will not now as a rule be valid or meaningful, they can often be established at least for certain subsets of the topological linear spaces under consideration. As before such identities will prove quite useful in the investigation of $M(X, Y)$. We shall pay special attention to the spaces $L_p(G)$, $1 \leqq p \leqq \infty$. It is easy to see that if $p = 1$ or if G is compact and $1 \leqq p < \infty$ then the definitions of a multiplier introduced in this chapter and that previously given for commutative Banach algebras are equivalent. This assertion for $p = 1$ was of course established in the introductory chapter, and a similar argument proves the equivalence when G is compact. Provided both definitions are meaningful, it is not generally known if the two notions are equivalent.

We shall give some indication in the following sections of the relationship between these several possible notions of a multiplier. As in the previous portion of the book we shall restrict our attention to the commutative cases, even though many of the results have valid analogues in the noncommutative context.

If G is a locally compact Abelian group we recall the following well-known facts. If $f \in L_1(G)$ and $g \in L_p(G)$, $1 \leqq p \leqq \infty$, then $f * g \in L_p(G)$ and $\|f * g\|_p \leqq \|f\|_1 \|g\|_p$. If $\mu \in M(G)$ then $\mu * g \in L_p(G)$ and $\|\mu * g\|_p \leqq \|\mu\| \|g\|_p$ (F.2). We shall make free use of these results in the next sections. If f is a function or μ is a measure on G we define $\tilde{f}$ and $\tilde{\mu}$ by $\tilde{f}(t) = f(t^{-1})$ and $\tilde{\mu}(E) = \mu(E^{-1})$.

3.1. $M(L_1(G), L_p(G))$, $1 \leqq p \leqq \infty$. The main result of this section is the following theorem.

Theorem 3.1.1. *Let G be a locally compact Abelian group and suppose T: $L_1(G) \to L_p(G)$ is a linear transformation where $1 \leqq p \leqq \infty$. Then the following are equivalent:*

(i) $T \in M(L_1(G), L_p(G))$.

(ii) *There exists a unique* μ *such that* $Tf = \mu * f$ *for each* $f \in L_1(G)$ *where* $\mu \in M(G)$ *if* $p=1$ *and* $\mu \in L_p(G)$ *if* $1 < p \leqq \infty$.

Moreover the correspondence between T *and* μ *defines an isometric linear isomorphism from* $M(L_1(G), L_p(G))$ *onto* $M(G)$ *if* $p=1$ *and onto* $L_p(G)$ *if* $1 < p < \infty$.

Proof. If $p=1$ then the result is contained in Theorem 0.1.1 and Corollary 0.1.1. Assume then that $1 < p \leqq \infty$.

If $\mu \in L_p(G)$ and we define $Tf = \mu * f$ for each $f \in L_1(G)$ then it is apparent by the remark at the end of the previous section that T is a bounded linear transformation from $L_1(G)$ to $L_p(G)$ which commutes with translations, that is, $T \in M(L_1(G), L_p(G))$.

Conversely, suppose $T \in M(L_1(G), L_p(G))$. Denote by $T^*: L_{p'}(G) \to L_\infty(G)$ the bounded linear transformation adjoint to T. Of course $1/p' + 1/p = 1$. Then if $f, g \in L_1(G)$ and $h \in L_{p'}(G)$ we have

$$\begin{aligned}\langle Tg*f, h\rangle &= \int_G Tg*f(t)\, h(t^{-1})\, d\lambda(t) = \int_G \left[\int_G Tg(t s^{-1}) f(s)\, d\lambda(s)\right] h(t^{-1})\, d\lambda(t)\\ &= \int_G f(s) \left[\int_G \tau_s Tg(t)\, h(t^{-1})\, d\lambda(t)\right] d\lambda(s)\\ &= \int_G f(s) \left[\int_G (T\tau_s g)(t)\, h(t^{-1})\, d\lambda(t)\right] d\lambda(s)\\ &= \int_G f(s) \left[\int_G \tau_s g(t)\, T^* \tilde{h}(t)\, d\lambda(t)\right] d\lambda(s)\\ &= \int_G \left[\int_G g(t s^{-1}) f(s)\, d\lambda(s)\right] T^* \tilde{h}(t)\, d\lambda(t)\\ &= \int_G g*f(t)\, T^* \tilde{h}(t)\, d\lambda(t) = \int_G T(g*f)(t)\, h(t^{-1})\, d\lambda(t)\\ &= \langle T(g*f), h\rangle.\end{aligned}$$

The applications of Fubini's Theorem (Theorem C.6.1) are valid because $g, f \in L_1(G)$ and $T^* h \in L_\infty(G)$.

Since this relation holds for all $h \in L_{p'}(G)$ we conclude that $T(g*f) = Tg*f$ for each $f, g \in L_1(G)$. Let $\{g_\alpha\} \subset L_1(G)$ be an approximate identity for $L_1(G)$ such that $\|g_\alpha\|_1 = 1$ for all α (F.7d). Then for each $f \in L_1(G)$ we have $\|Tf - Tg_\alpha * f\|_p = \|Tf - T(g_\alpha * f)\|_p \leqq \|T\|\, \|f - g_\alpha * f\|_1$, which shows that $Tf = \lim_\alpha Tg_\alpha * f$ in $L_p(G)$. Moreover $\|Tg_\alpha\|_p \leqq \|T\|\, \|g_\alpha\|_1 = \|T\|$ implies that $\{Tg_\alpha\}$ lies in a norm bounded subset of $L_p(G) = L_{p'}(G)^*$. Thus from Alaoglu's Theorem (Theorem D.4.3) and the reflexivity of $L_p(G)$ we conclude that there exists a subnet $\{Tg_\beta\}$ of $\{Tg_\alpha\}$ and a $\mu \in L_p(G)$ such that $\{Tg_\beta\}$ converges weak* to μ, that is, for each $h \in L_{p'}(G)$ we have $\lim_\beta \langle Tg_\beta, h\rangle = \langle \mu, h\rangle$.

Furthermore, suppose that $f, g \in C_c(G)$. Then, since $\lim_\beta \|Tg_\beta * f - Tf\|_p = 0$, we conclude that $\lim_\beta \langle Tg_\beta * f, g\rangle = \langle Tf, g\rangle$ as $C_c(G) \subset L_{p'}(G) \subset L_p(G)^*$. However $f * g \in L_{p'}(G)$ and $\{Tg_\beta\}$ converges weak* in $L_p(G)$ to μ, so we also have that

$$\begin{aligned}\langle Tf, g\rangle &= \lim_\beta \langle Tg_\beta * f, g\rangle = \lim_\beta (Tg_\beta * f) * g(e) = \lim_\beta Tg_\beta * (f * g)(e)\\ &= \lim_\beta \langle Tg_\beta, f * g\rangle = \langle \mu, f * g\rangle = \langle \mu * f, g\rangle.\end{aligned}$$

Since $C_c(G)$ is norm dense in $L_{p'}(G)$ we deduce that $Tf = \mu * f$ for each $f \in C_c(G)$.

Thus $Tf = \mu * f$ for each $f \in L_1(G)$ as $C_c(G)$ is norm dense in $L_1(G)$ and T and convolution with μ are continuous linear transformations from $L_1(G)$ into $L_p(G)$.

Suppose $\mu * f = 0$ for all $f \in L_1(G)$. In particular then for each $f \in C_c(G)$ the continuous function $\mu * f$ is identically zero. Hence $\langle \mu, f\rangle = 0$ for each $f \in C_c(G)$, which implies that $\mu = 0$ since $C_c(G)$ is norm dense in $L_{p'}(G)$ and $L_{p'}(G) = L_p(G)^*$.

Therefore μ is unique and the equivalence of (i) and (ii) is proved.

Clearly $\|Tf\|_p = \|\mu * f\|_p \leqq \|\mu\|_p \|f\|_1$ shows that $\|T\| \leqq \|\mu\|_p$. But given $\varepsilon > 0$ there exists an $f \in L_1(G)$, $\|f\|_1 \leqq 1$ such that $\|\mu * f\|_p > \|\mu\|_p - \varepsilon$ (F.2) provided $p < \infty$. Hence $\|T\| = \|\mu\|_p$ for $1 < p < \infty$. This establishes the isometric linear isomorphism between $M(L_1(G), L_p(G))$ and $L_p(G)$, $1 < p < \infty$. □

Corollary 3.1.1. *Let G be a locally compact Abelian group and $T: L_1(G) \to L_p(G)$ be a linear transformation. If G is compact and $1 \leqq p \leqq \infty$ or G is noncompact and $1 \leqq p \leqq 2$ then the following are equivalent:*

(i) $T \in M(L_1(G), L_p(G))$.

(ii) *There exists a μ such that $Tf = \mu * f$ for each $f \in L_1(G)$ where $\mu \in M(G)$ if $p = 1$ and $\mu \in L_p(G)$ if $p \neq 1$.*

(iii) *There exists a function φ on $\hat{G}$ such that $(Tf)\hat{} = \varphi \hat{f}$ for each $f \in L_1(G)$ where $\varphi \in L_\infty(\hat{G})$ if G is compact and $1 \leqq p \leqq \infty$, and $\varphi \in L_{p'}(\hat{G})$, $1/p + 1/p' = 1$, if G is noncompact and $1 \leqq p \leqq 2$.*

Proof. The implications (i) implies (ii) implies (iii) are immediate from the preceding theorem and the properties of the Fourier transform, its extensions in the case that G is noncompact and $1 < p \leqq 2$, and the Hausdorff-Young Theorem (F.8). It remains then only to establish that (iii) implies (i). Clearly the equation $(Tf)\hat{} = \varphi \hat{f}$ for each $f \in L_1(G)$ shows that T commutes with translation. Moreover, suppose $\lim_n \|f_n - f\|_1 = 0$ and

$\lim_n \|Tf_n - g\|_p = 0$. If G is noncompact and $1 \leq p \leq 2$ then

$$\begin{aligned}\|(Tf)\hat{} - \hat{g}\|_{p'} &\leq \|(Tf)\hat{} - (Tf_n)\hat{}\|_{p'} + \|(Tf_n)\hat{} - \hat{g}\|_{p'}\\ &\leq \|\varphi(\hat{f} - \hat{f}_n)\|_{p'} + \|Tf_n - g\|_p.\end{aligned}$$

But $\lim_n \|f_n - f\|_1 = 0$ implies that $\lim_n \|\hat{f}_n - \hat{f}\|_\infty = 0$, and an application of the Lebesgue Dominated Convergence Theorem (Theorem C.5.2) reveals that $\lim_n \|\varphi(\hat{f} - \hat{f}_n)\|_{p'} = 0$. Combining this with $\lim_n \|Tf_n - g\|_p = 0$ we conclude that $\|(Tf)\hat{} - \hat{g}\|_{p'} = 0$, that is, $(Tf)\hat{} = \hat{g}$. Thus the transformation T is closed and hence, by the Closed Graph Theorem (Theorem D.6.1), it is continuous. For compact G a similar argument can be made.

Therefore $T \in M(L_1(G), L_p(G))$ and the equivalence of (i), (ii) and (iii) is established. ∎

The equivalence between (i) and (iii) when G is noncompact and $p > 2$ is more delicate due to the difficulty of defining the Fourier transform when $p > 2$. We shall return to this problem in a general treatment of $M(L_p(G), L_q(G))$.

3.2. $M(M_w(G))$ and $M(M(G))$. For a locally compact Abelian group G the space $M(G)$ is a semi-simple commutative Banach algebra with identity (F.7b). Hence, as indicated in Chapter 1, every multiplier of $M(G)$, in the sense employed when discussing Banach algebras, corresponds to a unique element of $M(G)$. Indeed if T is such a multiplier then $T(\nu * \mu) = T\nu * \mu$ implies that $T\mu = T\delta * \mu$ where δ is the unit point mass concentrated at the identity of G. Clearly such a transformation also commutes with translations. However the converse assertion is not generally valid. That is, there may exist continuous linear mappings T from $M(G)$ to itself which commute with translation and can not be expressed as convolution with an element of $M(G)$. For example, suppose G is a nondiscrete locally compact Abelian group, and for each $\mu \in M(G)$ let $\mu = \mu_a + \mu_s$ be the Lebesgue decomposition with respect to Haar measure (Theorem C.1.3) of μ into its absolutely continuous and singular parts μ_a and μ_s, respectively. Let $f \in L_1(G)$, $f \neq 0$, and define the mapping T of $M(G)$ into itself by setting $T\mu = f * \mu_a$ for each $\mu \in M(G)$. It is easily seen that T is a continuous linear mapping on $M(G)$ which commutes with translations. But T cannot be expressed as convolution with an element of $M(G)$. As suppose there exists some $\nu \in M(G)$ such that $T\mu = \nu * \mu$ for each $\mu \in M(G)$. In particular, then we would have $\nu = \nu * \delta = f * \delta_a = 0$ as G is nondiscrete and so $T\mu = f * \mu_a = 0$ for all $\mu \in M(G)$. However this is clearly impossible as $f \neq 0$.

Thus we see for $M(G)$ and nondiscrete G that the Banach algebra definition of a multiplier as a continuous linear operator which commutes with convolution and the definition of a multiplier as a continuous linear operator which commutes with translations are not equivalent. This contrasts strongly of course with the algebra $L_1(G)$ as shown in Theorem 0.1.1.

If we consider $M(G)$ as the dual space of $C_0(G)$ (Theorem D.9.4) and endow it with the weak* topology induced by $C_0(G)$ then we obtain a topological linear space which we denote by $M_w(G)$. The space of continuous linear functionals on $M_w(G)$ can of course be identified with $C_0(G)$ (Theorem D.4.1) by the formula

$$\langle\mu, h\rangle = \int_G h(t^{-1})\, d\mu(t) \qquad (\mu\in M(G)).$$

One can give a fairly complete description of $M(M_w(G))$.

Theorem 3.2.1. *Let G be a locally compact Abelian group and T: $M_w(G)\to M_w(G)$ be a linear transformation. Then the following are equivalent:*

(i) *$T\in M(M_w(G))$.*

(ii) *There exists a unique $\omega\in M(G)$ such that $T\mu=\omega*\mu$ for each $\mu\in M(G)$.*

(iii) *There exists a unique bounded continuous function φ on $\hat{G}$ such that $(T\mu)\hat{} = \varphi\,\hat{\mu}$ for each $\mu\in M(G)$.*

Moreover the correspondence between T and ω defines a linear isomorphism from $M(M_w(G))$ onto $M(G)$.

Proof. Let $T\in M(M_w(G))$. Since T commutes with translations and the dual space of $M_w(G)$ is $C_0(G)$, essentially the same argument as that given in the proof of Theorem 3.1.1 shows that $\langle T(v*\mu), h\rangle = \langle Tv*\mu, h\rangle$ for each $h\in C_0(G)$, and hence $(Tv*\mu)=Tv*\mu$ for each $\mu, v\in M(G)$. Thus $T\mu=\omega*\mu$ for each $\mu\in M(G)$ where $\omega=T\delta$. Clearly ω is unique. Therefore (i) implies (ii).

It is evident that (ii) implies (iii). If (iii) holds then for each $\mu, v\in M(G)$ we have

$$T(v*\mu)\hat{} = \varphi(v*\mu)\hat{} = \varphi\,\hat{v}\,\hat{\mu} = (Tv*\mu)\hat{}\,.$$

Thus $T(v*\mu)=Tv*\mu$, and T defines a multiplier for the Banach algebra $M(G)$. From the remarks preceding the theorem we conclude that $T\mu=\omega*\mu$. Hence (ii) and (iii) are equivalent.

Finally, suppose (ii) holds. Obviously T: $M_w(G)\to M_w(G)$ and commutes with translations. It only remains to show that T is continuous. But let $\{\mu_\alpha\}\subset M_w(G)$ converge to μ, that is, $\lim_\alpha \langle\mu_\alpha, h\rangle = \langle\mu, h\rangle$ for each $h\in C_0(G)$. Then for each $h\in C_0(G)$

$$\langle T\mu_\alpha, h\rangle = \langle\omega*\mu_\alpha, h\rangle = \langle\mu_\alpha, \omega*h\rangle.$$

Since $\omega \in M(G)$ and $h \in C_0(G)$ it follows that $\omega * h \in C_0(G)$ and hence

$$\langle T\mu, h\rangle = \langle \omega * \mu, h\rangle = \langle \mu, \omega * h\rangle = \lim_\alpha \langle \mu_\alpha, \omega * h\rangle = \lim_\alpha \langle T\mu_\alpha, h\rangle .$$

That is, $\{T\mu_\alpha\}$ converges to $T\mu$ in $M_w(G)$.

Therefore T is continuous and belongs to $M(M_w(G))$.

The final assertion of the theorem is now apparent. ▯

It is evident that $\varphi = \hat{\mu}$.

The difficulty in extending these arguments to $M(G)$ lies in the fact that the dual space of $M(G)$ cannot be identified with $C_0(G)$ but only with some larger space. However one does have the following result. $M(M(G))$ here denotes the space of continuous linear transformations $T\colon M(G) \to M(G)$ such that $T\tau_s = \tau_s T$ for each $s \in G$, and not the space of multipliers for the Banach algebra $M(G)$. The reader should refer again to the comments preceding Theorem 3.2.1.

Theorem 3.2.2. *Let G be a locally compact Abelian group. If $T \in M(M(G))$ then there exists a unique $\omega \in M(G)$ such that $T\mu = \omega * \mu$ for each $\mu \in L_1(G)$. Moreover there exists a continuous homomorphism of $M(M(G))$ onto $M(G)$.*

Proof. Let $T \in M(M(G))$. Since T commutes with translation we see for each $\mu \in L_1(G)$ and each $s \in G$ that

$$\|\tau_s T\mu - T\mu\| = \|T\tau_s \mu - T\mu\| \leqq \|T\| \, \|\tau_s \mu - \mu\| .$$

But an element $\nu \in M(G)$ belongs to $L_1(G)$ whenever $\|\tau_s \nu - \nu\|$ is a continuous function of s (F.7a). Hence the preceding estimate immediately reveals that $T\mu \in L_1(G)$ for each $\mu \in L_1(G)$.

Clearly then the restriction T_1 of T to $L_1(G)$ is an element of $M(L_1(G))$, and thus by Theorem 0.1.1 or 3.1.1 there exists a unique $\omega \in M(G)$ such that $T\mu = T_1\mu = \omega * \mu$ for each $\mu \in L_1(G)$.

Furthermore, we have

$$\begin{aligned}\|\omega\| &= \sup\{\|T_1\mu\|_1 \mid \mu \in L_1(G),\ \|\mu\|_1 \leqq 1\} = \sup\{\|T\mu\| \mid \mu \in L_1(G),\ \|\mu\| \leqq 1\}\\ &\leqq \sup\{\|T\mu\| \mid \mu \in M(G),\ \|\mu\| \leqq 1\} = \|T\| .\end{aligned}$$

Hence it is apparent that the previous construction defines a continuous homomorphism of $M(M(G))$ into $M(G)$. However if $\omega \in M(G)$ then setting $T_\omega \mu = \omega * \mu$ for each $\mu \in M(G)$ clearly defines an element $T_\omega \in M(M(G))$. Moreover $T_\omega \mu = \omega * \mu$ for each $\mu \in L_1(G)$ and so the homomorphism is onto. ▯

Since $L_1(G)$ is not norm dense in $M(G)$ when G is not discrete, we cannot deduce that $T = T_\omega$ where $T\mu = \omega * \mu$ for each $\mu \in L_1(G)$. Thus we can only conclude the mapping of the theorem is a homomorphism rather than an isomorphism.

3.3. The Adjoint Argument: $M(L_{p'}(G), L_\infty(G))$, $1 \leqq p' < \infty$, and $M(C_0(G))$. We employed the notion of the adjoint of a linear transformation while discussing the spaces $M(L_1(G), L_p(G))$. A characterization of the elements in $M(L_{p'}(G), L_\infty(G))$ and $M(C_0(G))$ can be readily obtained by examining the adjoints of these multipliers in the light of the results of the previous two sections.

Theorem 3.3.1. *Let G be a locally compact Abelian group and suppose T: $L_{p'}(G) \to L_\infty(G)$ is a linear transformation where $1 \leqq p' < \infty$. Then the following are equivalent:*

(i) $T \in M(L_{p'}(G), L_\infty(G))$.

(ii) *There exists a unique $g \in L_p(G)$, $1/p + 1/p' = 1$, such that $Tf = g * f$ for each $f \in L_{p'}(G)$.*

Moreover the correspondence between T and g defines an isometric linear isomorphism from $M(L_{p'}(G), L_\infty(G))$ onto $L_p(G)$, $1/p + 1/p' = 1$.

Proof. If $g \in L_p(G)$ and we set $Tf = g * f$ for each $f \in L_{p'}(G)$ then a simple application of Hölder's inequality and the properties of convolution show that T: $L_{p'}(G) \to L_\infty(G)$ and that T is continuous and commutes with translations. Thus (ii) implies (i). Suppose $T \in M(L_{p'}(G), L_\infty(G))$ and denote by T^* the adjoint of T, that is, the continuous linear transformation from $L_\infty(G)^*$ to $L_{p'}(G)^* = L_p(G)$, $1/p + 1/p' = 1$. Restricting our attention to $L_1(G) \subset L_\infty(G)^*$ we have $\langle Tf, h\rangle = \langle f, T^* h\rangle$ for each $f \in L_{p'}(G)$, $h \in L_1(G)$. Moreover, for each $s \in G$ we have

$$\begin{aligned}\langle f, T^* \tau_s h\rangle &= \langle Tf, \tau_s h\rangle = \langle \tau_s Tf, h\rangle = \langle T \tau_s f, h\rangle = \langle \tau_s f, T^* h\rangle \\ &= \langle f, \tau_s T^* h\rangle,\end{aligned}$$

where $f \in L_{p'}(G)$, $h \in L_1(G)$. Consequently T^* restricted to $L_1(G)$ defines a multiplier for the pair $(L_1(G), L_p(G))$. Consequently by Theorem 3.1.1 there exists a unique $g \in L_p(G)$ such that $T^* h = g * h$ for each $h \in L_1(G)$. An elementary computation reveals for each $f \in L_{p'}(G)$ and $h \in L_1(G)$ that

$$\langle Tf, h\rangle = \langle f, T^* h\rangle = \langle f, g * h\rangle = \langle g * f, h\rangle.$$

Therefore $Tf = g * f$ for each $f \in L_{p'}(G)$.

From the form of T it is evident that $\|T\| \leqq \|g\|_p$. On the other hand from Theorem 3.1.1 we see that $\|g\|_p = \|T^*\|_{L_1(G)} \leqq \|T^*\| = \|T\|$ where $\|T^*\|_{L_1(G)}$ denotes the norm of T^* restricted to $L_1(G)$. Thus $\|g\|_p = \|T\|$ and the isometric isomorphism between $M(L_{p'}(G), L_\infty(G))$ and $L_p(G)$ is established. $\square$

It should be noted that for $p' > 1$ if $T \in M(L_{p'}(G), L_\infty(G))$ then $Tf \in C_0(G)$ for each $f \in L_{p'}(G)$, as $Tf = g * f \in C_0(G)$ since $g \in L_p(G)$, $f \in L_{p'}(G)$ and $1 < p' < \infty$ (F.3).

Employing a similar approach we can characterize $M(C_0(G))$.

Theorem 3.3.2. *Let G be a locally compact Abelian group and suppose $T\colon C_0(G)\to C_0(G)$ is a linear transformation. Then the following are equivalent:*

(i) $T\in M(C_0(G))$.

(ii) *There exists a unique $\omega\in M(G)$ such that $Tf=\omega*f$ for each $f\in C_0(G)$.*

Moreover the correspondence between T and ω defines an isometric algebra isomorphism from $M(C_0(G))$ onto $M(G)$.

Proof. The proof is quite similar to that of the previous theorem so we shall only give a brief sketch. Obviously (ii) implies (i) and if $T\in M(C_0(G))$ then the adjoint $T^*\colon M(G)\to M(G)$ satisfies $\langle Tf,\mu\rangle=\langle f, T^*\mu\rangle$ for each $f\in C_0(G)$, $\mu\in M(G)$. This identity shows immediately that T^* commutes with translation and is weak* continuous. Hence $T^*\in M(M_w(G))$. Thus by Theorem 3.2.1 there is a unique $\omega\in M(G)$ such that $T^*\mu=\omega*\mu$. Elementary calculations again reveal that

$$\langle Tf,\mu\rangle=\langle f, T^*\mu\rangle=\langle f,\omega*\mu\rangle=\langle \omega*f,\mu\rangle,$$

from which we conclude that $Tf=\omega*f$. The isometry is easily established. ▯

3.4. $M(L_\infty(G))$ and $M(L^w_\infty(G))$. In the previous sections we have avoided discussing $M(L_p(G), L_q(G))$ when $p=q=\infty$. This case presents difficulties similar to those encountered in the investigation of $M(M(G))$. In the latter case we saw that it was advantageous to replace $M(G)$ by $M_w(G)$. A similar approach is also fruitful in the present context. We shall denote by $L^w_\infty(G)$ the space $L_\infty(G)$ considered with the weak* topology induced by the elements of $L_1(G)$. As is well known (Theorem D.4.1) the space of continuous linear functionals on $L^w_\infty(G)$ can then be identified with $L_1(G)$ by the formula

$$\langle f,h\rangle=\int_G f(t)\,h(t^{-1})\,d\lambda(t) \qquad (f\in L_\infty(G))$$

where $h\in L_1(G)$.

Using adjoints as in the preceding section and the characterization of $M(L_1(G))$ we can examine $M(L^w_\infty(G))$.

Theorem 3.4.1. *Let G be a locally compact Abelian group and suppose $T\colon L^w_\infty(G)\to L^w_\infty(G)$ is a linear transformation. Then the following are equivalent:*

(i) $T\in M(L^w_\infty(G))$.

(ii) *There exists a unique $\mu\in M(G)$ such that $Tf=\mu*f$ for each $f\in L^w_\infty(G)$.*

Moreover the correspondence between T and μ defines a linear isomorphism from $M(L^w_\infty(G))$ onto $M(G)$.

Proof. Let $T \in M(L_\infty^w(G))$. Then for $f \in L_\infty(G)$, $h \in L_1(G)$ the pairing $\langle Tf, h\rangle = \langle f, T^* h\rangle$ defines the adjoint of T as a linear transformation T^* from $L_1(G)$ to $L_1(G)$. It follows as in the proof of Theorem 3.3.1 that for $s \in G$ we have $\langle f, \tau_s T^* h\rangle = \langle f, T^* \tau_s h\rangle$ for each $f \in L_\infty(G)$, $h \in L_1(G)$. Hence T^* commutes with translations. Moreover T^* is continuous on $L_1(G)$. Indeed, let h_n, h, $g \in L_1(G)$ be such that $\lim_n \|h_n - h\|_1 = 0$ and $\lim_n \|T^* h_n - g\|_1 = 0$. Then for each $f \in L_\infty(G)$ we have

$$
\begin{aligned}
|\langle f, T^* h - g\rangle| &\leqq |\langle f, T^* h - T^* h_n\rangle| + |\langle f, T^* h_n - g\rangle| \\
&= |\langle Tf, h - h_n\rangle| + |\langle f, T^* h_n - g\rangle| \\
&\leqq \|Tf\|_\infty \|h - h_n\|_1 + \|f\|_\infty \|T^* h_n - g\|_1 .
\end{aligned}
$$

Consequently $\langle f, T^* h - g\rangle = 0$ for each $f \in L_\infty(G)$, and hence $T^* h = g$. Thus T^* is a closed transformation and so, by the Closed Graph Theorem (Theorem D.6.1), it is continuous. Therefore $T^* \in M(L_1(G))$.

By Theorems 0.1.1 or 3.1.1 there exists a unique $\mu \in M(G)$ such that $T^* h = \mu * h$ for each $h \in L_1(G)$. But then we see once more that for each $f \in L_\infty(G)$ and $h \in L_1(G)$ we have

$$\langle Tf, h\rangle = \langle f, T^* h\rangle = \langle f, \mu * h\rangle = \langle f * \mu, h\rangle .$$

From this it follows at once that $Tf = \mu * f$ for each $f \in L_\infty(G)$. Thus (i) implies (ii).

Conversely, define $Tf = \mu * f$ for $f \in L_\infty(G)$ and some $\mu \in M(G)$. Clearly T defines a linear transformation from $L_\infty(G)$ to $L_\infty(G)$ which commutes with translations. Moreover, suppose $\{f_\alpha\} \subset L_\infty(G)$ converges in the weak* sense to $f \in L_\infty(G)$, that is $\lim_\alpha \langle f_\alpha, h\rangle = \langle f, h\rangle$ for each $h \in L_1(G)$. Then

$$
\begin{aligned}
\lim_\alpha \langle Tf_\alpha, h\rangle &= \lim_\alpha \langle \mu * f_\alpha, h\rangle = \lim_\alpha \langle f_\alpha, \mu * h\rangle = \langle f, \mu * h\rangle \\
&= \langle \mu * f, h\rangle = \langle Tf, h\rangle
\end{aligned}
$$

for each $h \in L_1(G)$ since $\mu * h \in L_1(G)$.

Therefore $T \in M(L_\infty^w(G))$. ☐

The chief difficulty which arises in attempting to carry over the proof of the preceding result to $M(L_\infty(G))$ is in trying to show that if $T \in M(L_\infty(G))$ then T is continuous with the weak* topology on $L_\infty(G)$. Such an attack will produce results however if we impose some additional restrictions on G and T. First we shall require that G be compact, and second that T be a linear transformation on $L_\infty(G)$ of the form $(Tf)\hat{} = \varphi \hat{f}$ where φ is some bounded function on $\hat{G}$. Since G is compact $L_\infty(G)$ is contained in $L_1(G)$ and the Fourier transform is well defined. Using the Closed Graph

Theorem it is easy to verify that any such linear transformation T is continuous and commutes with translations, that is, any such T is an element of $M(L_\infty(G))$.

Having made these preliminary remarks we now state and prove the next theorem.

Theorem 3.4.2. *Let G be a compact Abelian group and suppose $T: L_\infty(G) \to L_\infty(G)$ is a linear transformation. Then $T \in M(L_\infty(G))$ if any of the following three equivalent statements is valid:*

(i) *There exists a bounded function φ on $\hat{G}$ such that $(Tf)\hat{} = \varphi \hat{f}$ for each $f \in L_\infty(G)$.*

(ii) *There exists a unique $\mu \in M(G)$ such that $(Tf)\hat{} = \hat{\mu}\hat{f}$ for each $f \in L_\infty(G)$.*

(iii) *There exists a unique $\mu \in M(G)$ such that $Tf = \mu * f$ for each $f \in L_\infty(G)$.*

Proof. It is evident that (ii) and (iii) are equivalent, that both imply (i) and that T so defined belongs to $M(L_\infty(G))$.

Suppose then that $(Tf)\hat{} = \varphi \hat{f}$ for each $f \in L_\infty(G)$ and some $\varphi \in L_\infty(\hat{G})$. Clearly T commutes with translation. Moreover, suppose $\lim_n \|f_n - f\|_\infty = 0$ and $\lim_n \|Tf_n - g\|_\infty = 0$. Then

$$\begin{aligned} \|(Tf)\hat{} - \hat{g}\|_\infty &\leqq \|(Tf)\hat{} - (Tf_n)\hat{}\|_\infty + \|(Tf_n)\hat{} - \hat{g}\|_\infty \\ &\leqq \|\varphi\|_\infty \, \|f - f_n\|_1 + \|Tf_n - g\|_1 \\ &\leqq \|\varphi\|_\infty \, \|f - f_n\|_\infty + \|Tf_n - g\|_\infty \end{aligned}$$

since G is compact. Consequently $(Tf)\hat{} = \hat{g}$, and so $Tf = g$. Thus by the Closed Graph Theorem (Theorem D.6.1) we conclude that $T \in M(L_\infty(G))$.

We wish to show that $T \in M(L_\infty^w(G))$, that is, to show that T is continuous with the weak* topology on $L_\infty(G)$. First we note that the graph of T in $L_\infty^w(G) \times L_\infty^w(G)$ is closed. Indeed, if $\{(f_\alpha, Tf_\alpha)\} \subset L_\infty^w(G) \times L_\infty^w(G)$ is a net such that $\lim_\alpha (f_\alpha, Tf_\alpha) = (f, g)$ in the weak* sense then $\lim_\alpha (\hat{f}_\alpha(\gamma), (Tf_\alpha)\hat{}(\gamma)) = (\hat{f}(\gamma), \hat{g}(\gamma))$ for each $\gamma \in \hat{G}$. This assertion is valid as G is compact and hence $\hat{G} \subset L_1(G)$. But then it is also evident that $\lim_\alpha (\hat{f}_\alpha(\gamma), (Tf_\alpha)\hat{}(\gamma)) = (\hat{f}(\gamma), \varphi \hat{f}(\gamma))$ for each $\gamma \in \hat{G}$. Thus $\hat{g} = \varphi \hat{f} = (Tf)\hat{}$ and so $g = Tf$. That is, the graph of T is closed in $L_\infty^w(G) \times L_\infty^w(G)$.

To establish the weak* continuity of T it is sufficient to do so on bounded subsets of $L_\infty(G)$ (Theorem D.4.2). So suppose $\{f_\alpha\} \subset L_\infty(G)$ is norm bounded and $\{f_\alpha\}$ converges weak* to f. If $\{Tf_\alpha\}$ did not converge weak* to Tf then there exists a weak* neighborhood U of Tf and a subnet $\{Tf_\beta\}$ of $\{Tf_\alpha\}$ such that $\{Tf_\beta\} \not\subset U$. However $\{Tf_\beta\}$ is a norm bounded net in $L_\infty(G)$ as $T \in M(L_\infty(G))$ and so, by Alaoglu's Theorem (Theorem D.4.3), there exists a $g \in L_\infty(G)$ and a subnet $\{Tf_{\beta'}\}$ of $\{Tf_\beta\}$ such that $\{Tf_{\beta'}\}$ converges weak* to g. But $\{Tf_{\beta'}\} \not\subset U$ so that $g \neq Tf$, which contradicts the fact that the graph of T in $L^w_\infty(G) \times L^w_\infty(G)$ is closed. Hence T is weak* continuous on norm bounded sets, and so weak* continuous on $L_\infty(G)$.

Consequently $T \in M(L^w_\infty(G))$. Thus by Theorem 3.4.1 there exists a unique $\mu \in M(G)$ such that $Tf = \mu * f$ for each $f \in L_\infty(G)$.

Therefore (i) implies (iii) and the proof is complete. □

If one abandons the attempt to write $Tf = \mu * f$ for all $f \in L_\infty(G)$ and some $\mu \in M(G)$ when $T \in M(L_\infty(G))$ then one can obtain a characterization of $M(L_\infty(G))$ which is valid for arbitrary locally compact Abelian groups.

Theorem 3.4.3. *Let G be a locally compact Abelian group. If $T \in M(L_\infty(G))$ then there exists a unique $\mu \in M(G)$ such that $Tf = \mu * f$ for each $f \in C_0(G)$. Moreover there exists a continuous homomorphism of $M(L_\infty(G))$ onto $M(G)$.*

Proof. Let $T \in M(L_\infty(G))$. Since T commutes with translation we see for each $f \in C_0(G)$ and $s \in G$ that

$$\|\tau_s Tf - Tf\|_\infty = \|T\tau_s f - Tf\|_\infty \leqq \|T\| \, \|\tau_s f - f\|_\infty .$$

Hence Tf is a uniformly continuous function on G for each $f \in C_0(G)$.

In particular, then we may meaningfully define the linear functional F on $C_0(G)$ by $F(f) = Tf(e)$ for each $f \in C_0(G)$. Moreover F is continuous since

$$|F(f)| = |Tf(e)| \leqq \|Tf\|_\infty \leqq \|T\| \, \|f\|_\infty \qquad (f \in C_0(G)).$$

Thus by the Riesz Representation Theorem (Theorem D.9.4) there exists a unique $\mu \in M(G)$ such that $F(f) = \langle f, \mu \rangle$ for each $f \in C_0(G)$. Clearly $\|\mu\| = \|F\| \leqq \|T\|$.

But then for each $f \in C_0(G)$ and each $s \in G$ we obtain

$$Tf(s) = \tau_{s^{-1}} Tf(e) = T\tau_{s^{-1}} f(e) = F(\tau_{s^{-1}} f) = \langle \tau_{s^{-1}} f, \mu \rangle = \mu * f(s).$$

Therefore $Tf = \mu * f$ for each $f \in C_0(G)$.

It is evident that the preceding construction defines a continuous homomorphism from $M(L_\infty(G))$ into $M(G)$. However if $\mu \in M(G)$ then $\mu * f$ defines an element of $L_\infty(G)$ for each $f \in L_\infty(G)$ and $\|\mu * f\|_\infty \leqq \|\mu\| \, \|f\|_\infty$ (F.2). Thus setting $T_\mu f = \mu * f$ for each $f \in L_\infty(G)$ defines a continuous linear transformation $T_\mu \colon L_\infty(G) \to L_\infty(G)$ which commutes with translation. Obviously $T_\mu f = \mu * f$ for each $f \in C_0(G)$ and the homomorphism is onto. □

In general it is not the case for infinite G that $T = T_\mu$ where $Tf = \mu * f$ for each $f \in C_0(G)$. Consequently we must content ourselves with a homomorphism rather than an isomorphism in the preceding theorem. This can be seen from the following example which shows that there may exist $T \in M(L_\infty(G))$ which cannot be written as convolution with an element of $M(G)$.

Indeed, consider $G = Z$, the additive group of the integers and let Y be the linear subspace of $L_\infty(Z)$ consisting of all the $f \in L_\infty(Z)$ such that $\lim_{|n| \to \infty} f(n)$ exists. For each $f \in Y$ define $Tf = \lim_{|n| \to \infty} f(n)$, that is, Tf is the function in $L_\infty(Z)$ which is constantly equal to $\lim_{|n| \to \infty} f(n)$. Clearly T is a continuous linear mapping from Y considered as a subspace of $L_\infty(Z)$ into $L_\infty(Z)$ which commutes with translations. Moreover $\tau_k Tf = T\tau_k f = Tf$ for each $k \in Z$ and each $f \in Y$. An extension of the Hahn-Banach Theorem (Theorem D.6.6) assures the existence of a continuous linear mapping on all of $L_\infty(Z)$ to itself which is also invariant under translations and coincides with T on Y. We shall also denote this extension by T. Clearly this $T \in M(L_\infty(Z))$ and $T \neq 0$. But we claim that there exists no $\mu \in M(Z)$ such that $Tf = \mu * f$ for all $f \in L_\infty(Z)$. Indeed suppose such a μ did exist. Then since $Tf = 0$ for $f \in C_0(Z) \subset Y$ we see that $\mu * f = 0$ for $f \in C_0(Z)$. Hence by the Riesz Representation Theorem (Theorem D.9.4) and a consequence of the Hahn-Banach Theorem (D.6.c) we conclude that $\mu = 0$, contradicting the fact that $T \neq 0$. Thus T cannot be written as convolution with an element of $M(Z)$.

The measure $\mu \in M(Z)$ such that $Tf = \mu * f$ for all $f \in C_0(Z)$, whose existence is guaranteed by Theorem 3.4.3, is of course $\mu = 0$.

3.5. $M(L_1(G) \cap L_p(G), L_1(G))$, $1 < p < \infty$, and $M(L_1(G) \cap C_0(G), L_1(G))$. If $1 < p < \infty$ then it is easily verified that the linear space $L_1(G) \cap L_p(G)$ is a Banach space with the norm

$$\|f\|_{1,p} = \|f\|_1 + \|f\|_p \qquad (f \in L_1(G) \cap L_p(G)).$$

Similarly, $L_1(G) \cap C_0(G)$ is a Banach space with the norm

$$\|f\|_{1,\infty} = \|f\|_1 + \|f\|_\infty \qquad (f \in L_1(G) \cap C_0(G)).$$

Whenever G is noncompact the next theorems provide a characterization of the continuous linear transformations from these Banach spaces to $L_1(G)$ which commute with translations.

First, however, we shall prove a lemma which will be utilized again in Chapter 5.

Lemma 3.5.1. *Let G be a noncompact locally compact Abelian group.*
(i) *If $f \in L_p(G)$, $1 \leqq p < \infty$, then* $\lim_{s \to +\infty} \|f + \tau_s f\|_p = 2^{1/p} \|f\|_p$.
(ii) *If $f \in C_0(G)$ then* $\lim_{s \to +\infty} \|f + \tau_s f\|_\infty = \|f\|_\infty$.

Proof. Suppose $g\in C_c(G)$ with compact support K. Since G is noncompact if $s\notin KK^{-1}$ then the supports of g and $\tau_s g$ are disjoint. Consequently, on the one hand, if $s\notin KK^{-1}$ we have

$$\begin{aligned}\|g+\tau_s g\|_p &= \Big(\int_G |g(t)+\tau_s g(t)|^p \, d\lambda(t)\Big)^{1/p} \\ &= \Big(\int_K |g(t)|^p \, d\lambda(t) + \int_{Ks} |\tau_s g(t)|^p \, d\lambda(t)\Big)^{1/p} \\ &= 2^{1/p}\, \|g\|_p,\end{aligned}$$

for $1\leqq p<\infty$, while on the other hand,

$$\|g+\tau_s g\|_\infty = \|g\|_\infty .$$

Thus if $f\in L_p(G)$, $1\leqq p<\infty$, and $\varepsilon>0$ choose $g\in C_c(G)$ such that $\|f-g\|_p<\varepsilon/4$. Let the support of g be K. Then if $s\notin KK^{-1}$ we have

$$\begin{aligned}&\big|\|f+\tau_s f\|_p - 2^{1/p}\|f\|_p\big| \\ &\leqq \big|\|f+\tau_s f\|_p - \|g+\tau_s g\|_p\big| + \big|\|g+\tau_s g\|_p - 2^{1/p}\|g\|_p\big| + \big|2^{1/p}\|g\|_p - 2^{1/p}\|f\|_p\big| \\ &\leqq \|f-g\|_p + \|\tau_s f-\tau_s g\|_p + 2^{1/p}\|f-g\|_p < \frac{\varepsilon}{4} + \frac{\varepsilon}{4} + 2^{1/p}\frac{\varepsilon}{4} \leqq \varepsilon .\end{aligned}$$

Therefore, $\lim_{s\to+\infty} \|f+\tau_s f\|_p = 2^{1/p}\|f\|_p$ for each $f\in L_p(G)$, $1\leqq p<\infty$.

The assertion for $f\in C_0(G)$ is deduced in essentially the same manner. ▯

Theorem 3.5.1. *Let G be a noncompact locally compact Abelian group and $1<p<\infty$. If $T\colon L_1(G)\cap L_p(G)\to L_1(G)$ is a linear transformation then the following are equivalent:*

(i) *$T\in M(L_1(G)\cap L_p(G), L^1(G))$.*

(ii) *There exists a unique measure $\mu\in M(G)$ such that $Tf=\mu*f$ for each $f\in L_1(G)\cap L_p(G)$.*

Moreover the correspondence between T and μ defines an isometric algebra isomorphism of $M(L_1(G)\cap L_p(G), L_1(G))$ onto $M(G)$.

Proof. If $\mu\in M(G)$ and $Tf=\mu*f$ for each $f\in L_1(G)\cap L_p(G)$ then clearly

$$\|Tf\|_1 = \|\mu*f\|_1 \leqq \|\mu\|\,\|f\|_1 \leqq \|\mu\|\,\|f\|_{1,p}.$$

It is then evident that $T\in M(L_1(G)\cap L_p(G), L_1(G))$ and $\|T\|\leqq\|\mu\|$.

Conversely, suppose that $T\in M(L_1(G)\cap L_p(G), L_1(G))$. Then for each $f\in L_1(G)\cap L_p(G)$ we have

$$\|Tf\|_1 \leqq \|T\|\,(\|f\|_1 + \|f\|_p).$$

Combining this estimate with Lemma 3.5.1(i) we deduce that

$$2\|Tf\|_1 = \lim_{s\to+\infty} \|Tf + \tau_s Tf\|_1 = \lim_{s\to+\infty} \|T(f+\tau_s f)\|_1$$
$$\leqq \lim_{s\to+\infty} \|T\|(\|f+\tau_s f\|_1 + \|f+\tau_s f\|_p) = \|T\|(2\|f\|_1 + 2^{1/p}\|f\|_p)$$

for each $f\in L_1(G)\cap L_p(G)$. Thus

$$\|Tf\|_1 \leqq \|T\|(\|f\|_1 + 2^{(1/p)-1}\|f\|_p) \qquad (f\in L_1(G)\cap L_p(G)).$$

Repeating this process n times we see that

$$\|Tf\|_1 \leqq \|T\|(\|f\|_1 + 2^{n((1/p)-1)}\|f\|_p) \qquad (f\in L_1(G)\cap L_p(G)).$$

Since $p>1$ we have $\lim_n 2^{n((1/p)-1)}=0$, and so we conclude that

$$\|Tf\|_1 \leqq \|T\|\,\|f\|_1 \qquad (f\in L_1(G)\cap L_p(G)).$$

Hence T defines a continuous linear transformation from $L_1(G)\cap L_p(G)$ considered as a subspace of $L_1(G)$ to $L_1(G)$ which commutes with translation. Thus, since $L_1(G)\cap L_p(G)$ is norm dense in $L_1(G)$, T determines a unique element T' of $M(L_1(G))$ and $\|T'\|\leqq\|T\|$. By Theorems 0.1.1 or 3.1.1 there exists a unique $\mu\in M(G)$ such that $T'f=\mu*f$ for each $f\in L_1(G)$ and $\|\mu\|=\|T'\|$. Consequently, $Tf=\mu*f$ for each $f\in L_1(G)\in L_p(G)$ and $\|\mu\|\leqq\|T\|$.

Therefore (i) and (ii) are equivalent.

It is evident that the correspondence between T and μ defines an isometric algebra isomorphism from $M(L_1(G)\cap L_p(G), L_1(G))$ onto $M(G)$. □

Utilizing the second portion of Lemma 3.5.1 we can prove by essentially the same arguments as those just given the analogous result for $M(L_1(G)\cap C_0(G), L_1(G))$.

Theorem 3.5.2. *Let G be a noncompact locally compact Abelian group. If $T\colon L_1(G)\cap C_0(G)\to L_1(G)$ is a linear transformation then the following are equivalent:*

(i) *$T\in M(L_1(G)\cap C_0(G), L_1(G))$.*

(ii) *There exists a unique measure $\mu\in M(G)$ such that $Tf=\mu*f$ for each $f\in L_1(G)\cap C_0(G)$.*

Moreover the correspondence between T and μ determines an isometric algebra isomorphism from $M(L_1(G)\cap C_0(G), L_1(G))$ onto $M(G)$.

When G is a compact Abelian group then the Banach spaces $L_1(G)\cap L_p(G)$, $1<p<\infty$, and $L_1(G)\cap C_0(G)$ are easily seen to be the same as the Banach spaces $L_p(G)$, $1<p<\infty$, and $C(G)$. The problem of describing $M(L_p(G), L_1(G))$ and $M(C(G), L_1(G))$ in this case is somewhat more difficult than the same problem for noncompact G. Some results for $M(L_p(G), L_1(G))$, $1<p<\infty$, are given by Theorem 5.2.3, while the question for $M(C(G), M(G))$ is settled by Theorem 5.1.6 and the comments following that result.

Finally we note that Theorems 3.5.1 and 3.5.2 also give a complete description of the spaces $M(L_1(G)\cap L_p(G))$, $1<p<\infty$, and $M(L_1(G)\cap C_0(G))$ for noncompact G.

Indeed, suppose $T\in M(L_1(G)\cap L_p(G))$ for some p, $1<p<\infty$. Then for each $f\in L_1(G)\cap L_p(G)$ we have

$$\|Tf\|_1 \leqq \|Tf\|_{1,p} \leqq |||T|||\ \|f\|_{1,p}$$

where $|||T|||$ denotes the norm of the operator T considered as an element of $M(L_1(G)\cap L_p(G))$. The foregoing estimate shows that T is also an element of $M(L_1(G)\cap L_p(G), L_1(G))$ and $\|T\|\leqq |||T|||$. Applying Theorem 3.5.1 we deduce the existence of a unique $\mu\in M(G)$ such that $Tf=\mu * f$ for each $f\in L_1(G)\cap L_p(G)$ and for which $\|\mu\|=\|T\|\leqq|||T|||$. Conversely, if $\mu\in M(G)$ is such that $Tf=\mu * f$ for each $f\in L_1(G)\cap L_p(G)$ then we conclude at once from the estimate

$$\|Tf\|_{1,p} = \|\mu * f\|_1 + \|\mu * f\|_p \leqq \|\mu\|\ (\|f\|_1 + \|f\|_p) = \|\mu\|\ \|f\|_{1,p}$$

that $T\in M(L_1(G)\cap L_p(G))$ and $|||T|||\leqq\|\mu\|$.

The results of this argument, and a similar one applied to $M(L_1(G)\cap C_0(G))$, can be summarized in the following corollaries.

Corollary 3.5.1. *Let G be a noncompact locally compact Abelian group and $1<p<\infty$. If T: $L_1(G)\cap L_p(G)\to L_1(G)\cap L_p(G)$ is a linear transformation then the following are equivalent:*

(i) *$T\in M(L_1(G)\cap L_p(G))$.*

(ii) *There exists a unique measure $\mu\in M(G)$ such that $Tf=\mu * f$ for each $f\in L_1(G)\cap L_p(G)$.*

Moreover the correspondence between T and μ defines an isometric algebra isomorphism from $M(L_1(G)\cap L_p(G))$ onto $M(G)$.

Corollary 3.5.2. *Let G be a noncompact locally compact Abelian group. If T: $L_1(G)\cap C_0(G)\to L_1(G)\cap C_0(G)$ is a linear transformation then the following are equivalent:*

(i) *$T\in M(L_1(G)\cap C_0(G))$.*

(ii) *There exists a unique measure $\mu\in M(G)$ such that $Tf=\mu * f$ for each $f\in L_1(G)\cap C_0(G)$.*

Moreover the correspondence between T and μ defines an isometric algebraic isomorphism from $M(L_1(G)\cap C_0(G))$ onto $M(G)$.

3.6. Positive Multipliers and Isomorphisms of $L_p(G)$, $1 \leqq p < \infty$. In this section we shall give a characterization for locally compact Abelian groups of the positive multipliers for $L_p(G)$, and then apply the result to the problem of describing the isomorphisms of $L_p(G)$ when G is compact. We say that a linear transformation $T: L_p(G) \to L_p(G)$, $1 \leqq p < \infty$, is *positive* if $f \geqq 0$ almost everywhere implies $Tf \geqq 0$ almost everywhere.

The following lemma will be useful. $L_p^+(G)$ denotes the space of $f \in L_p(G)$ such that $f \geqq 0$ almost everywhere.

Lemma 3.6.1. *Let G be a locally compact Abelian group and $1 \leqq p < \infty$. If $0 < \varepsilon < 1$ then for each compact symmetric set K containing the identity of G there exists a $g_K \in L_\infty^+(G) \cap L_1^+(G)$ such that:*

(i) $\|g_K\|_p = 1$.

(ii) $\|\tau_s g_K - g_K\|_p \leqq (2\varepsilon)^{1/p}$ *for all* $s \in K$.

Proof. Given ε and K there exists (Theorem C.4.3) a Borel set V with compact closure such that $\lambda(VK) < (1+\varepsilon)\lambda(V)$. Let f be the characteristic function of V. Then for each $s \in G$ it is apparent $\tau_s f$ is the characteristic function of Vs, and $|\tau_s f - f|$ is the characteristic function of $Vs \Delta V = (Vs \sim V) \cup (V \sim Vs)$. Thus by appealing to the symmetry of K and the fact that $V \subset VK$ we conclude for each $s \in K$ that

$$\begin{aligned}
\|\tau_s f - f\|_p &= [\lambda(Vs \Delta V)]^{1/p} = [\lambda(Vs \sim V) + \lambda(V \sim Vs)]^{1/p} \\
&= [\lambda(Vs \sim V) + \lambda((V \sim Vs)\, s^{-1})]^{1/p} \\
&= [\lambda(Vs \sim V) + \lambda(Vs^{-1} \sim V)]^{1/p} \\
&\leqq [\lambda(VK \sim V) + \lambda(VK^{-1} \sim V)]^{1/p} = [2\lambda(VK \sim V)]^{1/p} \\
&= [2(\lambda(VK) - \lambda(V))]^{1/p} = [2\varepsilon\, \lambda(V)]^{1/p}.
\end{aligned}$$

The last inequality is valid due to the choice of V.

However $\|f\|_p = [\lambda(V)]^{1/p}$ and hence

$$\|\tau_s f - f\|_p \leqq (2\varepsilon)^{1/p}\, \|f\|_p \qquad (s \in K).$$

The conclusion of the lemma is now evident upon setting $g_K = f/\|f\|_p$ and noting that $\|f\|_p \neq 0$. ▯

The construction shows that g_K is just a scalar multiple of the characteristic function of some Borel set with compact closure.

Theorem 3.6.1. *Let G be a locally compact Abelian group and $1 \leqq p < \infty$. If $T: L_p(G) \to L_p(G)$ is a linear transformation then the following are equivalent:*

(i) *T is positive and $\tau_s T = T\tau_s$ for each $s \in G$.*
(ii) *T is positive and $T \in M(L_p(G))$.*
(iii) *There exists a unique nonnegative $\mu \in M(G)$ such that $Tf = \mu * f$ for each $f \in L_p(G)$.*

Proof. It is obvious that (iii) implies (ii) implies (i). Assume T satisfies (i) and suppose that T is not continuous. Then for each positive integer n there exists $g_n \in L_p(G)$ such that $\|g_n\|_p \leqq 1$ and $\|Tg_n\|_p \geqq 2n^3$. Since $\|Tg_n\|_p \leqq \|T\mathrm{Re}(g_n)\|_p + \|T\mathrm{Im}(g_n)\|_p$ it follows at once that there also exist $f_n \in L_p^R(G)$ for which $\|f_n\|_p \leqq 1$ and $\|Tf_n\|_p \geqq n^3$. Moreover, we may even assume, without loss of generality, that such f_n belong to $L_p^+(G)$. Indeed, since T is positive $-|f_n| \leqq f_n \leqq |f_n|$ almost everywhere implies $-T|f_n| \leqq Tf_n \leqq T|f_n|$, and so $|Tf_n| \leqq T|f_n|$, almost everywhere. Hence $\|T|f_n|\|_p \geqq n^3$ and $|f_n| \in L_p^+(G)$.

Thus assuming each $f_n \in L_p^+(G)$ it is apparent that $\sum_{n=1} n^{-2} f_n$ converges in $L_p(G)$ to some $f \in L_p^+(G)$. Appealing again to the positivity of T we conclude for each n that $0 \leqq Tf_n \leqq n^2\, Tf$ almost everywhere, since evidently $0 \leqq f_n \leqq n^2 f$ almost everywhere. But then for each n we obtain $n^2 \|Tf\|_p \geqq \|Tf_n\|_p \geqq n^3$, and hence $\|Tf\|_p \geqq n$, contradicting the fact that $T\colon L_p(G) \to L_p(G)$.

Therefore T is continuous and (i) implies (ii).

Now suppose that (ii) holds. Then, by essentially the same argument as given in the proof of Theorem 3.1.1, we see that $T(f*g) = Tf*g$ for each $f \in L_p(G)$ and $g \in L_1(G)$. Thus

$$\|Tf*g\|_p \leqq \|T\|\,\|f*g\|_p \leqq \|T\|\,\|f\|_1\,\|g\|_p \quad (f, g \in L_p(G) \cap L_1(G)).$$

Let $f \in L_p^+(G) \cap L_1^+(G)$, $\|f\|_1 = 1$, and set $h = Tf$. For $\varepsilon = 2^{-(p+1)}$ and K any compact symmetric subset of G containing the identity let $g_K \in L_\infty^+(G) \cap L_1^+(G)$ satisfy the conclusion of Lemma 3.6.1, that is, $\|g_K\|_p = 1$ and $\|\tau_s g_K - g_K\|_p \leqq (2\varepsilon)^{1/p} = 2^{-1}$ for each $s \in K$. Set $h_K = h\chi_K$, where χ_K denotes the characteristic function of K, and $m_K = \|h_K\|_1$. Now $h = Tf \geqq 0$ almost everywhere as T is positive so that $h_K \geqq 0$ almost everywhere. Since $g_K \geqq 0$ almost everywhere we see at once that $0 \leqq h_K * g_K \leqq h * g_K$, and hence

$$\|h_K * g_K\|_p \leqq \|h * g_K\|_p = \|Tf * g_K\|_p \leqq \|T\|\,\|f\|_1\,\|g_K\|_p = \|T\|,$$

since $f, g_K \in L_p^+(G) \cap L_1^+(G)$.

Furthermore, let $1/p + 1/p' = 1$. Evidently

$$h_K * g_K(t) - m_K\, g_K(t) = \int_K [\tau_s g_K(t) - g_K(t)]\, h_K(s)\, d\lambda(s).$$

So applying Hölder's inequalities where the integrations are taken with respect to the measure $h_K\,d\lambda$, we conclude that

$$|h_K * g_K(t) - m_K\, g_K(t)| \leqq \Big(\int_K |\tau_s\, g_K(t) - g_K(t)|^p\, h_K(s)\, d\lambda(s)\Big)^{1/p} \Big(\int_K h_K(s)\, d\lambda(s)\Big)^{1/p'}.$$

Hence

$$\begin{aligned}(\|h_K * g_K - m_K\, g_K\|_p)^p &\leqq \Big[\int_G \Big(\int_K |\tau_s\, g_K(t) - g_K(t)|^p\, h_K(s)\, d\lambda(s)\Big)\, d\lambda(t)\Big] \Big[\int_K h_K(s)\, d\lambda(s)\Big]^{p/p'}\\ &= \Big[\int_K \Big(\int_G |\tau_s\, g_K(t) - g_K(t)|^p\, d\lambda(t)\Big)\, h_K(s)\, d\lambda(s)\Big]\, m_K^{p/p'}\\ &= \Big[\int_K (\|\tau_s\, g_K - g_K\|_p)^p\, h_K(s)\, d\lambda(s)\Big]\, m_K^{p/p'} \leqq 2^{-p}\, m_K^{1+p/p'}.\end{aligned}$$

Consequently,

$$\|h_K * g_K - m_K\, g_K\|_p \leqq 2^{-1}\, m_K^{1/p+1/p'} = 2^{-1}\, m_K.$$

Combining this last inequality with the estimate of $\|h_K * g_K\|_p$, we immediately obtain that

$$m_K = m_K\, \|g_K\|_p = \|m_K\, g_K\|_p \leqq \|h_K * g_K\|_p + 2^{-1}\, m_K \leqq \|T\| + 2^{-1}\, m_K.$$

It is then obvious that

$$\int_K Tf(t)\, d\lambda(t) = \int_G h_K(t)\, d\lambda(t) = m_K \leqq 2\, \|T\|.$$

Since K was an arbitrary compact symmetric subset of G which contained the identity we conclude that $\|Tf\|_1 \leqq 2\, \|T\|$.

But this estimate is valid for every $f \in L_p^+(G) \cap L_1^+(G)$ for which $\|f\|_1 = 1$. Thus $\|Tf\|_1 \leqq 2\, \|T\|\, \|f\|_1$ for each $f \in L_p^+(G) \cap L_1^+(G)$. Moreover, as we indicated above, for any real valued $f \in L_p^+(G) \cap L_1^+(G)$ we have $|Tf| \leqq T|f|$ almost everywhere, and hence for such f,

$$\|Tf\|_1 \leqq \|T|f|\|_1 \leqq 2\, \|T\|\, \|f\|_1.$$

But then for each $f \in L_p(G) \cap L_1(G)$ we see that

$$\begin{aligned}\|Tf\|_1 &\leqq \|T\,\mathrm{Re}(f)\|_1 + \|T\,\mathrm{Im}(f)\|_1 \leqq 2\, \|T\|\, \|\mathrm{Re}(f)\|_1 + 2\, \|T\|\, \|\mathrm{Im}(f)\|_1\\ &\leqq 4\, \|T\|\, \|f\|_1.\end{aligned}$$

Therefore T defines a bounded linear transformation from the norm dense subspace $L_p(G) \cap L_1(G)$ into $L_1(G)$ which commutes with translation. Thus T can be uniquely extended to an element $T' \in M(L_1(G))$. Consequently, by either Theorem 0.1.1 or 3.1.1, there exists a unique $\mu \in M(G)$ such that $T' f = \mu * f$ for each $f \in L_1(G)$ and $Tf = \mu * f$ for each

$f \in L_p(G) \cap L_1(G)$. Moreover, since T is positive on the norm dense subset $L_p^+(G) \cap L_1^+(G)$ of $L_1^+(G)$, it follows from the continuity of T' that T' is positive. From this an easy argument shows that μ is a nonnegative measure.

Finally, if $f \in L_p^+(G)$ then let $\{f_n\} \subset L_p^+(G) \cap L_1^+(G)$ be an almost everywhere monotone increasing sequence which converges almost everywhere to f. Then by the Monotone Convergence Theorem (Theorem C.5.1) we have $\lim_n \|f_n - f\|_p = 0$ which implies via the estimate $\|Tf_n - Tf\|_p \leqq \|T\| \, \|f_n - f\|_p$, that $\lim_n \|Tf_n - Tf\|_p = 0$. However, since μ is nonnegative, the Monotone Convergence Theorem also can be applied to conclude that $\{Tf_n\} = \{\mu * f_n\}$ converges almost everywhere to $\mu * f$. Thus $Tf = \mu * f$ for each $f \in L_p^+(G)$, and hence for each $f \in L_p(G)$.

Therefore (ii) implies (iii), and the equivalence of (i), (ii) and (iii) is established. □

It should be noted that if $p > 1$ and G is infinite, compact and Abelian then there always exist $T \in M(L_p(G))$ which are not of the form $Tf = \mu * f$. Indeed, let $E \subset \hat{G}$ be any infinite Sidon set (F.11) and let φ be any bounded function on $\hat{G}$ which vanishes off E. If $1 < p \leqq 2$ then it is possible to show, using the properties of Sidon sets, that the equation $(Tf)\hat{} = \varphi \hat{f}$ for each $f \in L_p(G)$ defines an element of $M(L_p(G))$ (F.11d). But such a function φ is a Fourier-Stieltje's transform if and only if $\varphi \in L_2(\hat{G})$.

Indeed, suppose $\varphi = \hat{\mu}$ for some $\mu \in M(G)$ and $\{u_\alpha\}$ is an approximate identity in $L_1(G)$ composed of trigonometric polynomials for which $\|u_\alpha\|_1 = 1$ (F.7d). Then, since each $\mu * u_\alpha$ is a trigonometric polynomial which vanishes off of E, we conclude (F.11b) for each α that

$$\|\mu * u_\alpha\|_2 \leqq 2B \, \|\mu * u_\alpha\|_1 \leqq 2B \, \|\mu\|,$$

where B is the constant determined by the Sidon set E. Hence by Alaoglu's Theorem (Theorem D.4.3) and the reflexivity of $L_2(G)$ we see that there exists a subnet $\{\mu * u_\beta\}$ which converges weakly to some f in $L_2(G)$. This however implies that $\{(\mu * u_\beta)\hat{}\} = \{\hat{\mu} \, \hat{u}_\beta\}$ converges pointwise to $\hat{f}$ since G is compact and so $\hat{G} \subset L_2(G)$. Hence $\hat{\mu} = \hat{f}$ as $\{\hat{u}_\beta\}$ converges pointwise to one, and $\varphi = \hat{\mu} = \hat{f} \in L_2(\hat{G})$ by the Plancherel Theorem (Theorem F.8.2). Conversely, if $\varphi \in L_2(\hat{G})$ then by the Plancherel Theorem there exists a $\mu \in L_2(G)$ such that $\varphi = \hat{\mu}$. But since G is compact we have $L_2(G) \subset L_1(G) \subset M(G)$.

Since E is infinite it is apparent that one can always construct φ such that $\sum_\gamma |\varphi(\gamma)|^2 = +\infty$ and hence T is not obtained by convolution with an element of $M(G)$. Appealing to the duality between $L_p(G)$ and $L_{p'}(G)$, $1/p + 1/p' = 1$, we see that the same construction defines multipliers for $L_{p'}(G)$, $2 < p' < \infty$, which are not convolutions with bounded measures. This follows immediately from the Theorem 4.1.2, or from the following argument in the case of compact groups.

Suppose $2<p'<\infty$ and $T\in M(L_p(G))$ is as constructed above. Let $T^*: L_{p'}(G)\to L_{p'}(G)$ be the bounded linear transformation adjoint to T. Since G is compact $\hat{G}\subset L_p(G)\cap L_{p'}(G)$, and so for each $\gamma, \gamma'\in\hat{G}$, considered alternately as points of $\hat{G}$, and as functions on G, we have

$$\begin{aligned}(T^*\gamma)\hat{}\,(\gamma') &= \int_G (t^{-1},\gamma')\, T^*\gamma(t)\, d\lambda(t) = \langle T^*\gamma, \gamma'\rangle = \langle \gamma, T\gamma'\rangle \\ &= (T\gamma')\hat{}\,(\gamma) = \varphi(\gamma)\,\hat{\gamma}'(\gamma) = \varphi(\gamma)\int_G (t^{-1},\gamma)(t,\gamma')\, d\lambda(t) \\ &= \varphi(\gamma)\int_G (t^{-1},\gamma')(t,\gamma)\, d\lambda(t) = \varphi(\gamma)\,\hat{\gamma}(\gamma').\end{aligned}$$

Thus for each trigonometric polynomial g on G we see that $(T^* g)\hat{} = \varphi\hat{g}$ as $\hat{\gamma}=\chi_{\{\gamma\}}$ for each $\gamma\in\hat{G}$. Since the trigonometric polynomials are dense in $L_{p'}(G)$ (F.7d), it follows that $(T^* f)\hat{} = \varphi\hat{f}$ for each $f\in L_{p'}(G)$, and $T^*\in M(L_{p'}(G))$.

Let G_1 and G_2 be locally compact Abelian topological groups. If $\alpha: G_1\to G_2$ is a topological isomorphism of G_1 onto G_2 then it is easy to see that the mapping S defined by $(Sf)[\alpha(t)]=f(t)$ defines an isomorphism of $L_p(G_1)$ onto $L_p(G_2)$, $1\leqq p<\infty$. As indicated in the notes to the zeroth chapter a partial converse of this observation has been established for $p=1$ when the algebra isomorphism is bipositive. We shall now extend this result to $1<p<\infty$ under the additional assumption that G be compact.

A linear transformation $S: L_p(G_1)\to L_p(G_2)$ will be called *bipositive* whenever it is the case that $Sf\geqq 0$ almost everywhere if and only if $f\geqq 0$ almost everywhere.

Theorem 3.6.2. *Let G_1 and G_2 be compact Abelian groups and $1\leqq p<\infty$. If $S: L_p(G_1)\to L_p(G_2)$ is a bipositive algebra isomorphism of $L_p(G_1)$ onto $L_p(G_2)$ then G_1 and G_2 are topologically isomorphic.*

Proof. Let $h\in L_1(G_2)$ be nonnegative and for each $f\in L_p(G_1)$ define $Tf=S^{-1}(h*Sf)$. Since S is a bipositive isomorphism it is evident that T is a positive linear transformation from $L_p(G_1)$ to itself. Moreover, since $L_p(G_1)$ is an algebra under convolution we see for each $f, g\in L_p(G_1)$ that

$$T(f*g)=S^{-1}(h*S(f*g))=S^{-1}(h*Sf*Sg)=S^{-1}(h*Sf)*g=Tf*g.$$

Thus since $L_p(G_1)$ is a semi-simple commutative Banach algebra we conclude from Theorem 1.1.1 that $T\in M(L_p(G_1))$. Consequently, by the previous theorem, there exists a unique nonnegative $\mu\in M(G_1)$ such that $\mu*f=Tf=S^{-1}(h*Sf)$ for each $f\in L_p(G_1)$. Clearly this correspondence between $h \in L_1(G_2)$ and μ in $M(G_1)$ defines a positive linear transformation U from $L_1(G_2)$ to $M(G_1)$.

If $h \in L_p(G_2) \subset L_1(G_2)$ then $Tf = S^{-1}(h * Sf) = S^{-1} h * f$ for each $f \in L_p(G_1)$ and so $Uh = S^{-1} h$. Consequently if $g, h \in L_p(G_2)$ then $U(g * h) = S^{-1}(g * h) = S^{-1} g * S^{-1} h = Ug * Uh$. But U is a positive linear, hence order preserving, transformation, and so U is continuous. Indeed, if U were not continuous then for each positive integer n there would be an $f_n \in L_1(G_2)$, $f_n \geqq 0$, such that $\|Uf_n\| \geqq n \|f_n\|_1$. Set $f = \sum_{n=1}^{\infty} (n^2 \|f_n\|_1)^{-1} f_n$. Clearly $f \in L_1(G_2)$ and $f \geqq 0$. Thus $Uf \in M(G_1)$ and $Uf \geqq 0$. Moreover, $Uf_n \geqq 0$ for each n and $f - \sum_{n=1}^{N} (n^2 \|f_n\|_1)^{-1} f_n \geqq 0$ almost everywhere implies

$$U\left(f - \sum_{n=1}^{N} (n^2 \|f_n\|_1)^{-1} f_n\right) \geqq 0$$

for each positive integer N. Hence for each N we have

$$\begin{aligned} \|Uf\| &\geqq \int_G d\left\{U\left[\sum_{n=1}^{N} (n^2 \|f_n\|_1)^{-1} f_n\right]\right\}(t) = \sum_{n=1}^{N} (n^2 \|f_n\|_1)^{-1} \int_G dU f_n(t) \\ &= \sum_{n=1}^{N} (n^2 \|f_n\|_1)^{-1} \|Uf_n\| \geqq \sum_{n=1}^{N} \frac{1}{n}, \end{aligned}$$

which contradicts the fact that $\|Uf\| < +\infty$. Consequently U is continuous.

Since $L_p(G_2)$ is norm dense in $L_1(G_2)$ it then follows that $U(g * h) = Ug * Uh$ for all $g, h \in L_1(G_2)$. Similarly, the continuity of U and the identity $Uh = S^{-1} h$ for $h \in L_p(G_2)$ shows that the image of $L_1(G_2)$ under U lies in $L_1(G_1)$.

Therefore U is a positive continuous isomorphism of $L_1(G_2)$ into $L_1(G_1)$.

Interchanging the roles of G_1 and G_2 we construct a positive continuous isomorphism U' of $L_1(G_1)$ into $L_1(G_2)$ such that $(U'h) * f = S(h * S^{-1} f)$ for each $h \in L_1(G_1)$ and $f \in L_p(G_2)$.

Moreover, if $h \in L_p(G_2)$ then for each $f \in L_p(G_2)$ we have

$$(U'Uh) * f = S(Uh * S^{-1} f) = S[S^{-1}(h * SS^{-1} f)] = h * f.$$

Thus $U'Uh = h$ for each $h \in L_p(G_2)$, that is, $U'U$ is the identity transformation on $L_1(G_2)$ as $L_p(G_2)$ is norm dense in $L_1(G_2)$. Similarly UU' is the identity transformation on $L_1(G_1)$. Consequently $U^{-1} = U'$ and $U'^{-1} = U$ exist and both U and U' define bipositive isomorphisms between $L_1(G_1)$ and $L_1(G_2)$.

Therefore by Theorem 0.3.3 we conclude that G_1 and G_2 are topologically isomorphic. □

3.7. Notes. The majority of the development in this chapter is based on results which can be found in Brainerd and Edwards [1_{I}], Edwards [2, 9] and Gaudry [8]. In particular proofs of the Theorems in 3.1, 3.2, 3.3 and 3.4 are all available in either Brainerd and Edwards [1_{I}] or Edwards [2], while the results of 3.5 can be found in Gaudry [8]. Similar proofs of the results in the latter section have also been given in Figà-Talamanca and Gaudry [2] and Pigno [2]. The argument is essentially due to Hörmander [1] and F. Forelli. The proof of Theorem 3.6.1 is taken from Brainerd and Edwards [1_{I}] in a form suggested by R.B. Burckel, while Theorem 3.6.2 comes from Edwards [9].

Theorems 3.1.1, 3.3.1 and 3.4.3 were also proved in Hörmander [1] for $G=R^n$, while Comisky [1] and Gulick, Liu and van Rooij [1_{I}] have given proofs of the first two results in the context of Banach modules. Boehme [1] obtains similar results for the L_p-spaces of a half-line. In general, the multipliers of Banach modules have been studied and utilized by a number of authors, among whom we mention Comisky [1], Gulick, Liu and van Rooij [1], Harte [1], Kitchen [1], Liu, van Rooij and Wang [1], Máté [8] and Rieffel [2].

Many of the results of the first four sections are also discussed in Edwards [14_{II}] 16.1, 16.2 and 16.3, where the case of the circle group is emphasized, and Gaudry [6], V.1. In particular, the examples given in 3.2 and 3.4 concerning the disparities which arise due to the various possible definitions of multipliers are taken from the latter reference.

Pigno [1, 2] has obtained descriptions of the multipliers for various intersection spaces. Some of his results are collected in the following theorem.

Theorem 3.7.1. *Let G be a locally compact Abelian group and* $1 \leqq p \leqq \infty$. *If* T: $L_1(G) \cap L_p(G) \to L_1(G) \cap C_0(G)$ *is a linear transformation then the following are equivalent:*

(i) $T \in M(L_1(G) \cap L_p(G), L_1(G) \cap C_0(G))$.

(ii) *There exists a unique* $g \in L_1(G)$ *such that* $Tf = f * g$ *for each* $f \in L_1(G) \cap L_p(G)$ *with the property that if* $1 \leqq p < \infty$ *then* $g \in L_1(G) \cap L_{p'}(G)$, $1/p + 1/p' = 1$.

Pigno [2] shows also that $M(L_1(G) \cap L_p(G), L_1(G) \cap L_\infty(G)) = M(L_1(G) \cap L_p(G), L_1(G) \cap C_0(G))$, $1 \leqq p < \infty$.

A number of these results have also been obtained by other authors for particular groups. We mention in this connection Doss [1], Edwards [2], [14_{II}], 16.3.5, Hewitt and Ross [2], 35, Verblunsky [1], Young [2] and Zygmund [1]. Some other results concerning multipliers for intersection spaces can be found in Pigno [3], Ryan [1] and Wells [2].

The structure of the Banach space $L_1(G) \cap L_p(G)$, $1 \leqq p \leqq \infty$, has been investigated fairly extensively by Liu and van Rooij [1], Liu and Wang [1], Reiter [1], Warner [1] and Yap [1]. [1]

Theorem 3.6.1 is valid in the non Abelian case in the sense that every suitably defined positive multiplier T for $L_p(G)$, $1 \leqq p < \infty$, is such that $Tf = \mu * f$ for each $f \in C_c(G)$ and where μ is some not necessarily bounded nonnegative regular Borel measure on G. The reader is referred to Brainerd and Edwards [1_{I}] for the details. Some other results about positive multipliers are available in Edwards [5, 9], [14_{II}], 16.3.10.

Analogs of Theorem 3.6.2 for both bipositive and isometric multipliers have been obtained by Gaudry [3], Parrott [1] and Strichartz [2]. A summary of their results is given in the following theorems where we assume the measure involved is right Haar measure.

Theorem 3.7.2. *Let G_1 and G_2 be locally compact groups, $1 \leqq p \leqq \infty$ and suppose S: $L_p(G_1) \to L_p(G_2)$ is a surjective algebra isomorphism. Then G_1 and G_2 are topologically isomorphic whenever any of the following conditions are satisfied.*

(i) *$p \neq 2$, $p \neq \infty$ and S is an isometry.*

(ii) *$p = \infty$, G_1 and G_2 are compact groups and S is either bipositive or an isometry.*

(iii) *$p \neq 2$, $p \neq \infty$, G_1 and G_2 are compact Abelian groups and $\|Sf\|_p \leqq \|f\|_p$ for all $f \in L_p(G_1)$.*

An isomorphism theorem as this fails when $p = 2$ as observed in Gaudry [3], Parrott [1] and Strichartz [1].

Edwards [9] has also shown that two locally compact groups G_1 and G_2 are topologically isomorphic whenever there exists a bipositive or isometric algebra isomorphism of $C_c(G_1)$ onto $C_c(G_2)$.

In much the same vein as the previous results are those in Gaudry [7] regarding the relation between isomorphisms of $M(L_p(G_1))$ and $M(L_p(G_2))$ and isomorphisms of G_1 and G_2. His results are given in the next theorem which should be compared with Theorems 0.3.2, 0.3.3 and 0.3.4.

Theorem 3.7.3. *Let G_1 and G_2 be locally compact groups, $1 \leqq p < \infty$ and suppose $S: M(L_p(G_1)) \to M(L_p(G_2))$ is a surjective algebra isomorphism. Then G_1 and G_2 are topologically isomorphic whenever either of the following conditions is satisfied.*

(i) *$p \neq 2$ and S is an isometry.*

(ii) *S is bipositive.*

The first portion of the theorem fails if $p = 2$.

---

[1] See also Cigler [1] and Reiter [2].

Similar problems for Euclidean groups have also been discussed in Hörmander [1].

In addition to the papers already cited here and in the previous chapters, numerous other articles have been written concerning the multipliers of various pairs of linear spaces. It is not possible here to provide a detailed description of even a portion of these works. Instead we offer the following incomplete sampling of papers which deal with such subjects, with the exception that those papers we shall discuss in the succeeding chapters are not listed: Bachelis and Rosenthal [1], Boehme [2], Brainerd and Edwards [1], Byrnes and Newman [1], DeVore [1], Edwards [3, 6, 7, 12], Gaudry [1], [6], V and VI, Johnson [1, 2, 5], Kadec and Pelczynski [1], König and Meixner [1], McGiveney and Ruckle [1], Merlo [1], Meyer [1, 3], Price [1], Rowlands [1], Rudin [6], Taibleson [2], Thorp [1], Volevich and Paneyakh [1], Wada [1], Wells [1], and Weston [1–5].[1]

---

[1] See also Powell [1], Rivière [1] and Taibleson [3].

Chapter 4

The Multipliers for $L_p(G)$

4.0. Introduction. In this chapter we shall investigate the multipliers for the pair $(L_p(G), L_p(G))$. We have proven some scattered results pertaining to $M(L_p(G))$ in the previous chapters. In particular, we have already discussed to some extent the cases when $p=1$ and $p=\infty$. Consequently we shall now restrict our attention primarily to the values of p such that $1<p<\infty$. We shall show in the following sections that the multipliers for $L_p(G)$ can, in a certain sense, be represented either as multiplication of the Fourier transform by a bounded function or as a convolution operator, in this instance convolution with a pseudomeasure. We shall also investigate the relationships between the spaces $M(L_p(G))$ for various values of p, obtain some results on the existence of bounded functions which do not determine multipliers for $L_p(G)$, and examine the notion of the derived space for $L_p(G)$. As usual we shall consider only the commutative case.

In the following sections we shall need to consider the Fourier transform of elements in $L_p(G)$, at least for $1<p\leqq 2$, and to use the Hausdorff-Young Theorem (Theorem F.8.4). If $1<p<2$ and $f\in L_1(G)\cap L_p(G)$ then the Hausdorff-Young Theorem shows that $\|\hat{f}\|_{p'}\leqq\|f\|_p$ where $1/p+1/p'=1$. This inequality then enables one to uniquely extend the Fourier transform defined on the dense subspace $L_1(G)\cap L_p(G)$ of $L_p(G)$ to all of $L_p(G)$ in such a way that one still has $\|\hat{f}\|_{p'}\leqq\|f\|_p$ for each $f\in L_p(G)$. The Fourier-Plancherel transform is of course obtained in this way when $p=2$. We shall use the notation $\hat{f}$ to denote the usual Fourier transform, the Hausdorff-Young extension for $L_p(G)$, $1<p<2$, and the Fourier-Plancherel transform. Generally the context will make clear which transform is under discussion. Similarly $\check{f}$ will denote the various extensions of the mapping defined by

$$\check{f}(\gamma)=\int_G (t,\gamma)\,f(t)\,d\lambda(t) \qquad (f\in C_c(G)),$$

that is, by the inverse Fourier transform.

4.1. The Multipliers for $L_p(G)$ as Bounded Functions, $1<p<\infty$. For the study of the multipliers of $L_p(G)$ we shall need a notation for the norm of $T\in M(L_p(G))$ which will indicate the dependence on the index p. Thus if $T\in M(L_p(G))$, $1<p<\infty$, we shall denote the operator norm of T by $\|T\|_p$.

We begin the section with a description of the multipliers for $L_2(G)$.

Theorem 4.1.1. *Let G be a locally compact Abelian group and suppose T: $L_2(G)\to L_2(G)$ is a linear transformation. Then the following are equivalent:*

(i) $T\in M(L_2(G))$.

(ii) *There exists a unique $\varphi\in L_\infty(\hat{G})$ such that $(Tf)^\wedge=\varphi\hat{f}$ for each $f\in L_2(G)$.*

Moreover the correspondence between T and φ defines an isometric algebra isomorphism from $M(L_2(G))$ onto $L_\infty(\hat{G})$.

The notation $\hat{f}$ of course denotes the Fourier-Plancherel transform of an element in $L_2(G)$.

Proof. If $\varphi\in L_\infty(\hat{G})$ then $\varphi\hat{f}\in L_2(\hat{G})$ for each $f\in L_2(G)$. Thus there exists a unique $Tf\in L_2(G)$ for which $(Tf)^\wedge=\varphi\hat{f}$. It is apparent that T so defined is a linear transformation from $L_2(G)$ to $L_2(G)$ which commutes with translations and is bounded. Thus $T\in M(L_2(G))$. Furthermore, by the Plancherel Theorem (Theorem F.8.2) for each $f\in L_2(G)$ we have

$$\|Tf\|_2=\|(Tf)^\wedge\|_2=\|\varphi\hat{f}\|_2\leqq\|\varphi\|_\infty\|\hat{f}\|_2=\|\varphi\|_\infty\|f\|_2.$$

Hence $\|T\|_2\leqq\|\varphi\|_\infty$.

Conversely, suppose $T\in M(L_2(G))$. If $f,g\in C_c(G)$ then $Tf*g$ and $T(f*g)\in L_2(G)$, and for any $h\in L_2(G)$ we have

$$\begin{aligned}\langle Tf*g,h\rangle&=\int_G Tf*g(t)\,h(t^{-1})\,d\lambda(t)\\&=\int_G\Big[\int_G Tf(ts^{-1})\,g(s)\,d\lambda(s)\Big]h(t^{-1})\,d\lambda(t)\\&=\int_G g(s)\Big[\int_G(\tau_s Tf)(t)\,h(t^{-1})\,d\lambda(t)\Big]d\lambda(s)\\&=\int_G g(s)\Big[\int_G(T\tau_s f)(t)\,h(t^{-1})\,d\lambda(t)\Big]d\lambda(s)\\&=\int_G g(s)\Big[\int_G \tau_s f(t)\,T^*\tilde{h}(t)\,d\lambda(t)\Big]d\lambda(s)\\&=\int_G\Big[\int_G \tau_s f(t)\,g(s)\,d\lambda(s)\Big]T^*\tilde{h}(t)\,d\lambda(t)\\&=\int_G f*g(t)\,T^*\tilde{h}(t)\,d\lambda(t)\\&=\int_G T(f*g)(t)\,h(t^{-1})\,d\lambda(t)=\langle T(f*g),h\rangle,\end{aligned}$$

where as usual T^* denotes the operator adjoint of T. The applications of Fubini's Theorem (Theorem C.6.1) are valid since $f, g \in C_c(G)$. Since this holds for all $h \in L_2(G)$, we conclude that $Tf*g = T(f*g)$ for $f, g \in C_c(G)$.

Now if $f, g \in L_1(G) \cap L_2(G)$ then $Tf*g$ and $T(f*g)$ are in $L_2(G)$. Let $\{f_n\}$ and $\{g_n\}$ be sequences in $C_c(G)$ such that $\lim_n \|f_n - f\|_2 = 0$ and $\lim_n \|g_n - g\|_1 = 0$. Then we have

$$\begin{aligned}
&\|T(f*g) - Tf*g\|_2 \\
&\quad \leqq \|T(f*g) - T(f_n*g_n)\|_2 + \|Tf_n*g_n - Tf_n*g\|_2 + \|Tf_n*g - Tf*g\|_2 \\
&\quad \leqq \|T\|_2 (\|f*g - f_n*g\|_2 + \|f_n*g - f_n*g_n\|_2) + \|Tf_n\|_2 \, \|g_n - g\|_1 \\
&\qquad + \|Tf_n - Tf\|_2 \, \|g\|_1 \\
&\quad \leqq \|T\|_2 (\|f - f_n\|_2 \, \|g\|_1 + \|f_n\|_2 \, \|g - g_n\|_1) + \|T\|_2 \, \|f_n\|_2 \, \|g_n - g\|_1 \\
&\qquad + \|T\|_2 \, \|f_n - f\|_2 \, \|g\|_1 .
\end{aligned}$$

Hence $T(f*g) = Tf*g$ for $f, g \in L_1(G) \cap L_2(G)$. Interchanging the roles of f and g it follows that $T(f*g) = Tf*g = f*Tg$ for $f, g \in L_1(G) \cap L_2(G)$. Consequently we also see that

$$(Tf)\hat{}\ \hat{g} = \hat{f}(Tg)\hat{} \qquad (f, g \in L_1(G) \cap L_2(G)).$$

If $\gamma \in \hat{G}$ let $f \in L_1(G) \cap L_2(G)$ be such that $\hat{f}(\gamma) \neq 0$ and define $\varphi(\gamma) = (Tf)\hat{}\,(\gamma)/\hat{f}(\gamma)$. The identity just obtained above shows that this definition is independent of the choice of f, and, moreover, that $(Tf)\hat{} = \varphi \hat{f}$ for each $f \in L_1(G) \cap L_2(G)$. Clearly φ defines an equivalence class of measurable functions on $\hat{G}$ which we shall again denote by φ. Furthermore φ belongs locally to $L_2(\hat{G})$. Indeed let $K \subset \hat{G}$ be compact and let $f \in L_1(G) \cap L_2(G)$ be such that $\hat{f} = 1$ almost everywhere on K, $0 \leqq \hat{f} \leqq 1$ and $\hat{f} = 0$ almost everywhere off an open set U with finite measure. Clearly such a function exists in $L_1(G)$ (F.7e) and hence by the Plancherel Theorem (Theorem F.8.2) in $L_1(G) \cap L_2(G)$. Then we have

$$\|\varphi \chi_K\|_2 \leqq \|\varphi \hat{f}\|_2 = \|(Tf)\hat{}\,\|_2 = \|Tf\|_2 \leqq \|T\|_2 \, \|f\|_2 = \|T\|_2 \, \|\hat{f}\|_2$$

where χ_K denotes the characteristic function of the set K. This shows that φ belongs locally to $L_2(\hat{G})$.

From this result we can easily deduce that $\|\varphi\|_\infty \leqq \|T\|_2$. As suppose, without loss of generality, that K is a compact set in $\hat{G}$ such that $\eta(K) > 0$ and $\operatorname{ess\,inf}_{\gamma \in K} |\varphi(\gamma)| - \|T\|_2 = 2\delta > 0$. Then on K, $|\varphi(\gamma)| > \|T\|_2 + \delta$ for almost all γ and thus

$$\|\varphi \chi_K\|_2 > (\|T\|_2 + \delta) \sqrt{\eta(K)}.$$

On the other hand, let $U \supset K$ be an open set such that $\eta(U) < (1+\delta/2\,\|T\|_2)\,\eta(K)$ and choose $f \in L_1(G) \cap L_2(G)$ as before. Then, once again,

$$\|\varphi\,\chi_K\|_2 \leqq \|T\|_2\,\|\hat{f}\|_2 \leqq \|T\|_2\,\sqrt{\eta(U)}.$$

Combining these two inequalities we conclude that

$$\|T\|_2\,\sqrt{\eta(U)} > (\|T\|_2 + \delta)\,\sqrt{\eta(K)}$$

and hence

$$\eta(U) > (1+\delta/\|T\|_2)^2\,\eta(K) > (1+\delta/\|T\|_2)\,\eta(K),$$

contrary to the choice of U.

Therefore $\varphi \in L_\infty(\hat{G})$ and $\|\varphi\|_\infty \leqq \|T\|_2$.

An easy argument using the denseness of $L_1(G) \cap L_2(G)$ in $L_2(G)$, the Plancherel Theorem and the boundedness of φ shows that $(Tf)\hat{} = \varphi\hat{f}$ for each $f \in L_2(G)$. Clearly φ is unique.

Thus (i) implies (ii).

It is obvious that the correspondence between T and φ defines an isomorphism from $M(L_2(G))$ onto $L_\infty(\hat{G})$. The isomorphism is an isometry since $\|\varphi\|_\infty \leqq \|T\|_2 \leqq \|\varphi\|_\infty$. □

Using this theorem it is easy to obtain a characterization of the closed translation invariant linear subspaces of $L_2(G)$, that is, a characterization of the closed linear subspaces $X \subset L_2(G)$ such that $f \in X$ implies $\tau_s f \in X$ for each $s \in G$.

Corollary 4.1.1. *Let G be a locally compact Abelian group. Then the following are equivalent:*

(i) *$X \subset L_2(G)$ is a closed translation invariant linear subspace.*

(ii) *There exists a Borel measurable subset E of $\hat{G}$ such that $X = \{f \mid f \in L_2(G), \hat{f} = 0$ almost everywhere off $E\}$, that is, $\hat{X} = \chi_E L_2(\hat{G})$.*

Proof. It is easily seen that (ii) implies (i) since $(\tau_s f)\hat{} = (s^{-1},\cdot)\hat{f}$ for each $f \in L_2(G)$ and $s \in G$. Conversely, suppose (i) holds and let T be the Hilbert space projection (D.5) of $L_2(G)$ onto the closed linear subspace X. T is a bounded linear operator of norm one. Moreover, if $f \in L_2(G)$ and $f = f_1 + f_2$ where $f_1 \in X$ and $f_2 \in X^\perp$, the orthogonal complement of X in $L_2(G)$ (D.5), then for each $s \in G$ we have $\tau_s f = \tau_s f_1 + \tau_s f_2$ where $\tau_s f_1 \in X$, $\tau_s f_2 \in X^\perp$, and hence

$$(T\tau_s)(f) = T(\tau_s f_1 + \tau_s f_2) = \tau_s f_1 = \tau_s\,Tf_1 = \tau_s(Tf_1 + Tf_2) = (\tau_s\,T)(f).$$

Thus T commutes with translation and so $T \in M(L_2(G))$.

Consequently, by the preceding theorem, there exists a $\varphi \in L_\infty(\hat{G})$ such that $(Tf)\hat{} = \varphi\hat{f}$ for each $f \in L_2(G)$. Since T is a projection we conclude that $\varphi^2\hat{f} = \varphi\hat{f}$ almost everywhere for each $f \in L_2(G)$, and hence that $\varphi^2=\varphi$ almost everywhere. Setting $E=\varphi^{-1}(1)$ we see that $\varphi=\chi_E$ almost everywhere, and the assertion (ii) is evident.

Therefore (i) and (ii) are equivalent. ∎

The conclusion and arguments of the preceding theorem and the Riesz-Thorin Convexity Theorem (D.11) provide some results about $M(L_p(G))$.

Theorem 4.1.2. *Let G be a locally compact Abelian group. If* $1<p<\infty$ *and* $1/p+1/p'=1$ *then there exists an isometric algebra isomorphism of* $M(L_p(G))$ *onto* $M(L_{p'}(G))$.

Proof. Let $T\in M(L_p(G))$. By the same argument as in Theorem 4.1.1 one can easily show for $f, g\in C_c(G)$ and $h\in L_{p'}(G)$ that $\langle Tf*g, h\rangle = \langle T(f*g), h\rangle = \langle f*Tg, h\rangle$. Thus $Tf*g=T(f*g)=f*Tg$ for $f, g\in C_c(G)$.

Now for each $g\in C_c(G)$ define $F_g(f)=\langle Tg, f\rangle$ for $f\in C_c(G)$. Clearly F_g is a linear functional on $C_c(G)$ and moreover we have

$$|F_g(f)|=|f*Tg(e)|=|Tf*g(e)|\leqq \|Tf\|_p\,\|g\|_{p'}\leqq \|T\|_p\,\|f\|_p\,\|g\|_{p'}.$$

Thus F_g defines a bounded linear functional on the norm dense subspace $C_c(G)$ of $L_p(G)$, and hence can be extended to such a functional on all of $L_p(G)$ without increasing the norm. It follows immediately from the definition of F_g and the duality between $L_p(G)$ and $L_{p'}(G)$ that $Tg\in L_{p'}(G)$ and

$$\|Tg\|_{p'}=\|F_g\|\leqq \|T\|_p\,\|g\|_{p'}.$$

Therefore $T\in M(L_p(G))$ when restricted to $C_c(G)$ defines a continuous linear transformation from $C_c(G)$ into $L_{p'}(G)$. Furthermore it is evident that this transformation commutes with translations. T restricted to $C_c(G)$ can then be extended uniquely to a continuous linear transformation from $L_{p'}(G)$ to $L_{p'}(G)$ which commutes with translation without increasing the norm because $C_c(G)$ is norm dense in $L_{p'}(G)$. We again denote this transformation by T. Clearly from the construction $T\in M(L_{p'}(G))$ and $\|T\|_{p'}\leqq \|T\|_p$.

Interchanging the roles of p and p' we see at once that the isomorphism constructed in the preceding paragraph from $M(L_p(G))$ to $M(L_{p'}(G))$ is surjective, and that $\|T\|_p\leqq \|T\|_{p'}$.

Therefore $M(L_p(G))$ and $M(L_{p'}(G))$ are isometrically isomorphic. ∎

Theorem 4.1.3. *Let G be a locally compact Abelian group. If* $1<p<\infty$ *then there exists a continuous algebra isomorphism from* $M(L_p(G))$ *into* $M(L_2(G))$.

Proof. Let $T\in M(L_p(G))$. By a form of the Riesz-Thorin Convexity Theorem (Theorem D.11.3) the function $\log \|T\|_{1/a}$ is convex on $0\leqq a\leqq 1$.

In particular, since $1/p \cdot p + 1/p' \cdot p' = 2$ and $1/p + 1/p' = 1$ we have

$$\log \|T\|_2 \leqq \frac{1}{p} \log \|T\|_p + \frac{1}{p'} \log \|T\|_{p'} = \left(\frac{1}{p} + \frac{1}{p'}\right) \log \|T\|_p = \log \|T\|_p$$

as by the previous theorem $\|T\|_p = \|T\|_{p'}$. Thus the restriction of $T \in M(L_p(G))$ to the integrable simple functions determines a unique continuous linear transformation on $L_2(G)$ which commutes with translations, that is, a unique element $T \in M(L_2(G))$, and $\|T\|_2 \leqq \|T\|_p = \|T\|_{p'}$. This correspondence obviously defines a continuous isomorphism of $M(L_p(G))$ into $M(L_2(G))$. ▯

Corollary 4.1.2. *Let G be a locally compact Abelian group and $1 < p < \infty$. If $T \in M(L_p(G))$ then there exists a unique $\varphi \in L_\infty(\hat{G})$ such that $(Tf)\hat{} = \varphi \hat{f}$ for each $f \in L_2(G) \cap L_p(G)$ and $\|\varphi\|_\infty = \|T\|_2 \leqq \|T\|_p$.*

Proof. If $T \in M(L_p(G))$ then by Theorems 4.1.3 and 4.1.1 there exists a unique $\varphi \in L_\infty(\hat{G})$ such that $(Tf)\hat{} = \varphi \hat{f}$ for each integrable simple function $f \in L_2(G) \cap L_p(G)$. To establish this identity for all $f \in L_2(G) \cap L_p(G)$ it is clearly sufficient to consider only nonnegative f. Thus let $f \in L_2(G) \cap L_p(G)$ be nonnegative almost everywhere and let $\{g_n\}$ be a monotone increasing sequence of nonnegative simple functions which converges almost everywhere to f. Then by the Lebesgue Dominated Convergence Theorem (Theorem C.5.2) we have that $\lim_n \|g_n - f\|_p = 0$ and $\lim_n \|g_n - f\|_2 = 0$.

Denote by $T_2 \in M(L_2(G))$ the unique continuous extension to all of $L_2(G)$ of the restriction of T to the integrable simple functions as done in the previous theorem. Clearly $\lim_n \|Tg_n - Tf\|_p = 0$ and $\lim_n \|T_2 g_n - T_2 f\|_2 = 0$. Moreover, there exists a subsequence $\{Tg_k\}$ of $\{Tg_n\}$ which converges almost everywhere to both Tf and $T_2 f$. Hence $Tf = T_2 f$.

But then, appealing to the Plancherel Theorem (Theorem F.8.2), we have, on the one hand, that

$$\lim_n \|(T_2 g_n)\hat{} - (Tf)\hat{}\|_2 = \lim_n \|(T_2 g_n)\hat{} - (T_2 f)\hat{}\|_2 = \lim_n \|T_2 g_n - T_2 f\|_2 = 0,$$

while on the other hand,

$$\|(T_2 g_n)\hat{} - \varphi \hat{f}\|_2 = \|\varphi(\hat{g}_n - \hat{f})\|_2 \leqq \|\varphi\|_\infty \|g_n - f\|_2,$$

which shows that $\lim_n \|(T_2 g_n)\hat{} - \varphi \hat{f}\|_2 = 0$.

Therefore $(Tf)\hat{} = \varphi \hat{f}$ for each nonnegative $f \in L_2(G) \cap L_p(G)$, and hence for all $f \in L_2(G) \cap L_p(G)$. ▯

$\hat{f}$ of course again denotes the Fourier-Plancherel transform.

Corollary 4.1.3. *Let G be a locally compact Abelian group and suppose* $1 \leqq p \leqq q \leqq 2$. *Then there exists a continuous algebra isomorphism of* $M(L_p(G))$ *into* $M(L_q(G))$.

Proof. Let $T \in M(L_p(G))$. Then by the preceding theorem we see that $T \in M(L_2(G))$ and $\|T\|_2 \leqq \|T\|_p$. But then, since $\log \|T\|_{1/a}$ is convex on $0 \leqq a \leqq 1$ (Theorem D.11.3), there exists a λ, $0 \leqq \lambda \leqq 1$, such that

$$\begin{aligned}\log \|T\|_q &\leqq (1-\lambda)\log\|T\|_p + \lambda \log\|T\|_2 \leqq (1-\lambda)\log\|T\|_p + \lambda\log\|T\|_p \\ &= \log\|T\|_p.\end{aligned}$$

Thus, as in the proof of Theorem 4.1.3, $T \in M(L_p(G))$ restricted to the integrable simple functions determines a unique element of $M(L_q(G))$ and $\|T\|_q \leqq \|T\|_p$. □

Corollary 4.1.4. *Let G be a locally compact Abelian group and suppose* $2 \leqq q \leqq p < \infty$. *Then there exists a continuous algebra isomorphism of* $M(L_p(G))$ *into* $M(L_q(G))$.

Proof. The corollary follows immediately from Theorem 4.1.2 and the previous corollary. □

4.2. Pseudomeasures. It is evident that every $\mu \in M(G)$ defines a multiplier for $L_p(G)$, $1 < p < \infty$. This is obvious from the fact that $\|\mu * f\|_p \leqq \|\mu\|\,\|f\|_p$, $f \in L_p(G)$ (F.2). Previously it has often been the case that every multiplier T could be characterized in this way, that is, $Tf = \mu * f$. This is generally no longer possible for the operators in $M(L_p(G))$. However by considering a certain collection of mathematical objects which properly contains $M(G)$, the space of pseudomeasures, it is possible to obtain a description of the multipliers for $L_p(G)$ as convolution operators. In this section we shall define pseudomeasures and establish those properties of pseudomeasures which will be needed in the study of $M(L_p(G))$.

If G is a locally compact Abelian group we shall denote by $A(G)$ the space of Fourier transforms of elements in $L_1(\hat{G})$. Since $L_1(\hat{G})$ is semisimple it is evident that $A(G)$ is isomorphic to $L_1(\hat{G})$. Moreover, it is well known and easily proven that under pointwise operations the algebra $A(G)$ is a Banach algebra if one defines the norm of $\hat{f} \in A(G)$ by $\|\hat{f}\|_A = \|f\|_1 = \int_{\hat{G}} |f(\gamma)|\, d\eta(\gamma)$ (E.2). The subspace of all elements of $A(G)$ which have compact support, denoted by $A_c(G)$, is dense in $A(G)$ (F.7c). The space of continuous linear functionals on $A(G)$ will be denoted by $P(G)$. The elements in $P(G)$ are called *pseudomeasures*. We shall use $\|\cdot\|_P$ to denote the norm of an element in $P(G)$, and $\langle \hat{f}, \sigma \rangle$ will denote the pairing between elements of $A(G)$ and $P(G)$.

Theorem 4.2.1. *Let G be a locally compact Abelian group. Then $M(G)\subset P(G)$.*

Proof. If $\mu\in M(G)$ then the conclusion is apparent from the inequalities

$$\begin{aligned}\Big|\int_G \hat f(t^{-1})\,d\mu(t)\Big| &= \Big|\int_G \Big[\int_{\hat G} (t,\gamma)\, f(\gamma)\, d\eta(\gamma)\Big]\, d\mu(t)\Big| \\ &= \Big|\int_{\hat G} f(\gamma)\Big[\int_{\hat G} (t,\gamma)\, d\mu(t)\Big]\, d\eta(\gamma)\Big| \\ &\leqq \|\check\mu\|_\infty \|f\|_1 = \|\hat\mu\|_\infty \|\hat f\|_A,\end{aligned}$$

which are valid for each $\hat f\in A(G)$. ▯

$\check\mu$ of course denotes $\int_G (t,\gamma)\,d\mu(t)$.

An examination of the Fourier transform for elements of $P(G)$ will show that $M(G)\neq P(G)$.

If $\sigma\in P(G)$, define the linear functional F_σ on $L_1(\hat G)$ by the formula $F_\sigma(f)=\langle \hat f,\sigma\rangle$ for each $f\in L_1(\hat G)$. Since

$$|F_\sigma(f)|=|\langle \hat f,\sigma\rangle|\leqq \|\sigma\|_P \|\hat f\|_A = \|\sigma\|_P \|f\|_1,$$

it is obvious that F_σ defines a bounded linear functional on $L_1(\hat G)$. Hence there exists a unique element $\hat\sigma\in L_\infty(\hat G)$ such that

$$\langle \hat f,\sigma\rangle = \int_{\hat G} f(\gamma)\,\hat\sigma(\gamma^{-1})\,d\eta(\gamma)=\langle f,\hat\sigma\rangle$$

for each $f\in L_1(\hat G)$. Clearly $\|\hat\sigma\|_\infty\leqq\|\sigma\|_P$.

Conversely, if $\varphi\in L_\infty(\hat G)$ and we define $G_\varphi(\hat f)=\langle f,\varphi\rangle$ for each $\hat f\in A(G)$ then the inequalities

$$|G_\varphi(\hat f)|=|\langle f,\varphi\rangle|\leqq \|\varphi\|_\infty \|f\|_1 = \|\varphi\|_\infty \|\hat f\|_A$$

reveal that G_φ defines a bounded linear functional on $A(G)$. Hence there is a unique pseudomeasure $\sigma\in P(G)$ such that $\langle \hat f,\sigma\rangle=\langle f,\varphi\rangle$ for each $\hat f\in A(G)$ and $\|\sigma\|_P\leqq\|\varphi\|_\infty$.

From the previous construction it is evident that $\varphi=\hat\sigma$. Thus $\|\sigma\|_P=\|\hat\sigma\|_\infty$.

Evidently then the formula $\langle \hat f,\sigma\rangle=\langle f,\hat\sigma\rangle$ defines a linear isometry from $P(G)$ onto $L_\infty(\hat G)$. We shall call $\hat\sigma$ the *Fourier transform* of the pseudomeasure σ. Clearly if $\mu\in M(G)$ then the Fourier transform of μ as a measure and as a pseudomeasure are identical.

In order that the Fourier transform of pseudomeasures be completely analogous to the usual Fourier transform it is necessary that the transform preserve products. So far however no notion of multiplication has been introduced in $P(G)$. We can simultaneously accomplish both of these tasks if for $\sigma_1,\sigma_2\in P(G)$ we define $\sigma_1 * \sigma_2$ to be that pseudomeasure for which

$(\sigma_1 * \sigma_2)\hat{} = \hat{\sigma}_1 \hat{\sigma}_2$. This definition is meaningful because the Fourier transform is a linear isometry of $P(G)$ onto $L_\infty(\hat{G})$.

We may summarize this development in the following theorem.

Theorem 4.2.2. *Let G be a locally compact Abelian group. Then the Fourier transform $\sigma \to \hat{\sigma}$ is an isometric algebra isomorphism of $P(G)$ onto $L_\infty(\hat{G})$.*

Corollary 4.2.1. *Let G be an infinite locally compact Abelian group. Then $M(G) \neq P(G)$.*

Proof. Since G is infinite we have $M(G)\hat{}$ is a proper subset of $L_\infty(\hat{G}) = P(G)\hat{}$ (F.7c). □

It should be noted that the norm of $\mu \in M(G)$ as a pseudomeasure is, in general, less than its norm as a measure because $\|\mu\|_P = \|\hat{\mu}\|_\infty \leqq \|\mu\|$.

Furthermore, we remark that if $f \in L_1(G)$ then it obviously can be considered as an element of $P(G)$ since $L_1(G) \subset M(G) \subset P(G)$. Moreover, if $\sigma \in P(G)$ then $\sigma * f$ is defined, belongs to $P(G)$ and $(\sigma * f)\hat{} = \hat{\sigma} \hat{f}$ where $\hat{f}$ denotes the usual Fourier transform for elements of $L_1(G)$.

We shall say that $\sigma \in P(G)$ belongs to $L_2(G)$ if there exists a $g \in L_2(G)$ such that $\langle \hat{f}, \sigma \rangle = \langle \hat{f}, g \rangle$ for each $\hat{f} \in A_c(G)$. Since $A_c(G)$ is dense in $A(G)$ and, as we shall see in the succeeding chapter, dense in $L_2(G)$, the g is uniquely determined. Moreover, an easy computation using the definition of the Fourier transform for pseudomeasures and Parseval's Formula (Theorem F.8.3) shows for each $\hat{f} \in A_c(G)$ that

$$\langle f, \hat{\sigma} \rangle = \langle \hat{f}, \sigma \rangle = \langle \hat{f}, g \rangle = \langle f, \hat{g} \rangle .$$

Hence appealing to the Plancherel Theorem (Theorem F.8.2) and the denseness of $A_c(G)$ in $L_2(G)$ we conclude that $\hat{\sigma} = \hat{g} \in L_2(\hat{G}) \cap L_\infty(\hat{G})$, and the Fourier transform of σ as a pseudomeasure and as an element of $L_2(G)$ agree.

Conversely, if $\sigma \in P(G)$ is such that $\hat{\sigma} \in L_2(\hat{G}) \cap L_\infty(\hat{G})$ then σ belongs to $L_2(G)$. Indeed, $F(\hat{f}) = \langle f, \hat{\sigma} \rangle$ for each $f \in L_2(\hat{G})$ clearly defines a linear functional on $L_2(G)$ which is continuous, as by the Plancherel Theorem (Theorem F.8.2) and the Cauchy-Schwarz inequality we have

$$|F(\hat{f})| = |\langle f, \hat{\sigma} \rangle| \leqq \|f\|_2 \, \|\hat{\sigma}\|_2 = \|\hat{f}\|_2 \, \|\hat{\sigma}\|_2 .$$

Thus there exists a unique $g \in L_2(G)$ such that $\langle f, \hat{\sigma} \rangle = F(\hat{f}) = \langle \hat{f}, g \rangle$ for each $\hat{f} \in L_2(G)$. Obviously $\langle \hat{f}, g \rangle = \langle \hat{f}, \sigma \rangle$ for each $\hat{f} \in A_c(G)$, and by Parseval's Formula (Theorem F.8.3) we see that $\langle f, \hat{\sigma} \rangle = \langle \hat{f}, g \rangle = \langle f, \hat{g} \rangle$ for each $f \in L_2(\hat{G})$. Consequently, σ is in $L_2(G)$ and the Fourier transform of σ coincides with the Fourier-Plancherel transform of g.

We summarize the latter discussion in the following corollary.

Corollary 4.2.2. *Let G be a locally compact Abelian group. If $\sigma \in P(G)$ then the following are equivalent:*

(i) $\sigma \in L_2(G)$.

(ii) $\hat{\sigma} \in L_2(\hat{G}) \cap L_\infty(\hat{G})$.

Moreover if $\sigma \in L_2(G)$ then the Fourier transform of σ as a pseudomeasure and the Fourier-Plancherel transform of σ are equal.

4.3. The Multipliers of $L_p(G)$ as Pseudomeasures, $1 < p < \infty$. If we combine the previously obtained results characterizing elements of $M(L_p(G))$ as bounded functions and the relationship between pseudomeasures and their Fourier transforms then we easily see that multipliers for $L_p(G)$ can be considered as convolution with pseudomeasures.

Theorem 4.3.1. *Let G be a locally compact Abelian group and suppose $T: L_2(G) \to L_2(G)$ is a continuous linear transformation. Then the following are equivalent:*

(i) $T \in M(L_2(G))$.

(ii) *There exists a unique pseudomeasure $\sigma \in P(G)$ such that $Tf = \sigma * f$ for each $f \in L_1(G) \cap L_2(G)$.*

Moreover the correspondence between T and σ defines an isometric algebra isomorphism from $M(L_2(G))$ onto $P(G)$.

Proof. Suppose $T \in M(L_2(G))$. Then by Theorem 4.1.1 there exists a unique $\varphi \in L_\infty(\hat{G})$ such that $(Tf)\hat{} = \varphi \hat{f}$ for each $f \in L_2(G)$ and $\|\varphi\|_\infty = \|T\|_2$. Let $\sigma \in P(G)$ be the unique pseudomeasure whose existence is assured by Theorem 4.2.2 for which $\hat{\sigma} = \varphi$. For each $f \in L_1(G) \cap L_2(G)$ it is apparent that $\sigma * f$ exists as a pseudomeasure and that

$$(\sigma * f)\hat{} = \hat{\sigma} \hat{f} = \varphi \hat{f} = (Tf)\hat{}.$$

However it does not follow immediately from this that $\sigma * f = Tf$. First we must show that $Tf \in L_2(G)$ can be considered as a pseudomeasure and then that the Fourier-Plancherel transform of Tf, which appears on the right hand side of the identity, coincides with the Fourier transform of Tf as a pseudomeasure. But $(Tf)\hat{} = \varphi \hat{f} \in L_2(\hat{G}) \cap L_\infty(\hat{G})$ as $f \in L_1(G) \cap L_2(G)$, and so Tf defines an element of $P(G)$, as discussed preceding Corollary 4.2.2, by the equation

$$\langle Tf, \hat{g} \rangle = \langle (Tf)\hat{}, g \rangle \qquad (\hat{g} \in A(G)),$$

where $(Tf)\hat{}$ is the Fourier-Plancherel transform of Tf. Furthermore this definition reveals that the Fourier transform of the pseudomeasure Tf is precisely $(Tf)\hat{} = \varphi \hat{f}$.

Consequently by the uniqueness of the Fourier transform of pseudomeasures we conclude that $\sigma * f = Tf$ for each $f \in L_1(G) \cap L_2(G)$, and (i) implies (ii).

Conversely, suppose that $Tf=\sigma*f$ for each $f\in L_1(G)\cap L_2(G)$. Then, evidently, for each $f\in L_1(G)\cap L_2(G)$ and $s\in G$,

$$\begin{aligned}(T\tau_s f)\hat{} &= (\sigma*\tau_s f)\hat{} = \hat{\sigma}(\tau_s f)\hat{} = (s^{-1},\cdot)\,\hat{\sigma}\,\hat{f} = (s^{-1},\cdot)\,(\sigma*f)\hat{}\\ &= (s^{-1},\cdot)\,(Tf)\hat{} = (\tau_s Tf)\hat{}.\end{aligned}$$

Appealing again to the uniqueness of the pseudomeasure Fourier transform we see that $T\tau_s f=\tau_s Tf$ for each $f\in L_1(G)\cap L_2(G)$ and each $s\in G$. Since T is continuous and $L_1(G)\cap L_2(G)$ is norm dense in $L_2(G)$ we have $T\tau_s=\tau_s T$, that is, $T\in M(L_2(G))$. Therefore (ii) implies (i).

The isometric nature of the isomorphism between $M(L_2(G))$ and $P(G)$ defined by the correspondence between T and σ is an immediate consequence of Theorems 4.1.1 and 4.2.2. □

By a similar argument utilizing Theorem 4.1.3, Corollary 4.1.2 and Theorem 4.2.2 one can establish the next theorem. We shall omit the proof.

Theorem 4.3.2. *Let G be a locally compact Abelian group and suppose $1<p<\infty$, $p\neq 2$. If $T\in M(L_p(G))$ then there exists a unique $\sigma\in P(G)$ such that $Tf=\sigma*f$ for each $f\in L_1(G)\cap L_2(G)\cap L_p(G)$. Moreover the correspondence between T and σ defines a continuous algebra isomorphism from $M(L_p(G))$ into $P(G)$.*

It should perhaps be explicitly noted that for any p, $1<p<\infty$, if $T\in M(L_p(G))$ then the formula $Tf=\sigma*f$ holds for all $f\in C_c(G)$, and the function φ in Theorem 4.1.1 and Corollary 4.1.2 is precisely $\hat{\sigma}$.

Finally, we wish to point out that, in general, for $1<p<\infty$ there may exist $\varphi\in L_\infty(\hat{G})\sim M(G)\hat{}$ which defines a multiplier for $L_p(G)$, and if $p\neq 2$ that there may exist $\varphi\in L_\infty(\hat{G})$ which corresponds to no multiplier for $L_p(G)$. An example to substantiate the first assertion when G is an infinite compact Abelian group appears after the proof of Theorem 3.6.1. The situation for noncompact Abelian groups is discussed in 4.7.

To establish the second assertion we may assume that $p>2$ since $M(L_p(G))=M(L_{p'}(G))$, $1/p+1/p'=1$, by Theorem 4.1.2. Let $G=\Gamma$, the circle group, and let $g\in L_2(\Gamma)$, $g\notin L_p(\Gamma)$. There exists a function φ on $\hat{\Gamma}=Z$ such that φ takes only the values $+1$ and -1 and $\varphi\,\hat{g}$ is the Fourier transform of some function $f\in L_p(\Gamma)$ (F.10d). Clearly $\hat{g}=\varphi\,\hat{f}$ and $g\notin L_p(\Gamma)$. Hence φ does not define a multiplier for $L_p(\Gamma)$ but $\varphi\in L_\infty(Z)$. The situation will be discussed more fully in 4.5.

In terms of measures and pseudomeasures the previous assertions are equivalent to saying that, in general, the injection of $M(G)$ into $M(L_p(G))$ is proper for $1<p<\infty$, and the injection of $M(L_p(G))$ into $P(G)$ is proper for $1<p<\infty$, $p\neq 2$.

4.4. Sets of Uniqueness. In the following section we shall examine in some detail the relationships between the spaces $M(L_p(G))$ for various values of p. To be able to carry out this investigation we need to know something about the existence of *sets of uniqueness* for $L_p(\hat{G})$, that is, about the existence of measurable sets $E \subset G$ with the property that if $f \in L_1(G)$, $f=0$ almost everywhere off of E and $\hat{f} \in L_p(\hat{G})$ then $f=0$.

Trivially every measurable subset of G with measure zero is a set of uniqueness for $L_p(\hat{G})$, $1 \leqq p \leqq \infty$. On the other hand it is easily seen that for some values of p these are the only sets of uniqueness which have finite Haar measure. Indeed if $p=2$ and E is a measurable subset of G such that $\lambda(E)<\infty$ then $f=\chi_E \in L_1(G) \cap L_2(G)$, $f=0$ everywhere off of E and $\hat{f} \in L_2(\hat{G})$. But $f=0$ if and only if $\lambda(E)=0$.

The main theorem of this section asserts that when G is a nondiscrete locally compact Abelian group then there always exists sets of uniqueness for $L_p(\hat{G})$, $1 \leqq p < 2$, which have finite positive Haar measure. We shall also use this result in the study of the derived spaces of $L_p(G)$.

The proof of the indicated theorem is rather intricate, and we shall approach it via a series of lemmas.

We begin by recalling a result from general measure theory (Theorem C.4.2) applied to topological groups. Let G be a nondiscrete locally compact Abelian group and suppose $F \subset G$ is a measurable set such that $0<\lambda(F)<\infty$. Then there exists a measurable set $F' \subset F$ for which $\lambda(F')=\lambda(F)/2$. Hence, given a measurable $F \subset G$ such that $0<\lambda(F)<\infty$ we can construct a sequence of partitions $\{\Pi_n\}$ of F in the following manner: $\Pi_0=\{F\}$. Suppose $\Pi_n=\{E_1, \ldots, E_{2^n}\}$ where the E_j are mutually disjoint measurable subsets of F such that $\bigcup_{j=1}^{2^n} E_j=F$ and $\lambda(E_j)=\lambda(F)/2^n$, $j=1,2,\ldots,2^n$. Then for each $j=1,2,\ldots,2^n$ we let $E_{ij} \subset E_j$, $i=1,2$, be measurable subsets of E_j such that $\lambda(E_{ij})=\lambda(E_j)/2$, $i=1,2$, and set $\Pi_{n+1}=\{E_{11}, E_{21}, E_{12}, E_{22}, \ldots, E_{12^n}, E_{22^n}\}$. Clearly Π_{n+1} partitions F into 2^{n+1} mutually disjoint measurable subsets each of measure $\lambda(F)/2^{n+1}$.

Having constructed such a sequence of partitions $\{\Pi_n\}$ we define a sequence of functions $\{r_n\}$ on G as follows. $r_0=\chi_E$. Let $\Pi_n=\{E_1, \ldots, E_{2^n}\}$ and $\Pi_{n+1}=\{E_{11}, E_{21}, E_{12}, E_{22}, \ldots, E_{12^n}, E_{22^n}\}$. Then r_{n+1} is zero off of F, and given a pair $E_{1j}, E_{2j}, j=1,2,\ldots,2^n$, r_{n+1} takes the values ± 1 on $E_j=E_{1j} \cup E_{2j}$, is constant on E_{ij}, $i=1,2$ and $\int_{E_j} r_{n+1}(t)\, d\lambda(t)=0$, $j=1,2,\ldots,2^n$. That is, for each pair E_{1j}, E_{2j}, r_{n+1} is constantly $+1$ on one of the sets and constantly -1 on the other. Evidently the choice of the functions r_n, which are clearly generalizations of the Rademacher functions on the unit interval (Edwards [14_{II}], 14.1), is not unique. In what follows we shall always assume that some one choice for the sequence of functions $\{r_n\}$ has been made relative to the given sequence of partitions $\{\Pi_n\}$.

An immediate consequence of the first lemma is that the family of all finite products of such functions is an orthogonal family in $L_2(G)$.

Lemma 4.4.1. *Let G be a nondiscrete locally compact Abelian group and suppose $F\subset G$ is a measurable set such that $0<\lambda(F)<\infty$. If $\alpha_1, \alpha_2, \ldots, \alpha_k$ are nonnegative integers and $n_1<n_2<\cdots<n_k$ then*

$$\int_G r_{n_1}^{\alpha_1}\ldots r_{n_k}^{\alpha_k}(t)\,d\lambda(t)=\begin{cases}\lambda(F), & \text{if } \alpha_1, \alpha_2, \ldots, \alpha_k \text{ are even,}\\ 0, & \text{otherwise.}\end{cases}$$

Proof. Since each of the functions r_{n_i}, $i=1,2,\ldots,k$, is zero off of F and takes only the values ± 1 on F, it is evident that if all the α_i are even then the integrand is precisely the characteristic function of F. Hence the value of the integral in this case is $\lambda(F)$.

On the other hand if any of the α_i are odd then, without loss of generality, we may assume that all the α_i are odd, since factors with even α_i contribute only the factor one to the product in the integrand. Moreover, in this case, since each of the r_{n_i} takes only the values ± 1 on F, it is obvious that $r_{n_i}^{\alpha_i}=r_{n_i}$, $i=1,2,\ldots,k$. Finally if $E\in\Pi_{n_k-1}$ then it follows from the construction of the r_n that each of the r_{n_i}, $i=1,2,\ldots,k-1$, is of constant sign on E. Hence for $t\in E$ we have $r_{n_1}^{\alpha_1}\ldots r_{n_k}^{\alpha_k}(t)=r_{n_1}\ldots r_{n_k}(t)=\pm r_{n_k}(t)$. Consequently we have

$$\int_G r_{n_1}^{\alpha_1}\ldots r_{n_k}^{\alpha_k}(t)\,d\lambda(t)=\sum_{E\in\Pi_{n_k-1}}\left[\pm\int_E r_{n_k}(t)\,d\lambda(t)\right]=0. \quad \square$$

Given a nondiscrete locally compact Abelian group G and a measurable $F\subset G$, $0<\lambda(F)<\infty$, we shall denote by W the family of all finite products of the functions r_n. The previous lemma shows that W is a countable orthogonal system in $L_2(G)$ such that if $w\in W$ then $\|w\|_2=[\lambda(F)]^{\frac{1}{2}}$.

Lemma 4.4.2. *Let G be a nondiscrete locally compact Abelian group and suppose $F\subset G$ is a measurable set such that $0<\lambda(F)<\infty$. Let $\varepsilon>0$, let N be a positive integer and suppose k is a positive integer such that $1/2^k<\varepsilon/N\leqq 1/2^{k-1}$. Then for each positive integer n, $1\leqq n\leqq N$, there exist functions $g_j\in L_1(G)$, $j=1,2,\ldots,n$, such that:*

(i) $g_j=0$ *off of* F, $j=1,2,\ldots,n$.

(ii) $\int_F |g_j(t)|\,d\lambda(t)\leqq 2\lambda(F)$, $j=1,2,\ldots,n$.

(iii) $\int_F g_j^2(t)\,d\lambda(t)\leqq 2^k\lambda(F)$, $j=1,2,\ldots,n$.

(iv) $\lambda(\{t|g_j(t)=1\})\geqq(1-\varepsilon/N)\,\lambda(F)$, $j=1,2,\ldots,n$.

(v) $\|\hat{g}_1+\cdots+\hat{g}_n\|_\infty\leqq(2+n/N)\,\lambda(F)$.

(vi) $\int_G g_i\,g_j(t)\,d\lambda(t)=0$, $i\neq j$, $i,j=1,2,\ldots,n$.

(vii) g_j *is a finite real linear combination of elements of* W, $j=1,2,\ldots,n$.

Proof. It is easily seen for $n=1$ that $g_1=\chi_F$ satisfies (i)–(vii). Assume that for some n, $1\leqq n<N$, we have constructed $g_1, \ldots, g_n \in L_1(G)$ which satisfy (i)–(vii). Since each $g_j \in L_1(G)$ there exists a compact set $K\subset\hat{G}$ such that $|\hat{g}_j(\gamma)|\leqq \lambda(F)/N$, $1\leqq j\leqq n$, for each $\gamma\notin K$.

The system W is countable so we may assume, without loss of generality, that it has been enumerated, say $W=\{w_i\}$. Then we conclude at once from Bessel's inequality (Theorem D.5.2) that for each $\gamma\in K$ and each positive integer m we have

$$\sum_{j=1}^{m} |(\gamma, w_j)|^2 = \sum_{j=1}^{m} |(\gamma\,\chi_F, w_j)|^2 \leqq \lambda(F)\,(\|\gamma\,\chi_F\|_2)^2$$

where $(f, g)=\int_G f(t)\,\overline{g(t)}\,d\lambda(t)$. It is evident from this estimate that for each m the sums $\sum_{j=1}^{m} |(\gamma, w_j)|^2$ and $\sum_{j=1}^{\infty} |(\gamma, w_j)|^2$ are continuous functions of γ on K. Consequently by Dini's Theorem (Theorem C.8.1) we see that the series $\sum_{j=1}^{\infty} |(\gamma, w_j)|^2$ converges uniformly on the compact set $K\subset\hat{G}$. In particular, there exists an m_0 such that if $m>m_0$ then $\sum_{j=m}^{\infty} |(\gamma, w_j)|^2 < (\lambda(F)/2^{k/2} N)^2$ for all $\gamma\in K$. Moreover, we may assume that m_0 is chosen so large that all of the w_j's in W which appear in the expansions of g_j, $j=1, 2, \ldots, n$, have indices less than or equal to m_0. Thus for $i>m_0$ we have $(g_j, w_i)=0$, $j=1, 2, \ldots, n$.

Recall that the elements of W are finite products of the r_n's and that by hypothesis $g_1, \ldots, g_n$ are finite real linear combinations of the elements of W. Hence for a fixed m_0 it is apparent for a sufficiently large positive integer r that the partition $\Pi_r=\{E_1, E_2, \ldots, E_{2^r}\}$ of F has the property that the functions g_j, $j=1, 2, \ldots, n$ and w_i, $i=1, 2, \ldots, m_0$, are constant on the sets E_j, $j=1, 2, \ldots, 2^r$. The partition Π_{r+k} of F then partitions each set in Π_r into 2^k subsets. That is, $E_j=\bigcup_{i=1}^{2^k} E_{ij}$, $j=1, 2, \ldots, 2^r$, where $E_{ij}\in\Pi_{r+k}$, $i=1, 2, \ldots, 2^k$, $j=1, 2, \ldots, 2^r$. Define g_{n+1} as follows,

$$g_{n+1}(t)=\begin{cases} 0, & t\notin F \\ 1-2^k, & t\in E_{1j}, \quad j=1, 2, \ldots, 2^r \\ 1, & t\in E_j\sim E_{1j}, \; j=1, 2, \ldots, 2^r. \end{cases}$$

Clearly $g_{n+1}\in L_1(G)$ and is zero off of F. Moreover, for each $j=1, 2, \ldots, 2^r$ we have

$$\int_{E_j} g_{n+1}(t)\,d\lambda(t) = \int_{E_{1j}} (1-2^k)\,d\lambda(t) + \sum_{i=2}^{2^k} \int_{E_{ij}} 1\,d\lambda(t)$$
$$=(1-2^k)\,\lambda(F)/2^{r+k}+(2^k-1)\,\lambda(F)/2^{r+k}=0.$$

Hence if $i=1,2,\ldots,m_0$ we have

$$(g_{n+1}, w_i) = \int_F g_{n+1}(t)\, w_i(t)\, d\lambda(t) = \sum_{j=1}^{2^r} \int_{E_j} g_{n+1}(t)\, w_i(t)\, d\lambda(t) = 0,$$

because w_i is constant on each of the E_j. In particular, since each g_j, $j=1,2,\ldots,n$, is a real linear combination of w_i, $i=1,2,\ldots,m_0$, we conclude that $(g_{n+1}, g_j)=0$, $j=1,2,\ldots,n$.

Furthermore suppose $i>m_0$. If $w_i=\chi_F$, as would be the case when $w_i=r_{n_1}^{\alpha_1}\ldots r_{n_k}^{\alpha_k}$, $\alpha_1,\ldots,\alpha_k$ even integers, then as before we see that $(g_{n+1}, w_i)=0$. If $w_i \neq \chi_F$ then for i sufficiently large we would have that the integral of w_i over each set in Π_{r+k} is zero. This follows immediately from the definition of the partition $\{\Pi_n\}$ and the functions in W. Since g_{n+1} is constant on the sets in Π_{r+k} we conclude that in this case also $(g_{n+1}, w_i)=0$. Thus there exist at most a finite number of $w_i \in W$ such that $(g_{n+1}, w_i) \neq 0$, and if $(g_{n+1}, w_i) \neq 0$ then $i>m_0$. Consequently there exists $m_0<m_1<m_2<\cdots<m_h$ and real numbers $a_1, a_2, \ldots, a_h$ such that $g_{n+1}=\sum_{i=1}^{n} a_i\, w_{m_i}$. The a_i's are real since g_{n+1} and the functions in W are real valued.

So far then we have defined g_{n+1} such that $g_1, g_2, \ldots, g_{n+1}$ satisfy (i), (vi) and (vii). To complete the proof we note first that

$$\begin{aligned}\int_F |g_{n+1}(t)|\, d\lambda(t) &= \sum_{j=1}^{2^r} \int_{E_{1i}} (2^k-1)\, d\lambda(t) + \sum_{j=1}^{2^r} \sum_{i=2}^{2^k} \int_{E_{ij}} d\lambda(t)\\ &= 2^r(2^k-1)\, 2^{-(r+k)}\, \lambda(F) + 2^r(2^k-1)\, 2^{-(r+k)}\, \lambda(F)\\ &= 2(1-2^k)\, \lambda(F) \leqq 2\lambda(F),\end{aligned}$$

and

$$\begin{aligned}\int_F g_{n+1}^2(t)\, d\lambda(t) &= \sum_{j=1}^{2^r} \int_{E_{1j}} (1-2^k)^2\, d\lambda(t) + \sum_{j=1}^{2^r} \sum_{i=2}^{2^k} \int_{E_{ij}} d\lambda(t)\\ &= 2^r(1-2^k)^2\, 2^{-(r+k)}\, \lambda(F) + 2^r(2^k-1)\, 2^{-(r+k)}\, \lambda(F)\\ &= (2^k-1)\, \lambda(F) \leqq 2^k\, \lambda(F).\end{aligned}$$

Moreover, we have from the definition of g_{n+1} that

$$\begin{aligned}\lambda(\{t \mid g_{n+1}(t)=1\}) &= \sum_{j=1}^{2^r} \sum_{i=2}^{2^k} \lambda(E_{ij}) = 2^r(2^k-1)\, 2^{-(r+k)}\, \lambda(F)\\ &= (1-2^{-k})\, \lambda(F) > (1-\varepsilon/N)\, \lambda(F)\end{aligned}$$

as $1/2^k < \varepsilon/N$.

Finally, if $\gamma\in\hat{G}$ and $\gamma\notin K$ then by the choice of the compact set K we have

$$|\hat{g}_1(\gamma)+\cdots+\hat{g}_n(\gamma)+\hat{g}_{n+1}(\gamma)|$$
$$\leqq n\lambda(F)/N+|\hat{g}_{n+1}(\gamma)|\leqq n\lambda(F)/N+\|g_{n+1}\|_1\leqq(2+n/N)\,\lambda(F)$$
$$<(2+n+1/N)\,\lambda(F),$$

by the first estimate established above. While if $\gamma\in K$ then using both the Cauchy-Schwarz and Bessel's inequalities we deduce that

$$|\hat{g}_{n+1}(\gamma)|=\left|\int_F(t^{-1},\gamma)\sum_{i=1}^h a_i\,w_{m_i}(t)\,d\lambda(t)\right|\leqq\sum_{i=1}^h|a_i|\left|\int_F(t^{-1},\gamma)\,w_{m_i}(t)\,d\lambda(t)\right|$$
$$\leqq\left(\sum_{i=1}^h|a_i|^2\right)^{\frac{1}{2}}\left(\sum_{i=1}^h|(\gamma,w_{m_i})|^2\right)^{\frac{1}{2}}$$
$$\leqq\lambda(F)^{-\frac{1}{2}}\|g_{n+1}\|_2\left(\sum_{j=m_1}^{\infty}|(\gamma_j\,w_j)|^2\right)^{\frac{1}{2}}$$
$$\leqq\lambda(F)^{-\frac{1}{2}}\,2^{k/2}\,\lambda(F)^{\frac{1}{2}}\,2^{-k/2}\,N^{-1}\,\lambda(F)=\lambda(F)/N.$$

The last inequality is valid because of the choice of m_0. Combining this estimate with the property (v) of $g_1,g_2,\ldots,g_n$ we see that for $\gamma\in K$,

$$|\hat{g}_1(\gamma)+\cdots\hat{g}_n(\gamma)+\hat{g}_{n+1}(\gamma)|\leqq(2+n/N)\,\lambda(F)+\lambda(F)/N$$
$$=(2+n+1/N)\,\lambda(F).$$

Consequently, $\|\hat{g}_1+\cdots+\hat{g}_{n+1}\|_\infty\leqq(2+n+1/N)\,\lambda(F)$, and the proof of the lemma is complete. ▯

One further lemma is necessary.

Lemma 4.4.3. *Let G be a nondiscrete locally compact Abelian group and suppose $F\subset G$ is a measurable set such that $0<\lambda(F)<\infty$. If $1\leqq p<2$ then for each $\varepsilon>0$ there exists a measurable set $E(\varepsilon,p)\subset F$ such that:*

(i) $\lambda(E(\varepsilon,p))\geqq(1-\varepsilon)\,\lambda(F)$.

(ii) *If $f\in L_1(G)$ and $f=0$ almost everywhere off of $E(\varepsilon,p)$ then $\|\hat{f}\|_\infty\leqq\varepsilon\|\hat{f}\|_p$.*

Proof. Let $\varepsilon>0$. For each positive integer N let $g_1,g_2,\ldots,g_N\in L_1(G)$ be functions which satisfy the conclusions of Lemma 4.4.2. Set $\Phi_N=\sum_{j=1}^N g_j/N$.

Utilizing Lemma 4.4.2 (iii) and (vi) and the fact that $\varepsilon/N < 1/2^{k-1}$ we deduce that

$$\|\Phi_N\|_2 = \left(\sum_{j=1}^{N} (\|g_j\|_2)^2\right)^{\frac{1}{2}} \Big/ N \leqq (N\, 2^k\, \lambda(F))^{\frac{1}{2}}/N = (2\, 2^{k-1}\, \lambda(F)/N)^{\frac{1}{2}}$$
$$< (2\lambda(F)/\varepsilon)^{\frac{1}{2}},$$

for every positive integer N. Moreover, by part (v) of Lemma 4.4.2 we have

$$\|\hat{\Phi}_N\|_\infty = \left\| \sum_{j=1}^{N} \hat{g}_j \right\|_\infty \Big/ N \leqq (2 + N/N)\, \lambda(F)/N = 3\lambda(F)/N.$$

Let $1/p + 1/p' = 1$. Then $2 < p' \leqq \infty$ and by the convexity of $\log \|\cdot\|_{1/a}$ as a function of a, $0 \leqq a \leqq 1$, (Theorem D.11.3), and the Plancherel Theorem (Theorem F.8.2) we conclude for each N that

$$\|\hat{\Phi}_N\|_{p'} \leqq (\|\hat{\Phi}_N\|_2)^{2/p'} (\|\hat{\Phi}_N\|_\infty)^{1-2/p'} = (\|\Phi_N\|_2)^{2/p'} (\|\hat{\Phi}_N\|_\infty)^{1-2/p'}$$
$$\leqq (2\lambda(F)/\varepsilon)^{1/p'} (3\lambda(F)/N)^{1-2/p'}.$$

Since $1 - 2/p' > 0$ it is apparent that we can choose N so large that $\|\hat{\Phi}_N\|_{p'} \leqq \varepsilon$.

Having made such a choice of N we write $\Phi = \Phi_N$ and set $E(\varepsilon, p) = \{t \mid \Phi(t) = 1\}$. Obviously $E(\varepsilon, p)$ is a measurable subset of F as the measurable function Φ vanishes off of F. Moreover $E(\varepsilon, p) \supset \bigcap_{j=1}^{N} \{t \mid g_j(t) = 1\}$. Thus by Lemma 4.4.2 (iv) we have

$$\lambda(E(\varepsilon, p)) \geqq \lambda\left(\bigcap_{j=1}^{N} \{t \mid g_j(t) = 1\}\right) = \lambda\left(F \sim \bigcup_{j=1}^{N} \{t \mid t \in F, g_j(t) \neq 1\}\right)$$
$$\geqq \lambda(F) - \sum_{j=1}^{N} \lambda(\{t \mid t \in F, g_j(t) \neq 1\})$$
$$\geqq \lambda(F) - N[\lambda(F) - (1 - \varepsilon/N)\, \lambda(F)] = (1 - \varepsilon)\, \lambda(F).$$

Finally, suppose that $f \in L_1(G)$ and f vanishes almost everywhere off of $E(\varepsilon, p)$. If $\|\hat{f}\|_p = \infty$ then the second conclusion of the lemma is trivially true. So assume that $\|\hat{f}\|_p < \infty$. Then by the Hausdorff-Young Theorem (Theorem F.8.4) and the injective character of the Fourier transform we conclude that $f \in L_1(G) \cap L_{p'}(G)$, $1/p + 1/p' = 1$. And once more appealing to the convexity of $\log \|\cdot\|_{1/a}$ we see that $f \in L_2(G)$. Thus by means of Parseval's Formula (Theorem F.8.3) we deduce for each $\gamma \in \hat{G}$ that

$$|\hat{f}(\gamma)| = \left|\int_G (t^{-1}, \gamma) f(t)\, \Phi(t)\, d\lambda(t)\right| = \left|\int_{\hat{G}} \hat{f}(\gamma\, \delta^{-1})\, \hat{\Phi}(\delta)\, d\eta(\delta)\right|$$
$$\leqq \|\hat{\Phi}\|_{p'}\, \|\hat{f}\|_p \leqq \varepsilon\, \|\hat{f}\|_p.$$

Consequently $\|\hat{f}\|_\infty \leqq \varepsilon\, \|\hat{f}\|_p$. □

Now we can state and prove the theorem which guarantees the existence of sets of uniqueness with positive measure for $L_p(\hat{G})$, $1 \leqq p < 2$, provided G is nondiscrete.

Theorem 4.4.1. *Let G be a nondiscrete locally compact Abelian group and suppose $F \subset G$ is a measurable set such that $0 < \lambda(F) < \infty$. If $\varepsilon > 0$ then there exists a measurable set $E \subset F$ such that:*

(i) $\lambda(E) > \lambda(F) - \varepsilon$.

(ii) *E is a set of uniqueness for $L_p(\hat{G})$, $1 \leqq p < 2$.*

Proof. Let n_0 be a positive integer such that $\lambda(F)/2 \cdot 3^{n_0} < \varepsilon$. Set $\varepsilon_n = 1/3^{n+n_0}$ and $p_n = 2 - \varepsilon_n$, $n = 1, 2, 3, \ldots$. Clearly $1 \leqq p_n < 2$ for each n. Let $\{E(\varepsilon_n, p_n)\}$ be a sequence of measurable subsets of F which satisfy the conclusions of Lemma 4.4.3. Set $E = \bigcap_{n=1}^{\infty} E(\varepsilon_n, p_n)$. Obviously E is a measurable subset of F and

$$\lambda(E) = \lambda\left[F \sim \left(G \sim \bigcap_{n=1}^{\infty} E(\varepsilon_n, p_n)\right) \cap F\right] \geqq \lambda(F) - \sum_{n=1}^{\infty} \lambda[(G \sim E(\varepsilon_n, p_n)) \cap F]$$

$$= \lambda(F) - \sum_{n=1}^{\infty} [\lambda(F) - \lambda(E(\varepsilon_n, p_n))] \geqq \lambda(F) - \sum_{n=1}^{\infty} [\lambda(F) - (\lambda(F) - \varepsilon_n \lambda(F))]$$

$$= \lambda(F) - \lambda(F) \sum_{n=1}^{\infty} \frac{1}{3^{n+n_0}} = \lambda(F) - \lambda(F)/2 \cdot 3^{n_0} \geqq \lambda(F) - \varepsilon.$$

Furthermore, suppose $f \in L_1(G)$, f vanishes almost everywhere off of E and $\|\hat{f}\|_p < \infty$ for some p, $1 \leqq p < 2$. By the same argument as used at the end of the proof of Lemma 4.4.3 we see that $f \in L_2(G)$. Hence by the Plancherel Theorem (Theorem F.8.2) we conclude that $\hat{f} \in L_2(\hat{G}) \cap L_p(\hat{G})$. The usual reasoning based on the convexity of $\log \|\cdot\|_{1/a}$ (Theorem D.11.3) then reveals that $\hat{f} \in L_r(\hat{G})$, $p \leqq r \leqq 2$. Thus since $\lim_n p_n = 2$ we see that there exists a positive integer N such that for $n \geqq N$ we have $p < p_n < 2$. Hence by Lemma 4.4.3 (ii) we have $\|\hat{f}\|_\infty \leqq \varepsilon_n \|\hat{f}\|_{p_n}$ for $n \geqq N$.

However since $p < p_n < 2$ for $n \geqq N$ we see that $1/p_n = \alpha_n \cdot \frac{1}{2} + (1 - \alpha_n) 1/p$ where $\alpha_n = 1/p_n - 1/p/(\frac{1}{2} - 1/p)$. Appealing one final time to the convexity of $\log \|\cdot\|_{1/a}$ we deduce that for $n \geqq N$

$$\|\hat{f}\|_{p_n} \leqq (\|\hat{f}\|_2)^{\alpha_n} (\|\hat{f}\|_p)^{1-\alpha_n}.$$

But $\lim_n \alpha_n = 1$. Thus the sequence $\{\|\hat{f}\|_{p_n}\}$ for $n \geqq N$ is bounded.

Then we obviously have $\lim_n \varepsilon_n \|\hat{f}\|_{p_n} = 0$, whence $\|\hat{f}\|_\infty = 0$.

Therefore $f = 0$ and E is a set of uniqueness for $L_p(\hat{G})$, $1 \leqq p < 2$. □

4.5. Inclusion Results. Theorems 4.1.2 and 4.1.3 provide some information about the relationship between the spaces of multipliers $M(L_p(G))$ for various values of p. In particular these theorems say that if $1<p<\infty$, $1/p+1/p'=1$, then there is an isometric isomorphism from $M(L_p(G))$ onto $M(L_{p'}(G))$, and a continuous isomorphism from $M(L_p(G))$ into $M(L_2(G))$. Thus, with a slight abuse of terminology, we can consider $M(L_p(G))$ and $M(L_{p'}(G))$ as identical spaces and $M(L_p(G))$ as a subspace of $M(L_2(G))$. We shall generally make such identifications throughout the following development.

In this section we shall further investigate such relationships among the spaces $M(L_p(G))$. Our main goal will be to show that if G is an infinite group and $|1/p-\frac{1}{2}|>|1/r-\frac{1}{2}|>0$ then $M(L_p(G))\subset M(L_r(G))\subset M(L_2(G))$ and all the inclusions are proper. We shall only prove this theorem at the end of the section. First we shall establish a number of auxiliary results, some of which are of independent interest.

On the basis of the development in 4.1, 4.2, 4.3 and Theorem 0.1.1 it is evident that we may define a Fourier transform for the elements of $M(L_p(G))$, $1\leqq p<\infty$. Indeed, if $T\in M(L_p(G))$ then we set $\hat{T}=\hat{\sigma}$ where σ is the unique pseudomeasure such that $(Tf)\hat{}=\hat{\sigma}\hat{f}$, $f\in L_1(G)\cap L_2(G)\cap L_p(G)$. Then for $1\leqq p<\infty$ we define $\mathscr{M}(L_p(G))=\{\varphi|\varphi\in L_\infty(\hat{G}), \varphi=\hat{T}, T\in M(L_p(G))\}$. Clearly $\mathscr{M}(L_p(G))$, $1\leqq p<\infty$, is a subalgebra of $L_\infty(\hat{G})$, and Theorems 0.1.1 and 4.1.1, respectively, reveal that $\mathscr{M}(L_1(G))=M(G)\hat{}$ and $\mathscr{M}(L_2(G))=L_\infty(\hat{G})$. Here we shall be primarily interested in $\mathscr{M}(L_p(G))$, $1<p<\infty$. The previous remarks show that $\mathscr{M}(L_p(G))\subset\mathscr{M}(L_2(G))$ and $\mathscr{M}(L_p(G))=\mathscr{M}(L_{p'}(G))$, $1/p+1/p'=1$.

We shall begin the discussion by considering some results for compact Abelian groups. First we need two preliminary lemmas

Lemma 4.5.1. *Let G be an infinite compact Abelian group and $1\leqq p<2$. Then $L_2(G)$ is a proper subspace of $L_p(G)$.*

Proof. Suppose not. Then $L_2(G)=L_p(G)$ and since $\|f\|_p\leqq\|f\|_2$, $f\in L_p(G)=L_2(G)$, we deduce from the Two Norm Theorem (Theorem D.6.3) that there exists a constant $B>0$ such that $\|f\|_2\leqq B\|f\|_p$, $f\in L_p(G)=L_2(G)$.

However G is nondiscrete and so $\lambda(\{e\})=0$. Thus, since Haar measure is regular, there exists a sequence of open neighborhoods U_n of the identity e such that $0<\lambda(U_n)<1/n$. Hence we conclude from

$$\|\chi_{U_n}\|_2\leqq B\|\chi_{U_n}\|_p$$

that

$$B\geqq\frac{\|\chi_{U_n}\|_2}{\|\chi_{U_n}\|_p}=\lambda(U_n)^{\frac{1}{2}-1/p}\geqq n^{1/p-\frac{1}{2}}$$

for each n because $1/p-\frac{1}{2}>0$. But such an estimate obviously leads to a contradiction.

Therefore $L_2(G)$ is a proper subset of $L_p(G)$. ▯

Lemma 4.5.2. *Let G be a compact Abelian group and suppose g is a function on $\hat{G}$. Then the following are equivalent:*

(i) $g\in L_2(\hat{G})$.

(ii) $\varphi g\in L_1(G)\hat{}$ *for each* $\varphi\in C_0(\hat{G})$.

Proof. Since the Fourier-Plancherel transform is an isometry from $L_2(G)$ onto $L_2(\hat{G})$ we deduce immediately from Theorem 4.1.1 that if $g\in L_2(\hat{g})$ then $\varphi g\in L_2(\hat{G})\subset L_1(G)\hat{}$ for each $\varphi\in C_0(\hat{G})$ since $L_2(G)\subset L_1(G)$.

Conversely, if $\varphi g\in L_1(G)\hat{}$ for each $\varphi\in C_0(\hat{G})$ then $g\in L_\infty(\hat{G})$. As if $g\notin L_\infty(\hat{G})$ then there exists a sequence $\{\gamma_n\}\subset\hat{G}$ such that $|g(\gamma_n)|\geqq n$. Let $\varphi(\gamma_n)=1/\sqrt{n}$, $n=1,2,3,\ldots$, and $\varphi(\gamma)=0$, if $\gamma\neq\gamma_n$. Then, since $\hat{G}$ is discrete, we see that $\varphi\in C_0(\hat{G})$ and $|\varphi g(\gamma_n)|\geqq\sqrt{n}$. Thus $\varphi g\notin L_1(G)\hat{}$ as $L_1(G)\hat{}\subset C_0(\hat{G})$, contrary to the hypotheses on g.

Since $g\in L_\infty(\hat{G})$ the formula $(Tf)\hat{}=g\hat{f}$, $f\in C(G)$, defines, by Theorem 4.1.1, a linear transformation T from $C(G)$ to $L_2(G)$ which commutes with translations. Moreover, suppose $f_n, f\in C(G)$, $h\in L_2(G)$ are such that $\lim_n\|f_n-f\|_1=0$ and $\lim_n\|Tf_n-h\|_2=0$. Then, since $\lim_n\|g\hat{f}_n-g\hat{f}\|_\infty=0$ and $\lim_n\|(Tf_n)\hat{}-\hat{h}\|_2=\lim_n\|g\hat{f}_n-\hat{h}\|_2=0$, it is apparent that $Tf=h$. Thus, by the Closed Graph Theorem (Theorem D.6.1), we conclude that T is continuous from $C(G)$ as a subspace of $L_1(G)$ to $L_2(G)$. Hence T defines an element of $M(L_1(G),L_2(G))$ as $C(G)$ is norm dense in $L_1(G)$, and so by Theorem 3.1.1 we conclude that $g\in L_2(\hat{G})$. ▯

Theorem 4.5.1. *Let G be an infinite compact Abelian group. Then for each $p\neq 2$, $1<p<\infty$, $\mathcal{M}(L_p(G))\cap C_0(\hat{G})$ is a proper subset of $C_0(\hat{G})$.*

Proof. Without loss of generality we may assume that $1<p<2$. Suppose that $C_0(\hat{G})\subset\mathcal{M}(L_p(G))$. Then for each $f\in L_p(G)$ we have that $\varphi\hat{f}\in L_p(G)\hat{}$ for each $\varphi\in C_0(\hat{G})$. Hence by Lemma 4.5.2 we conclude that $\hat{f}\in L_2(\hat{G})$ for each $f\in L_p(G)$, that is, $L_p(G)\subset L_2(G)$. However this implies that $L_p(G)=L_2(G)$ as G is compact and $1<p<2$, thereby contradicting Lemma 4.5.1.

Therefore $\mathcal{M}(L_p(G))\cap C_0(\hat{G})$ is a proper subset of $C_0(\hat{G})$. ▯

Combining the preceding result with Theorem 4.4.1 we can now show that $M(L_p(G))$ is a proper subset of $M(L_2(G))$ for $p\neq 2$ and all infinite groups.

Theorem 4.5.2. *Let G be an infinite locally compact Abelian group. For each $p\neq 2$, $1<p<\infty$, $M(L_p(G))$ is a proper subspace of $M(L_2(G))$.*

Proof. If G is an infinite compact Abelian group then by Theorem 4.5.1 we have that $\mathscr{M}(L_p(G)) \cap C_0(\hat{G})$ is a proper subset of $C_0(\hat{G}) \subset L_\infty(\hat{G}) = \mathscr{M}(L_2(G))$. Hence $M(L_p(G))$ is a proper subspace of $M(L_2(G))$.

Suppose then that G is an infinite noncompact group, and assume, without loss of generality, that $1<p<2$. Let $F \subset \hat{G}$ be a compact set for which $\eta(F)>0$, and let $E \subset F$ be a set of uniqueness for $L_p(G)$ such that $\eta(E)>0$. The existence of E is guaranteed by Theorem 4.4.1 since $\hat{G}$ is nondiscrete. By Theorem 4.1.1 the characteristic function χ_E defines a multiplier for $L_2(G)$. However, χ_E does not define a multiplier for $L_p(G)$, $1<p<2$.

As if it did then by Theorem 4.3.2 we would have $\chi_E \hat{f} \in L_p(G)\hat{}$ for each $f \in L_1(G) \cap L_2(G) \cap L_p(G)$. In particular if $f \in L_1(G) \cap L_2(G) \cap L_p(G)$ is such that $\hat{f} \in C_c(\hat{G})$ and $\hat{f}(\gamma)=1$, $\gamma \in F$ (F.7e) then $\chi_E \hat{f} = \chi_E \in L_p(G)\hat{}$. Thus there exists a unique $g \in L_p(G)$ such that $\check{g} = \chi_E$. Since $\chi_E \in L_2(\hat{G}) \cap L_1(\hat{G})$ it follows immediately that $\hat{\chi}_E = g \in L_p(G)$. But this contradicts the choice of E as a set of uniqueness for $L_p(G)$.

Therefore χ_E does not define a multiplier for $L_p(G)$ and $M(L_p(G))$ is a proper subspace of $M(L_2(G))$. □

The result analogous to Theorem 4.5.1 is also valid for noncompact groups.

Lemma 4.5.3. *Let G be an infinite noncompact locally compact Abelian group and $1<p<2$. Suppose $F \subset \hat{G}$ is compact, $\eta(F)>0$, and let $E \subset F$ be a set of uniqueness for $L_p(G)$ such that $\eta(E)>0$. Then there exists a $g \in L_1(\hat{G})$ such that $g * \chi_E$ does not define a multiplier for $L_p(G)$.*

Proof. Suppose $g * \chi_E$ defines a multiplier for $L_p(G)$ for each $g \in L_1(\hat{G})$. Denote by T_g the element of $M(L_p(G))$ so determined, that is, $(T_g f)\hat{} = (g * \chi_E)\hat{f}$, $f \in L_1(G) \cap L_2(G) \cap L_p(G) = L_1(G) \cap L_2(G)$. The latter equality follows from the convexity of $\log \|\cdot\|_{1/a}$ (Theorem D.11.3) and the assumption that $1<p<2$. Define the mapping $S\colon L_1(\hat{G}) \to M(L_p(G))$ by $S(g) = T_g$. Clearly S is linear. Moreover, suppose $g_n, g \in L_1(\hat{G})$ and $T \in M(L_p(G))$ are such that $\lim_n \|g_n - g\|_1 = 0$ and $\lim_n \|T_{g_n} - T\|_p = 0$.

Then by Corollary 4.1.2 we see that

$$\begin{aligned}\|\hat{T} - g * \chi_E\|_\infty &\leqq \|\hat{T} - \hat{T}_{g_n}\|_\infty + \|\hat{T}_{g_n} - g * \chi_E\|_\infty \\ &\leqq \|T - T_{g_n}\|_p + \|g_n * \chi_E - g * \chi_E\|_\infty \leqq \|T - T_{g_n}\|_p + \|g_n - g\|_1.\end{aligned}$$

Hence $\hat{T} = g * \chi_E$ and so $T = S(g) = T_g$. Thus from the Closed Graph Theorem (Theorem D.6.1) we conclude that there exists a constant $B>0$ such that

$$\|T_g\|_p \leqq B\|g\|_1 \qquad (g \in L_1(\hat{G})).$$

Let $f\in L_1(G)\cap L_2(G)$ be such that $\hat{f}\in C_c(\hat{G})$ and $\hat{f}(\gamma)=1$, $\gamma\in F$ (F.7e). And let $\{g_\alpha\}\subset L_1(\hat{G})$ be an approximate identity for $L_1(\hat{G})$ such that $\|g_\alpha\|_1\leqq 1$. Then for each α

$$\|T_{g_\alpha} f\|_p\leqq\|T_{g_\alpha}\|_p\,\|f\|_p\leqq B\,\|g_\alpha\|_1\,\|f\|_p\leqq B\,\|f\|_p.$$

Hence $\{T_{g_\alpha} f\}$ is a uniformly bounded net in $L_p(G)$ and so by Alaoglu's Theorem (Theorem D.4.3) and the reflexivity of $L_p(G)$ there exists a subnet $\{T_{g_\beta} f\}$ of $\{T_{g_\alpha} f\}$ and a $g\in L_p(G)$ such that

$$\langle g, h\rangle=\lim_\beta\langle T_{g_\beta} f, h\rangle \qquad \big(h\in L_{p'}(G)\big),$$

where as usual $1/p+1/p'=1$.

In particular, for each $h\in A_c(G)\subset L_{p'}(G)$ we have

$$\begin{aligned}\langle g, h\rangle&=\lim_\beta\langle T_{g_\beta} f, h\rangle=\lim_\beta\int_G T_{g_\beta} f(t)\,h(t^{-1})\,d\lambda(t)\\&=\lim_\beta\int_{\hat{G}}(T_{g_\beta} f)\hat{\ }(\gamma)\,\hat{h}(\gamma)\,d\eta(\gamma)\\&=\lim_\beta\int_{\hat{G}} g_\beta*\chi_E(\gamma)\,\hat{f}(\gamma)\,\hat{h}(\gamma)\,d\eta(\gamma).\end{aligned}$$

The application of Parseval's Formula (Theorem F.8.3) is valid because $A_c(G)\subset C_c(G)\subset L_2(G)$ and $\{T_{g_\beta} f\}\subset L_2(G)$ as $M\big(L_p(G)\big)\subset M\big(L_2(G)\big)$ and $f\in L_2(G)$. Consequently, since $\{g_\beta\}$ is an approximate identity for $L_1(\hat{G})$ and $\hat{f}\,\hat{h}\in L_\infty(\hat{G})$ we conclude for each $h\in A_c(G)$ that

$$\begin{aligned}\langle g, h\rangle&=\lim_\beta\int_{\hat{G}} g_\beta*\chi_E(\gamma)\,\hat{f}(\gamma)\,\hat{h}(\gamma)\,d\eta(\gamma)=\int_{\hat{G}}\chi_E(\gamma)\,\hat{f}(\gamma)\,\hat{h}(\gamma)\,d\eta(\gamma)\\&=\int_{\hat{G}}\chi_E(\gamma)\,\hat{h}(\gamma)\,d\eta(\gamma)=\int_G\hat{\chi}_E(t)\,h(t^{-1})\,d\lambda(t)=\langle\hat{\chi}_E, h\rangle.\end{aligned}$$

Hence $\hat{\chi}_E=g\in L_p(G)$ as $A_c(G)$ is norm dense in $L_p(G)$, contradicting the choice of E as a set of uniqueness for $L_p(G)$.

Therefore there exists some $g\in L_1(\hat{G})$ for which $g*\chi_E$ does not define a multiplier for $L_p(G)$. □

Theorem 4.5.3. *Let G be an infinite noncompact locally compact Abelian group. For each $p\neq 2$, $1<p<\infty$, $\mathscr{M}\big(L_p(G)\big)\cap C_0(\hat{G})$ is a proper subset of $C_0(\hat{G})$.*

Proof. As usual we may assume that $1<p<2$. Let $E\subset\hat{G}$ be a set of uniqueness for $L_p(G)$ as in the preceding lemma, and let $g\in L_1(\hat{G})$ be such that $g*\chi_E$ does not define a multiplier for $L_p(G)$, that is, such that $g*\chi_E\notin\mathscr{M}\big(L_p(G)\big)$. But since $g\in L_1(\hat{G})$ and $\chi_E\in L_\infty(\hat{G})$ we see that $g*\chi_E\in C_0(\hat{G})$. Thus $\mathscr{M}\big(L_p(G)\big)\cap C_0(\hat{G})$ is a proper subset of $C_0(\hat{G})$. □

We can carry the study of $\mathscr{M}(L_p(G)) \cap C_0(\hat{G})$ even further as is evidenced by the next result.

Theorem 4.5.4. *Let G be an infinite locally compact Abelian group. If $|1/p-\frac{1}{2}| \neq |1/r-\frac{1}{2}|$, $1<p$, $r<\infty$, then $\mathscr{M}(L_p(G)) \cap C_0(\hat{G}) \neq \mathscr{M}(L_r(G)) \cap C_0(\hat{G})$.*

Proof. We assume without loss of generality that $1<p<r \leqq 2$. If $r=2$ then $\mathscr{M}(L_r(G)) \cap C_0(\hat{G})) = L_\infty(\hat{G}) \cap C_0(\hat{G}) = C_0(\hat{G})$ and we conclude from Theorems 4.5.1 and 4.5.3 that $\mathscr{M}(L_p(G)) \cap C_0(\hat{G}) \neq \mathscr{M}(L_r(G)) \cap C_0(\hat{G})$.

Suppose that $1<p<r<2$ and that

$$\mathscr{M}(L_p(G)) \cap C_0(\hat{G}) = \mathscr{M}(L_r(G)) \cap C_0(\hat{G}).$$

If $\varphi \in \mathscr{M}(L_p(G)) \cap C_0(\hat{G})$ then there exists a unique $T \in M(L_p(G))$ such that $\hat{T} = \varphi$. We set $\|\varphi\| = \|T\|_p$. Since $\|\varphi\|_\infty \leqq \|T\|_p$ by Corollary 4.1.2 we see that $\|\cdot\|$ is a norm on the linear space $\mathscr{M}(L_p(G)) \cap C_0(\hat{G})$. Let $\{\varphi_n\} = \{\hat{T}_n\}$ be a Cauchy sequence in $\mathscr{M}(L_p(G)) \cap C_0(\hat{G})$ with respect to this norm. Then $\{T_n\}$ is a Cauchy sequence in $M(L_p(G))$ and so there exists a $T \in M(L_p(G))$ for which $\lim_n \|T_n - T\|_p = 0$. Moreover, we deduce from the inequalities

$$\|\hat{T} - \varphi_n\|_\infty = \|\hat{T} - \hat{T}_n\|_\infty \leqq \|T - T_n\|_p$$

that $\varphi = \hat{T} \in C_0(\hat{G})$. Hence $\mathscr{M}(L_p(G)) \cap C_0(\hat{G})$ is a Banach space under the norm $\|\cdot\|$.

Similarly we see that $\mathscr{M}(L_r(G)) \cap C_0(G)$ is a Banach space under the norm induced on it by the norm in $M(L_r(G))$ as above. Since by assumption $\mathscr{M}(L_p(G)) \cap C_0(\hat{G}) = \mathscr{M}(L_r(G)) \cap C_0(\hat{G})$ we conclude immediately via the Closed Graph Theorem (Theorem D.6.1) that the two norms so constructed are equivalent. Thus there exists a constant $B>0$ such that for each $\hat{T} \in \mathscr{M}(L_p(G)) \cap C_0(\hat{G}) = \mathscr{M}(L_r(G)) \cap C_0(\hat{G})$ we have $\|T\|_p \leqq B\|T\|_r$.

Now, as indicated previously, if $f \in L_1(G)$ then the formula $T_f g = f * g$, $g \in L_p(G)$, defines an element of $M(L_p(G))$. Moreover $\hat{T}_f = \hat{f} \in C_0(\hat{G})$ so that $\hat{T}_f \in \mathscr{M}(L_p(G)) \cap C_0(\hat{G})$. In this way we can consider $L_1(G)\hat{}$ as a subspace of $\mathscr{M}(L_p(G)) \cap C_0(\hat{G})$. Similarly $L_1(G)\hat{}$ may also be considered as a subspace of $\mathscr{M}(L_2(G)) \cap C_0(\hat{G}) = C_0(\hat{G})$. However, the norm of the Banach space $\mathscr{M}(L_p(G)) \cap C_0(\hat{G})$ restricted to $L_1(G)\hat{}$ defines a topology on $L_1(G)\hat{}$ which is not equivalent to the topology so induced on $L_1(G)\hat{}$ considered as a subspace of $C_0(\hat{G})$ with the usual supremum norm topology. This is so because $L_1(G)\hat{}$ is supremum norm dense in $C_0(\hat{G})$ but $\mathscr{M}(L_p(G)) \cap C_0(\hat{G}) \neq C_0(\hat{G})$ by Theorems 4.5.1 and 4.5.3. Consequently there exists a sequence $\{f_n\} \subset L_1(G)$ such that $\|T_{f_n}\|_p = 1$ but $\lim_n \|\hat{T}_{f_n}\|_\infty = \lim_n \|T_{f_n}\|_2 = 0$.

But then, on the one hand, since $\frac{1}{2}<1/r<1/p$ we have by a form of the Riesz-Thorin Convexity Theorem (Theorem D.11.3) that

$$\|T_{f_n}\|_r \leqq (\|T_{f_n}\|_2)^\alpha (\|T_{f_n}\|_p)^{1-\alpha} = (\|T_{f_n}\|_2)^\alpha$$

where $1/r=\alpha\cdot\frac{1}{2}+(1-\alpha)\,1/p$, $0<\alpha<1$. From this it is immediate that $\lim_n \|T_{f_n}\|_r=0$.

While on the other hand, for each n we have

$$1=\|T_{f_n}\|_p \leqq B\,\|T_{f_n}\|_r,$$

contradicting the previous assertion.

Therefore $\mathscr{M}(L_p(G))\cap C_0(\hat{G}) \neq \mathscr{M}(L_r(G))\cap C_0(\hat{G})$. $\square$

Finally, we prove the result indicated at the beginning of this section.

Theorem 4.5.5. *Let G be an infinite locally compact Abelian group. If $|1/p-\frac{1}{2}|>|1/r-\frac{1}{2}|>0$, $1<p$, $r<\infty$, then $M(L_p(G))\subset M(L_r(G))\subset M(L_2(G))$ and the inclusions are proper.*

Proof. The inclusions here are meant in the sense that there exists a continuous algebra isomorphism of the one space into the other.

Theorems 4.1.3 and 4.5.2 show that $M(L_p(G))\subset M(L_2(G))$ and that the inclusion is proper.

To establish the existence of the other isomorphism we assume as usual that $1<p<2$. Then $|1/r-\frac{1}{2}|<1/p-\frac{1}{2}$ and an elementary computation reveals that $1/p'<1/r<1/p$ where $1/p+1/p'=1$. If $T\in M(L_p(G))=M(L_{p'}(G))$ then we may apply a form of the Riesz-Thorin Convexity Theorem (Theorem D.11.3) to conclude by Theorem 4.1.2 that

$$\|T\|_r \leqq (\|T\|_p)^\alpha (\|T\|_{p'})^{1-\alpha} = \|T\|_p,$$

where $1/r=\alpha\cdot 1/p+(1-\alpha)\cdot 1/p'$, $0<\alpha<1$. Thus the restriction of $T\in M(L_p(G))$ to the integrable simple functions determines a unique continuous linear transformation on $L_r(G)$ which commutes with translations, that is, a unique element of $M(L_r(G))$. Clearly this correspondence defines a continuous algebra isomorphism from $M(L_p(G))$ into $M(L_r(G))$. Hence $M(L_p(G))\subset M(L_r(G))$.

That the inclusion is proper follows immediately from Theorem 4.5.4. $\square$

A few final remarks are perhaps in order. First the concerns of this section are obviously trivial if G is finite. In this case $M(L_p(G))=M(L_2(G))$, $1\leqq p\leqq\infty$. Secondly, by Theorem 3.1.1, we see for infinite groups that $M(L_1(G))=M(G)$ and $\mathscr{M}(L_1(G))\cap C_0(\hat{G})=M(\hat{G})\cap C_0(\hat{G})$.

From the remarks at the end of 4.3 we see that $M(L_1(G)) \subset M(L_p(G))$, $1 < p < \infty$, and the inclusion is proper. Moreover, it is well known that $M(G)\hat{} \cap C_0(\hat{G}) \neq C_0(\hat{G})$ (F.7c), so that $\mathscr{M}(L_1(G)) \cap C_0(\hat{G}) \neq C_0(\hat{G})$.

4.6. The Derived Space for $L_p(G)$, $1 \leqq p < \infty$. We discussed in Chapter 1 the notion of the derived algebra of a commutative Banach algebra. In particular, we examined in some detail the derived algebra of $L_p(G)$, $1 \leqq p < \infty$, when G is a compact Abelian group. If G is noncompact and $p > 1$ then of course $L_p(G)$ is no longer a Banach algebra and so the concept of the derived *algebra* is meaningless. However, recalling the content of Lemma 1.8.1 we see that there exists a rather natural analog of the derived algebra for the spaces $L_p(G)$, $1 < p < \infty$. More precisely, if G is a locally compact Abelian group and $1 \leqq p < \infty$ then for $f \in L_p(G)$ we set

$$\|f\|_0 = \sup\{\|h * f\|_p \,|\, h \in L_1(G),\ \|\hat{h}\|_\infty \leqq 1\}.$$

The *derived space* of $L_p(G)$ is the linear subspace of $L_p(G)$ consisting of all those $f \in L_p(G)$ such that $\|f\|_0 < \infty$. We shall denote the derived space of $L_p(G)$ by $L_p^0(G)$.

When $p = 1$ we see at once from Lemma 1.8.1 and the fact that $L_1(G)$ is a Banach algebra that $L_1^0(G)$ coincides with the derived algebra $(L_1(G))_0$. Similarly if G is compact then it is easily seen that the definition of the derived space is also equivalent to that of the derived algebra for $L_p(G)$, $1 < p < \infty$.

For certain values of p it is also possible to carry over the definition of the derived algebra directly to the present situation. In particular if G is a locally compact Abelian group and $1 \leqq p \leqq 2$ then we can once again set $(L_p(G))_0 = \{f \,|\, f \in L_p(G),\ \varphi \hat{f} \in L_p(G)\hat{}\ ,\ \varphi \in C_0(\hat{G})\}$. Obviously $(L_p(G))_0$ is a linear subspace of $L_p(G)$ and it is identical to the derived algebra whenever $L_p(G)$ is a Banach algebra. Thus $(L_1(G))_0 = L_1^0(G)$. Moreover, since by Theorem 4.1.1 every bounded function on $\hat{G}$ defines a multiplier for $L_2(G)$, we see at once that $(L_2(G))_0 = L_2(G)$. Some information about the relationship between $(L_p(G))_0$ and $L_p^0(G)$ is given in the next theorem.

Theorem 4.6.1. *Let G be a locally compact Abelian group and suppose $1 \leqq p \leqq 2$. Then $(L_p(G))_0 \subset L_p^0(G)$.*

Proof. Let $f \in (L_p(G))_0$. For $\varphi \in C_0(\hat{G})$ define $T\varphi \in L_p(G)$ by $(T\varphi)\hat{}\ = \varphi \hat{f}$. Clearly T is a linear transformation from $C_0(\hat{G})$ to $L_p(G)$. Suppose $\lim_n \|\varphi_n - \varphi\| = 0$ and $\lim_n \|T\varphi_n - g\|_p = 0$. Then

$$\begin{aligned}
\|\varphi \hat{f} - \hat{g}\|_{p'} &\leqq \|\varphi \hat{f} - (T\varphi_n)\hat{}\ \|_{p'} + \|(T\varphi_n)\hat{}\ - \hat{g}\|_{p'} \\
&\leqq \|\varphi \hat{f} - \varphi_n \hat{f}\|_{p'} + \|(T\varphi_n)\hat{}\ - \hat{g}\|_{p'} \\
&\leqq \|\varphi - \varphi_n\|_\infty \|f\|_p + \|T\varphi_n - g\|_p
\end{aligned}$$

where $1/p+1/p'=1$. An appeal to the Hausdorff-Young Theorem (Theorem F.8.4) is of course necessary when $1<p<2$. Hence $\varphi\hat{f}=\hat{g}$, that is, $T\varphi=g$, and the transformation T is closed. Consequently by the Closed Graph Theorem (Theorem D.6.1) the transformation is continuous. Thus, since $(h*f)\hat{}=\hat{h}\hat{f}$ for each $h\in L_1(G)$, we see that

$$\|h*f\|_p=\|T\hat{h}\|_p\leqq\|T\|\,\|\hat{h}\|_\infty \qquad (h\in L_1(G)).$$

Therefore $\|f\|_0<\infty$ and $f\in L_p^0(G)$. □

From the remarks preceding the theorem it is apparent that the inclusion is an equality whenever G is compact or G is noncompact and $p=1$ or $p=2$. In general for noncompact G and $1<p<2$ it is not clear whether $L_p^0(G)$ and $(L_p(G))_0$ are identical. We shall show below, however, that if G is noncompact and connected or infinite and discrete then both spaces consist only of the zero element, and hence, they coincide. We shall show first for noncompact G and $1\leqq p<2$ that $(L_p(G))_0=\{0\}$. This is an immediate corollary of the following theorem.

Theorem 4.6.2. *Let G be a noncompact locally compact Abelian group and $1\leqq p<2$. If f is a function on $\hat{G}$ such that $\varphi f\in L_p(G)\hat{}$ for each $\varphi\in C_0(\hat{G})$ then $f=0$ locally almost everywhere.*

We recall that $f=0$ *locally almost everywhere* means that $f=0$ almost everywhere on each compact set.

Proof. Suppose f does not vanish locally almost everywhere. Then there exists some compact set $K\subset\hat{G}$ such that $\eta(K)>0$ and f does not vanish almost everywhere on K. Let $\psi\in C_c(\hat{G})$ be such that $\psi\equiv1$ on K. Then ψf does not vanish locally almost everywhere as $\psi f=f$ on K. Moreover, since $\varphi\psi\in C_0(\hat{G})$ for each $\varphi\in C_0(\hat{G})$, it follows that $\varphi\psi f\in L_p(G)\hat{}$ for each $\varphi\in C_0(\hat{G})$. Thus, without loss of generality, we may assume that f vanishes off of some compact set F, $\eta(F)>0$.

Now if $p=1$ then $\varphi f\in L_1(G)\hat{}$ for each $\varphi\in C_0(\hat{G})$. Thus for each φ there exists a unique $g\in L_1(G)$ such that $\hat{g}=\varphi f$. Since f vanishes off of a compact set we conclude that $\varphi f\in C_c(\hat{G})\subset L_2(\hat{G})$. Hence by the Plancherel Theorem (Theorem F.8.2) we have $g\in L_1(G)\cap L_2(G)$. Since $1<p<2$ it then follows from the convexity of $\log\|\cdot\|_{1/a}$ (Theorem D.11.3) that $g\in L_p(G)$, $1<p<2$. Consequently, $\varphi f\in L_p(G)\hat{}$, $1<p<2$, for each $\varphi\in C_0(\hat{G})$. Thus we may further assume, without loss of generality, that $\varphi f\in L_p(G)\hat{}$ for each $\varphi\in C_0(\hat{G})$ and some fixed p, $1<p<2$.

In this case we see, upon choosing $\varphi\in C_0(\hat{G})$ which is identically one on F, that $f\in L_p(G)\hat{}\subset L_{p'}(\hat{G})$, $1/p+1/p'=1$ (Theorem F.8.4). Thus the estimate

$$\int_{\hat{G}} |f(\gamma)|^2 \, d\eta(\gamma) = \int_{\hat{G}} \chi_F(\gamma) |f(\gamma)|^2 \, d\eta(\gamma)$$

$$\leqq \Big(\int_{\hat{G}} |f(\gamma)|^{p'} \, d\eta(\gamma)\Big)^{2/p'} \Big(\int_{\hat{G}} \chi_F(\gamma) \, d\eta(\gamma)\Big)^{1-2/p'}$$

shows that $f \in L_2(\hat{G})$.

Now for each $\varphi \in C_0(\hat{G})$ let $T\varphi$ denote the unique element of $L_p(G)$ such that $(T\varphi)^{\check{}} = \varphi f$. Since $f \in L_2(\hat{G})$ it is also apparent that $T\varphi = (\varphi f)^{\hat{}} \in L_2(G)$. In this way we define a linear mapping T from $C_0(\hat{G})$ to $L_p(G)$. Moreover, by the same argument *mutatis mutandis* as used in the proof of Theorem 4.6.1, we deduce that T is continuous. Thus there exists a constant $B > 0$ such that

$$\|T\varphi\|_p \leqq B \|\varphi\|_\infty \qquad \big(\varphi \in C_0(\hat{G})\big).$$

Let $E \subset \hat{G}$ be any compact set. We wish to show that $(\chi_E f)^{\hat{}} \in L_p(G)$. Let $\{\varphi_n\} \subset C_c(\hat{G})$ be such that $0 \leqq \varphi_n(\gamma) \leqq 1, \gamma \in \hat{G}, \varphi_n(\gamma) = 1, \gamma \in E$, the support of φ_n is E_n, $E_n \supset E_{n+1}$ and $\bigcap_{n=1}^{\infty} E_n = E$. Clearly $\{\varphi_n\}$ converges pointwise to χ_E. Furthermore, $\|T\varphi_n\|_p \leqq B \|\varphi_n\|_\infty \leqq B$ shows that $\{T\varphi_n\}$ is a bounded sequence in $L_p(G)$. Hence (Theorem D.4.4) the sequence $\{T\varphi_n\}$ has a weakly convergence subsequence $\{T\varphi_k\}$. Let $g \in L_p(G)$ be such that

$$\lim_k \langle T\varphi_k, h \rangle = \langle g, h \rangle \qquad \big(h \in L_{p'}(G)\big).$$

Now the sequence $\{\varphi_k f\}$ clearly converges pointwise to $\chi_E f$ and $|\varphi_k f(\gamma)| \leqq |f(\gamma)|$, $\gamma \in \hat{G}$, so we may appeal to the Lebesgue Dominated Convergence Theorem (Theorem C.5.2) to deduce for each $h \in C_c(G)$ that

$$\langle g, h \rangle = \lim_k \langle T\varphi_k, h \rangle = \lim_k \int_G T\varphi_k(t) \, h(t^{-1}) \, d\lambda(t)$$

$$= \lim_k \int_{\hat{G}} (T\varphi_k)^{\hat{}}(\gamma) \, \hat{h}(\gamma) \, d\eta(\gamma)$$

$$= \lim_k \int_{\hat{G}} \varphi_k f(\gamma) \, \hat{h}(\gamma) \, d\eta(\gamma) = \int_{\hat{G}} \chi_E f(\gamma) \, \hat{h}(\gamma) \, d\eta(\gamma)$$

$$= \int_G (\chi_E f)^{\hat{}}(t) \, h(t^{-1}) \, d\lambda(t) = \langle (\chi_E f)^{\hat{}}, h \rangle.$$

The applications of Parseval's Formula (Theorem F.8.3) are valid since $f \in L_2(\hat{G})$. Since $C_c(G)$ is norm dense in $L_{p'}(G)$ we conclude that $(\chi_E f)^{\hat{}} = g$ almost everywhere, and so $(\chi_E f)^{\hat{}} \in L_p(G)$.

But now by Theorem 4.4.1 there exists a measurable subset $E \subset F$ such that f does not vanish almost everywhere on E and E is a set of uniqueness for $L_p(G)$. Since Haar measure is regular and a measurable subset of a set of uniqueness is again a set of uniqueness, it is apparent that we may assume that E is compact. From the argument of the preceding paragraph we have $(\chi_E f)\hat{} \in L_p(G)$, while $\chi_E f$ does not vanish almost everywhere, thereby contradicting the choice of E as a set of uniqueness for $L_p(G)$.

Therefore $f=0$ locally almost everywhere and the proof is complete. □

Corollary 4.6.1. *Let G be a noncompact locally compact Abelian group and $1 \leqq p < 2$. Then $(L_p(G))_0 = \{0\}$.*

In particular, the corollary shows that the derived algebra of $L_1(G)$ for noncompact groups G consists only of the zero element. As seen in Theorem 1.9.1 this is in sharp contrast to the situation for compact groups where $(L_p(G))_0 = L_2(G)$, $1 \leqq p \leqq 2$.

The counterpart of the second portion of Theorem 1.9.1 takes the following form.

Corollary 4.6.2. *Let G be a noncompact locally compact Abelian group and $1<p<\infty$, $p \neq 2$. Suppose $\varphi \in \mathcal{M}(L_p(G))$ has the property that if $|\psi(\gamma)| \leqq |\varphi(\gamma)|$ for almost all $\gamma \in \hat{G}$ then $\psi \in \mathcal{M}(L_p(G))$. Then $\varphi = 0$ locally almost everywhere.*

Proof. Without loss of generality we may assume that $1<p<2$. By Corollary 4.1.2 we see that $\varphi \hat{f} \in L_p(G)\hat{}$ for each $f \in L_2(G) \cap L_p(G)$. If $\psi \in C_0(\hat{G})$ then $|\psi\, \varphi(\gamma)|/\|\psi\|_\infty \leqq |\varphi(\gamma)|$ for almost all $\gamma \in \hat{G}$ and so $\psi\varphi/\|\psi\|_\infty \in \mathcal{M}(L_p(G))$. Hence $\psi\varphi \in \mathcal{M}(L_p(G))$ for each $\psi \in C_0(\hat{G})$. In particular then we have $\psi(\varphi \hat{f}) \in L_p(G)\hat{}$ for each $\psi \in C_0(\hat{G})$ and $f \in L_2(G) \cap L_p(G)$. Consequently we conclude by Theorem 4.6.2 that $\varphi \hat{f} = 0$ locally almost everywhere for each $f \in L_2(G) \cap L_p(G)$. It follows immediately that $\varphi = 0$ locally almost everywhere. □

Next we wish to turn our attention to the somewhat more intricate investigation of $L_p^0(G)$. First we shall obtain an alternative description of these spaces.

Theorem 4.6.3. *Let G be a locally compact Abelian group and suppose $1<p<\infty$ and $1/p + 1/p' = 1$. Then the following are equivalent:*

(i) *$f \in L_p^0(G)$.*

(ii) *For each $g \in L_{p'}(G)$ there exists a unique $\mu \in M(\hat{G})$ such that $f * g = \hat{\mu}$.*

Proof. Let $f \in L_p^0(G)$ and $g \in L_{p'}(G)$. If $h \in L_1(G)$ we set $F(\hat{h}) = h * f * g(e)$. Clearly this defines a linear functional on $L_1(G)\hat{}$ since $f * g \in C_0(G)$. Moreover, F is continuous on $L_1(G)\hat{}$ with the supremum norm since by Hölder's inequality we have

$$|F(\hat{h})| = |h * f * g(e)| \leqq \|h * f\|_p \|g\|_{p'} \leqq \|\hat{h}\|_\infty \|f\|_0 \|g\|_{p'},$$

as $f \in L_p^0(G)$. Consequently F has a unique extension to a continuous linear functional on $C_0(\hat{G})$ and $\|F\| \leqq \|f\|_0 \|g\|_{p'}$. Let $\mu \in M(\hat{G})$ be the unique measure associated with this functional by the Riesz Representation Theorem (Theorem D.9.4). Then for each $h \in L_1(G)$ we have

$$h*f*g(e) = \int_{\hat{G}} \hat{h}(\gamma^{-1})\, d\mu(\gamma)$$

Moreover, for each $h \in L_1(G)$, we have

$$\begin{aligned}\langle f*g, h\rangle = h*f*g(e) &= \int_{\hat{G}} \hat{h}(\gamma^{-1})\, d\mu(\gamma) = \int_{\hat{G}} \Big[\int_G (t,\gamma)\, h(t)\, d\lambda(t)\Big]\, d\mu(\gamma)\\ &= \int_G h(t) \Big[\int_{\hat{G}} (t,\gamma)\, d\mu(\gamma)\Big]\, d\lambda(t) = \int_G h(t)\, \hat{\mu}(t^{-1})\, d\lambda(t)\\ &= \int_G h(t^{-1})\, \hat{\mu}(t)\, d\lambda(t) = \langle \hat{\mu}, h\rangle.\end{aligned}$$

Therefore $f*g = \hat{\mu}$, and μ is unique from the uniqueness of the Fourier-Stieltjes transform (F.7b).

Suppose now that $f \in L_p(G)$ and for each $g \in L_{p'}(G)$ there is a unique $\mu \in M(\hat{G})$ such that $f*g = \hat{\mu}$. Clearly this equation defines a linear transformation S: $L_{p'}(G) \to M(\hat{G})$. Moreover, if $\lim_n \|g_n - g\|_{p'} = 0$ and $\lim_n \|S g_n - v\| = 0$ then

$$\begin{aligned}\|(S g)\hat{} - \hat{v}\|_\infty &\leqq \|(S g)\hat{} - (S g_n)\hat{}\|_\infty + \|(S g_n)\hat{} - \hat{v}\|_\infty\\ &\leqq \|f*g - f*g_n\|_\infty + \|S g_n - v\| \leqq \|f\|_p \|g - g_n\|_{p'} + \|S g_n - v\|.\end{aligned}$$

Hence $S g = v$ and S is a closed linear transformation. Thus by the Closed Graph Theorem (Theorem D.6.1) S is continuous.

If $h \in L_1(G)$ and $S g = \mu$ where $f*g = \hat{\mu}$ then we have

$$\begin{aligned}|h*f*g(e)| = |h*\hat{\mu}(e)| &= \Big|\int_G h(t^{-1})\, \hat{\mu}(t)\, d\lambda(t)\Big|\\ &= \Big|\int_G h(t^{-1}) \Big[\int_{\hat{G}} (t^{-1},\gamma)\, d\mu(\gamma)\Big]\, d\lambda(t)\Big|\\ &= \Big|\int_{\hat{G}} \int_G (t^{-1},\gamma)\, h(t^{-1})\, d\lambda(t)\, d\mu(\gamma)\Big| = \Big|\int_{\hat{G}} \hat{h}(\gamma^{-1})\, d\mu(\gamma)\Big|\\ &= |\langle \hat{h}, \mu\rangle| \leqq \|\hat{h}\|_\infty \|\mu\| = \|\hat{h}\|_\infty \|S g\| \leqq \|\hat{h}\|_\infty \|S\|\, \|g\|_{p'}.\end{aligned}$$

Consequently, by a corollary to the Hahn-Banach Theorem (D.6d), we conclude that

$$\begin{aligned}\|h*f\|_p &= \sup\{|h*f*g(e)| \mid \|g\|_{p'} \leqq 1\}\\ &\leqq \sup\{\|\hat{h}\|_\infty \|S\|\, \|g\|_{p'} \mid \|g\|_{p'} \leqq 1\} \leqq \|\hat{h}\|_\infty \|S\|.\end{aligned}$$

Therefore $\|f\|_0 \leqq \|S\| < \infty$ and $f \in L_p^0(G)$. ☐

It is apparent from this result that if $f \in L_p^0(G)$ then f defines a continuous linear transformation $S: L_{p'}(G) \to M(\hat{G})$ by means of the equation $(Sg)\hat{} = \hat{\mu} = f * g$, and that $\|f\|_0 \leqq \|S\|$. It is elementary to verify that S is such that $S(\tau_s g) = (s, \cdot) S g$ for each $s \in G$. The next theorem shows that the converse of this observation is also valid.

Theorem 4.6.4. *Let G be a locally compact Abelian group and suppose $1 < p < \infty$ and $1/p + 1/p' = 1$. If $S: L_{p'}(G) \to M(\hat{G})$ is a linear transformation then the following are equivalent:*

(i) *There exists a unique $f \in L_p^0(G)$ such that $(Sg)\hat{} = f * g$ for each $g \in L_{p'}(G)$.*

(ii) *S is continuous and $S(\tau_s g) = (s, \cdot) S g$ for each $s \in G$ and $g \in L_{p'}(G)$. Moreover the correspondence between S and f defines a linear isometry from $L_p^0(G)$ onto the Banach space of all continuous linear transformations $S: L_{p'}(G) \to M(\hat{G})$ such that $S(\tau_s g) = (s, \cdot) S g$.*

Proof. In view of the remark preceding the theorem it is apparent that (i) implies (ii).

Conversely, suppose (ii) holds. If $g \in L_{p'}(G)$ then define $F(g) = \int_{\hat{G}} d\tilde{\mu}(\gamma)$ where $Sg = \mu \in M(\hat{G})$ and $\tilde{\mu}(E) = \mu(E^{-1})$. Clearly F is a linear functional on $L_{p'}(G)$, and furthermore it is continuous since

$$|F(g)| \leqq \|\tilde{\mu}\| = \|S g\| \leqq \|S\| \, \|g\|_{p'}$$

for each $g \in L_{p'}(G)$. Let $f \in L_p(G)$ be the unique element such that $F(g) = \langle f, g \rangle = f * g(e)$ for each $g \in L_{p'}(G)$. Then we have for each $s \in G$,

$$\begin{aligned} f * g(s) &= f * \tau_{s^{-1}} g(e) = F(\tau_{s^{-1}} g) = \int_{\hat{G}} d(S \tau_{s^{-1}} g)\tilde{}\,(\gamma) = \int_{\hat{G}} d((s^{-1}, \cdot) S g)\tilde{}\,(\gamma) \\ &= \int_{\hat{G}} (s, \gamma) \, d\tilde{\mu}(\gamma) = \int_{\hat{G}} (s^{-1}, \gamma) \, d\mu(\gamma) = \hat{\mu}(s). \end{aligned}$$

Thus for each $g \in L_{p'}(G)$ we have $f * g = \hat{\mu} = (Sg)\hat{}$, and $f \in L_p^0(G)$ by Theorem 4.6.3.

It is evident that the correspondence established by the equivalence of (i) and (ii) defines a linear mapping from $L_p^0(G)$ onto the Banach space of all continuous linear transformations $S: L_{p'}(G) \to M(\hat{G})$ for which $S(\tau_s g) = (s, \cdot) S g$. Moreover, if $(Sg)\hat{} = \hat{\mu} = f * g$ for each $g \in L_{p'}(G)$ then from the proof of Theorem 4.6.3 we know that $\|f\|_0 \leqq \|S\|$. On the other hand, for each $g \in L_{p'}(G)$ we have from the proof of Theorem 4.6.3 that $h * f * g(e) = \int_{\hat{G}} \hat{h}(\gamma^{-1}) \, d\mu(\gamma)$ for each $h \in L_1(G)$. Hence since $L_1(G)\hat{}$ is norm dense in $C_0(\hat{G})$ we can conclude that

$$\begin{aligned}\|S g\| = \|\mu\| &= \sup\{|\int_{\hat{G}} \hat{h}(\gamma^{-1})\, d\mu(\gamma)|\,|\,\hat{h}\in C_0(\hat{G}),\ \|\hat{h}\|_\infty\leqq 1\}\\ &= \sup\{|\int_{\hat{G}} \hat{h}(\gamma^{-1})\, d\mu(\gamma)|\,|\,h\in L_1(G),\ \|\hat{h}\|_\infty\leqq 1\}\\ &= \sup\{|h*f*g(e)|\,|\,h\in L_1(G),\ \|\hat{h}\|_\infty\leqq 1\}\\ &\leqq \sup\{\|h*f\|_p\,\|g\|_{p'}\,|\,h\in L_1(G),\ \|\hat{h}\|_\infty\leqq 1\} = \|f\|_0\,\|g\|_{p'}.\end{aligned}$$

Thus $\|S\|\leqq\|f\|_0$ and the mapping is an isometry. □

Corollary 4.6.3. *Let G be a locally compact Abelian group and $1<p<\infty$. Then $L_p^0(G)$ is a Banach space.*

These general results will now enable us to obtain conditions under which $L_p^0(G)=(L_p(G))_0$, $1<p<2$, and G noncompact. What we shall show is that whenever G is either noncompact and connected or infinite and discrete then $L_p^0(G)=\{0\}$, $1<p<2$. This, combined with Corollary 4.6.1, immediately yields the indicated assertion. We require one further preliminary result.

Lemma 4.6.1. *Let G be a locally compact Abelian group and $1<p<2$. If $f\in L_p^0(G)$ and $h\in L_1(\hat{G})$, $\|h\|_1=1$, then $\hat{h}f\in L_p^0(G)$.*

Proof. Given $f\in L_p^0(G)$ and $g\in L_1(G)$ an elementary computation reveals that for each $\gamma\in\hat{G}$ we have

$$\begin{aligned}\|g*[(\cdot,\gamma)f]\|_p/\|\hat{g}\|_\infty &= \|(\cdot,\gamma)\{[(\cdot,\gamma^{-1})\,g]*f\}\|_p/\|\hat{g}\|_\infty\\ &= \|[(\cdot,\gamma^{-1})\,g]*f\|_p/\|[(\cdot,\gamma^{-1})\,g]^\wedge\|_\infty.\end{aligned}$$

We then conclude from the definition of $L_p^0(G)$ that $(\cdot,\gamma)f\in L_p^0(G)$ for each $\gamma\in\hat{G}$ and $\|(\cdot,\gamma)f\|_0=\|f\|_0$.

Let $h\in L_1(\hat{G})$ and, without loss of generality, assume that $\|h\|_1=1$.

To show that $\hat{h}f\in L_p^0(G)$ we shall construct a net of trigonometric polynomials p_β such that if S_β: $L_{p'}(G)\to M(\hat{G})$ are the operators corresponding to $p_\beta f\in L_p^0(G)$ as in Theorem 4.6.4, then $\{S_\beta\}$ will converge in the strong operator topology to a continuous operator S from $L_{p'}(G)$ to $M(\hat{G})$ for which $(S g)^\wedge=\hat{h}f*g$ for each $g\in L_{p'}(G)$. An application of Theorem 4.6.3 will then imply that $\hat{h}f\in L_p^0(G)$.

First consider h as an element in the unit sphere of $M(\hat{G})$. Then by the Krein-Milman and Alaoglu Theorems (Theorems D.7.2 and D.4.3) there exists a net of convex linear combinations of point masses in $M(\hat{G})$ which converges in the weak* topology to h. Let these measures be $\mu_\alpha=\sum_j a_j(\alpha)\, b_j(\alpha)\, \delta_{\gamma_j(\alpha)}$ where

$$a_j(\alpha)\geqq 0,\qquad \sum_j a_j(\alpha)=1,\qquad b_j(\alpha)\in C,\qquad \sum_j a_j(\alpha)\,|b_j(\alpha)|=1,$$

and $\delta_{\gamma_j(\alpha)}$ are unit point masses concentrated at $\gamma_j(\alpha)$. For each α and $g \in L_{p'}(\hat{G})$ define $L_\alpha g = \mu_\alpha * g$. Clearly each L_α is a continuous linear transformation from $L_{p'}(\hat{G})$ to $L_{p'}(\hat{G}$ since $\|\mu_\alpha * g\|_{p'} \leqq \|\mu_\alpha\| \, \|g\|_{p'}$ (F.2).

If $g \in L_{p'}(\hat{G})$ then for each $k \in L_p(\hat{G})$ we have

$$\begin{aligned}\lim_\alpha \int_{\hat{G}} L_\alpha g(\gamma)\, k(\gamma^{-1})\, d\eta(\gamma) &= \lim_\alpha L_\alpha g * k(e) = \lim_\alpha \mu_\alpha * g * k(e) \\ &= \lim_\alpha \int_{\hat{G}} g * k(\gamma^{-1})\, d\mu_\alpha(\gamma) \\ &= \int_{\hat{G}} g * k(\gamma^{-1})\, h(\gamma)\, d\eta(\gamma) = h * g * k(e) \\ &= \int_{\hat{G}} h * g(\gamma)\, k(\gamma^{-1})\, d\eta(\gamma),\end{aligned}$$

since $g * k \in C_0(\hat{G})$ and $\{\mu_\alpha\}$ converges in the weak* topology to h. Thus $\{L_\alpha g\}$ converges weakly to $h * g$ for each $g \in L_{p'}(\hat{G})$. In particular, since $f \in L_p^0(G) \subset L_p(G)$, we see that $\{L_\alpha \check{f}\}$ converges weakly to $h * \check{f} = (\hat{h} f)\check{}$ as $\check{f} \in L_{p'}(\hat{G})$. But the weak and norm topologies in $L_{p'}(\hat{G})$ have the same closed convex sets (Theorem D.7.3). Thus there exists a net of convex linear combinations of the elements $\{L_\alpha \check{f}\} = \{\mu_\alpha * \check{f}\}$ which converges in norm to $h * \check{f}$. Recalling the definition of the μ_α it is obvious that we obtain in this way a net of measures $\{\nu_\beta\} \subset M(\hat{G})$ of the form $\nu_\beta = \sum_j c_j(\beta)\, \delta_{\omega_j(\beta)}$, where $\sum_j |c_j(\beta)| = 1$ and $\omega_j(\beta)$ are points in $\hat{G}$, such that $\{\nu_\beta * \check{f}\}$ converges in the norm of $L_{p'}(\hat{G})$ to $h * \check{f}$. Obviously, $\hat{\nu}_\beta(t) = \sum_j c_j(\beta)\,(t, \omega_j(\beta)^{-1})$ and so $\hat{\nu}_\beta f = f_\beta \in L_p^0(G)$. Let S_β be the continuous linear transformation from $L_{p'}(G)$ to $M(\hat{G})$ defined by the equation $(S_\beta g)\hat{} = f_\beta * g$ for each $g \in L_{p'}(G)$. Theorem 4.6.4 assures that the transformations S_β are meaningful.

We shall now prove that $\{S_\beta\}$ converges in the strong operator topology to a continuous linear operator S on $L_{p'}(G)$ to $M(\hat{G})$. From the last portion of Theorem 4.6.4 and the relation $\|(\cdot, \gamma) f\|_0 = \|f\|_0$ we see that

$$\|S_\beta\| = \|f_\beta\|_0 = \|\hat{\nu}_\beta f\|_0 = \Big\|\sum_j c_j(\beta)\big(\cdot, \omega_j(\beta)^{-1}\big) f\Big\|_0 \leqq \sum_j |c_j(\beta)|\, \|f\|_0 = \|f\|_0$$

for each β. Thus the net $\{S_\beta\}$ is uniformly bounded. Let $X = L_p(\hat{G})\hat{} \subset L_{p'}(G)$. Then X is norm dense in $L_{p'}(G)$. If $g \in X$ then there exists a $k \in L_p(\hat{G})$ such that $g = \hat{k}$ and $(S_\beta g)\hat{} = f_\beta * g = (\check{f}_\beta k)\hat{}$. Thus $S_\beta g = \check{f}_\beta k \in L_1(\hat{G})$ for each $g \in X$ since $\check{f}_\beta \in L_{p'}(\hat{G})$ and $k \in L_p(\hat{G})$. Furthermore

$$\|S_\beta g - (h * \check{f})\, k\|_1 = \|\check{f}_\beta k - (h * \check{f})\, k\|_1 \leqq \|\nu_\beta * \check{f} - h * \check{f}\|_{p'} \, \|k\|_p,$$

from which it follows that $\lim_\beta \|S_\beta g - (h * \check{f})\, k\|_1 = 0$ for each $g \in X$ because $\lim_\beta \|\nu_\beta * \check{f} - h * \check{f}\|_{p'} = 0$.

But then $\{S_\beta\}$ is a uniformly bounded net of continuous linear transformations from $L_{p'}(G)$ to $M(\hat{G})$ such that for each g in the norm dense subset X of $L_{p'}(G)$ the net $\{S_\beta g\}$ converges in $M(\hat{G})$. Consequently, by the Uniform Boundedness Theorem (Theorem D.6.4), we conclude that there exists a bounded linear transformation S from $L_{p'}(G)$ to $M(\hat{G})$ such that $\lim_\beta \|S_\beta g - S g\| = 0$ for each $g \in L_{p'}(G)$. In particular, if $g \in X$ then $S g = (\hat{h} * \check{f})\, k$ where $\hat{k} = g$, that is, $(S g)\hat{} = \hat{h} f * g$ for each $g \in X$. Since X is norm dense in $L_{p'}(G)$ and S is continuous, it follows that $(S g)\hat{} = \hat{h} f * g$ for each $g \in L_{p'}(G)$. Thus $\hat{h} f * g \in M(\hat{G})$ for each $g \in L_{p'}(G)$.

Therefore by Theorem 4.6.3 we conclude that $\hat{h} f \in L_p^0(G)$. ▯

The assertion of the lemma is trivially true if $p = 2$.

Theorem 4.6.5. *Let G be a noncompact locally compact connected Abelian group. If $1 < p < 2$ then $L_p^0(G) = \{0\}$.*

Proof. Suppose $f \in L_p^0(G)$ and $f \neq 0$. By Lemma 4.6.1 we can assume that $f = 0$ almost everywhere off of some compact set K. Without loss of generality we may assume that K contains the identity element of G. Let $0 < \varepsilon < \|f\|_p/2$. Choose a compact symmetric neighborhood W of the identity in G such that if $h \in L_p^+(G) \cap L_1^+(G)$, $h = 0$ off of W and $\|h\|_1 = 1$ then $\|f * h - f\|_p < \varepsilon$. Since KW is compact and G is noncompact and connected, there exists an element $d \in G$ for which the sets $\{K W d^n\}$, $n = 0, \pm 1, \pm 2, \ldots$, are pairwise disjoint. Denote by $F(\varepsilon)$ all the elements $g \in L_p(G) \cap L_1(G)$ which are of the form

$$g(t) = \sum_{n \in Z}{}' b_n \tau_{d^n} h(t)$$

where $\{b_n\}$ are complex numbers, $\sum'$ indicates a finite sum and h is chosen as above.

The fact that K contains the identity and the pairwise disjointness of the sets $\{K W d^n\}$ imply that the sets $\{W d^n\}$, $n = 0, \pm 1, \pm 2, \ldots$, are also pairwise disjoint. Thus, if $g \in F(\varepsilon)$ it is apparent that $\|g\|_p = (\sum_{n \in Z}' |b_n|^p)^{1/p}$. Moreover, if $g \in F(\varepsilon)$ then

$$\begin{aligned}
\big|\|f\|_p \|g\|_p - \|f * g\|_p\big| &= \Big|\|f\|_p \|g\|_p - \Big\|\sum_{n \in Z}{}' b_n f * \tau_{d^n} h\Big\|_p\Big| \\
&= \Big|\|f\|_p \Big(\sum_{n \in Z}{}' |b_n|^p\Big)^{1/p} - \Big(\sum_{n \in Z}{}' |b_n|^p (\|f * \tau_{d^n} h\|_p)^p\Big)^{1/p}\Big| \\
&= \Big|(\|f\|_p - \|f * h\|_p)\Big(\sum_{n \in Z}{}' |b_n|^p\Big)^{1/p}\Big| \\
&\leqq \|f - f * h\|_p \Big(\sum_{n \in Z}{}' |b_n|^p\Big)^{1/p} < \varepsilon \|g\|_p,
\end{aligned}$$

since $K W d^n$ are pairwise disjoint and $f * h$ is zero off of KW.

However, since $f\in L_p^0(G)$, we have $\|f*g\|_p\leqq\|f\|_0\,\|\hat{g}\|_\infty$. Thus for each $g\in F(\varepsilon)$

$$\|f\|_0\,\|\hat{g}\|_\infty\geqq\|f*g\|_p>\|f\|_p\,\|g\|_p-\varepsilon\,\|g\|_p>(2\varepsilon-\varepsilon)\,\|g\|_p=\varepsilon\,\|g\|_p.$$

Consequently, if $g(t)=\sum_{n\in Z}' b_n\,\tau_{d^n}\,h(t)$ then

$$\begin{aligned}\Big(\sum_{n\in Z}'|b_n|^p\Big)^{1/p}&=\|g\|_p\leqq C(\varepsilon)\,\|\hat{g}\|_\infty=C(\varepsilon)\,\Big\|\sum_{n\in Z}' b_n(\tau_{d^n}h)^\wedge\Big\|_\infty\\&=C(\varepsilon)\,\Big\|\sum_{n\in Z}' b_n(d^{-n},\cdot)\,\hat{h}\Big\|_\infty\leqq C(\varepsilon)\,\Big\|\sum_{n\in Z}' b_n(d^{-n},\cdot)\Big\|_\infty\,\|\hat{h}\|_\infty\\&\leqq C(\varepsilon)\,\Big\|\sum_{n\in Z}' b_n(d^{-n},\cdot)\Big\|_\infty\end{aligned}$$

where $C(\varepsilon)=\|f\|_0/\varepsilon$.

Let $D=\{d^n\,|\,n=0,\pm1,\pm2,\ldots\}$ be considered as a discrete topological group. Then $\hat{D}$ is compact and the continuous characters on D are precisely the elements of $\hat{G}$ restricted to D (B.2). If $\varphi=\sum_{n\in Z}' b_n\,\chi_{\{d^{-n}\}}$ then $\hat{\varphi}(\gamma)=\sum_{n\in Z}' b_n(d^{-n},\gamma)$ and the previous inequality shows that $\|\varphi\|_p\leqq C(\varepsilon)\,\|\hat{\varphi}\|_\infty$.

Let $\psi\in C(\hat{D})$. Since $\hat{D}$ is compact there exists a sequence of trigonometric polynomials $\{\hat{\varphi}_n\}$ on $\hat{D}$ such that $\lim_n\|\hat{\varphi}_n-\psi\|_\infty=0$. Thus

$$\|\varphi_n-\varphi_m\|_p\leqq C(\varepsilon)\,\|\hat{\varphi}_n-\hat{\varphi}_m\|_\infty$$

shows that $\{\varphi_n\}$ is a Cauchy sequence in $L_p(D)$. Hence there exists some $\varphi\in L_p(D)$ for which $\lim_n\|\varphi_n-\varphi\|_p=0$. Clearly $\hat{\varphi}=\psi$. Thus $C(\hat{D})\subset L_p(D)^\wedge$, which is known to be false unless D is finite (F.10c).

Since D is infinite we conclude that $f=0$. □

The preceding result is also valid for infinite discrete G.

Theorem 4.6.6. *Let G be an infinite discrete Abelian group. If $1<p<2$ then $L_p^0(G)=\{0\}$.*

Proof. Suppose $f\in L_p^0(G)$ and $f\neq0$. Clearly $L_p^0(G)$ is invariant under translation and so we may assume, without loss of generality, that $f(e)\neq0$. Let h be the function which is identically one on $\hat{G}$. Since $\hat{G}$ is compact we have $h\in L_1(\hat{G})$ and $\|h\|_1=1$. From Lemma 4.6.1 we conclude that $\hat{h}f=f(e)\,\chi_{\{e\}}\in L_p^0(G)$ where $\chi_{\{e\}}$ is the characteristic function of the set consisting only of the identity of G. But then Theorem 4.6.3 implies that $\chi_{\{e\}}*g=g\in M(\hat{G})^\wedge$ for each $g\in L_{p'}(G)$, $1/p+1/p'=1$, which is a contradiction because $L_{p'}(G)$ is not contained in $M(\hat{G})^\wedge$ for $p'>2$ (F.10b).

Therefore $L_p^0(G)=\{0\}$. □

It should be noted that $L_p^0(G) \neq \{0\}$ when $p > 2$. Indeed, we claim that in this case $L_{p'}(\hat{G})^\wedge \subset L_p^0(G)$ where $1/p + 1/p' = 1$. Clearly by the Hausdorff-Young Theorem (Theorem F.8.4) we see that $L_{p'}(\hat{G})^\wedge \subset L_p(G)$. The $\wedge$ here of course denotes the Hausdorff-Young transform. Moreover, if $h \in L_1(G)$ and $f = \hat{g} \in L_{p'}(\hat{G})^\wedge$ then $\hat{h}\, g \in L_{p'}(\hat{G})$ and $(\hat{h}\, g)^\wedge = h * f$. Hence by the Hausdorff-Young Theorem we have

$$\|h * f\|_p \leqq \|\hat{h}\, g\|_{p'} \leqq \|\hat{h}\|_\infty \, \|g\|_{p'},$$

from which we conclude that $\|f\|_0 \leqq \|g\|_{p'}$. Thus $L_{p'}(\hat{G})^\wedge \subset L_p^0(G)$.

4.7. Some Further Results Concerning $\mathcal{M}(L_p(G))$, $1 \leqq p < \infty$. Previously we noted that $\mathcal{M}(L_1(G)) = M(G)^\wedge$ and $\mathcal{M}(L_2(G)) = L_\infty(\hat{G})$. Thus it is apparent that $L_p(\hat{G}) * L_{p'}(\hat{G}) \subset \mathcal{M}(L_2(G))$, $1 \leqq p < \infty$, $1/p + 1/p' = 1$, and that $L_2(\hat{G}) * L_2(\hat{G}) \subset \mathcal{M}(L_p(G))$, $1 \leqq p < \infty$, where $L_p(\hat{G}) * L_{p'}(\hat{G}) = \{f * g \mid f \in L_p(\hat{G}),\ g \in L_{p'}(\hat{G})\}$, $1 \leqq p < \infty$, $1/p + 1/p' = 1$. We now wish to discuss a similar sort of inclusion result and then apply it to show that for noncompact groups $M(G)^\wedge$ is always a proper subset of $\mathcal{M}(L_p(G))$, $1 < p < \infty$.

First we prove a lemma. The linear space of integrable simple functions on a locally compact Abelian group G will be denoted by $\mathcal{S}(G)$.

Lemma 4.7.1. *Let G be a locally compact Abelian group and $1 < p < \infty$. Then for each $\varphi \in L_\infty(\hat{G})$ the following are equivalent:*

(i) $\varphi \in \mathcal{M}(L_p(G))$.

(ii) *There exists a constant $K > 0$ such that*

$$\Big|\int_{\hat{G}} \varphi(\gamma)\, \hat{f}(\gamma)\, \hat{g}(\gamma)\, d\eta(\gamma)\Big| \leqq K \, \|f\|_p \, \|g\|_{p'} \qquad (f, g \in \mathcal{S}(G)),$$

where $1/p + 1/p' = 1$.

Proof. If $\varphi \in \mathcal{M}(L_p(G))$ then from $L_p(G)^* = L_{p'}(G)$ and the norm density of $\mathcal{S}(G)$ in both $L_p(G)$ and $L_{p'}(G)$ we conclude that

$$\begin{aligned}
\|T\|_p &= \sup_f \{\|Tf\|_p \mid \|f\|_p \leqq 1, f \in \mathcal{S}(G)\} \\
&= \sup_{f,g} \Big\{\Big|\int_G Tf(t)\, g(t^{-1})\, d\lambda(t)\Big| \;\Big|\; \|f\|_p \leqq 1,\ \|g\|_{p'} \leqq 1, f, g \in \mathcal{S}(G)\Big\} \\
&= \sup_{f,g} \Big\{\Big|\int_{\hat{G}} \varphi(\gamma)\, \hat{f}(\gamma)\, \hat{g}(\gamma)\, d\eta(\gamma)\Big| \;\Big|\; \|f\|_p \leqq 1,\ \|g\|_{p'} \leqq 1, f, g \in \mathcal{S}(G)\Big\}
\end{aligned}$$

where $\hat{T} = \varphi$ and $T \in M(L_p(G))$. The last equality follows from Parseval's Formula (Theorem F.8.3) and Theorem 4.1.3.

Thus (i) implies (ii) with $K = \|T\|_p$.

Conversely, suppose (ii) holds. Then by the Plancherel Theorem (Theorem F.8.2) $\varphi \hat{f} \in L_2(\hat{G})$ for each $f \in \mathcal{S}(G)$, and we denote by Tf the unique element in $L_2(G)$ for which $(Tf)\hat{} = \varphi \hat{f}, f \in \mathscr{S}(G)$. Hence for f, $g \in \mathscr{S}(G)$, $\|f\|_p \leqq 1$, $\|g\|_{p'} \leqq 1$, we can again apply Parseval's Formula to see that

$$|\int_G Tf(t)\, g(t^{-1})\, d\lambda(t)| = |\int_{\hat{G}} \varphi(\gamma)\, \hat{f}(\gamma)\, \hat{g}(\gamma)\, d\eta(\gamma)| \leqq K.$$

It follows easily from this and the density of $\mathscr{S}(G)$ in $L_p(G)$ and $L_{p'}(G)$ that T defines a bounded linear operator on $L_p(G)$. From the definition of T it is obvious that T commutes with translation and so $T \in M(L_p(G))$.

Therefore $\varphi \in \mathscr{M}(L_p(G))$, and (ii) implies (i). □

Theorem 4.7.1. *Let G be a locally compact Abelian group, $1 < r < 2$ and $1/r + 1/r' = 1$. Then $L_r(\hat{G}) * L_{r'}(\hat{G}) \subset \mathscr{M}(L_p(G))$ for $2r/3r - 2 \leqq p \leqq 2r/2 - r$.*

Proof. Consider the multilinear mapping $U\colon \mathscr{S}(\hat{G}) \times \mathscr{S}(\hat{G}) \times \mathscr{S}(G) \times \mathscr{S}(G) \to C$ defined by

$$U(\psi, \omega, f, g) = \int_{\hat{G}} \psi * \omega(\gamma)\, \hat{f}(\gamma)\, \hat{g}(\gamma)\, d\eta(\gamma) \qquad (\psi, \omega \in \mathscr{S}(\hat{G}), f, g \in \mathscr{S}(G)).$$

Then appealing to the Cauchy-Schwarz inequality, the Plancherel Theorem and Parseval's Formula (Theorems F.8.2 and F.8.3) we conclude for $\psi, \omega \in \mathscr{S}(\hat{G}), f, g \in \mathscr{S}(G)$ that, on the one hand,

$$\begin{aligned} |\int_{\hat{G}} \psi * \omega(\gamma)\, \hat{f}(\gamma)\, \hat{g}(\gamma)\, d\eta(\gamma)| &\leqq \|\psi * \omega\|_\infty \int_{\hat{G}} |\hat{f}(\gamma)\, \hat{g}(\gamma)|\, d\eta(\gamma) \\ &\leqq \|\psi\|_1\, \|\omega\|_\infty\, \|\hat{f}\|_2\, \|\hat{g}\|_2 \\ &= \|\psi\|_1\, \|\omega\|_\infty\, \|f\|_2\, \|g\|_2, \end{aligned}$$

while on the other hand,

$$\begin{aligned} |\int_{\hat{G}} \psi * \omega(\gamma)\, \hat{f}(\gamma)\, \hat{g}(\gamma)\, d\eta(\gamma)| &= |\int_G \hat{\psi}(t)\, \hat{\omega}(t)\, f * g(t^{-1})\, d\lambda(t)| \\ &\leqq \|f\|_1\, \|g\|_\infty\, \|\psi\|_2\, \|\omega\|_2. \end{aligned}$$

Thus

$$|U(\psi, \omega, f, g)| \leqq \|\psi\|_r\, \|\omega\|_{r'}\, \|f\|_p\, \|g\|_{p'} \qquad (\psi, \omega \in \mathscr{S}(\hat{G}), f, g \in \mathscr{S}(G))$$

whenever either $r = 1, p = 2$ or $r = 2$, $p = 1$ and $1/r + 1/r' = 1$, $1/p + 1/p' = 1$, that is, the multilinear mapping $U\colon \mathscr{S}(\hat{G}) \times \mathscr{S}(\hat{G}) \times \mathscr{S}(G) \times \mathscr{S}(G) \to C$ is continuous when the components of the product space are considered with the topologies indicated above.

Appealing to the Multilinear Riesz-Thorin Theorem (Theorem D.11.1), we conclude that

$$|\int_{\hat{G}} \psi * \omega(\gamma)\, \hat{f}(\gamma)\, \hat{g}(\gamma)\, d\eta(\gamma)| \leqq \|\psi\|_p\, \|\omega\|_q\, \|f\|_r\, \|g\|_s$$

$$(\psi, \omega \in \mathscr{S}(\hat{G}), f, g \in \mathscr{S}(G))$$

whenever

$$1/r = (1-\alpha)\cdot 1 + \alpha \cdot \tfrac{1}{2}, \qquad 1/r + 1/r' = 1$$

$$1/p = (1-\alpha)\cdot \tfrac{1}{2} + \alpha \cdot 1, \qquad 1/p + 1/p' + 1,$$

where $0 \leqq \alpha \leqq 1$. Solving for r and applying Lemma 4.7.1 with $\varphi = \psi * \omega$, it follows at once that $\mathcal{S}(\hat{G}) * \mathcal{S}(\hat{G}) \subset \mathcal{M}(L_p(G))$, $p = 2r/3r-2$. Hence $L_r(\hat{G}) * L_{r'}(\hat{G}) \subset \mathcal{M}(L_p(G))$, $p = 2r/3r-2$, $1 < r < 2$. Furthermore, we note that $1 < p = 2r/3r - 2 < 2 < p' = 2r/2 - r$ for $1 < r < 2$ and $1/p + 1/p' = 1$. Consequently Theorem 4.5.5 shows that $L_r(\hat{G}) * L_{r'}(\hat{G}) \subset \mathcal{M}(L_p(G))$, $2r/3r - 2 \leqq p \leqq 2r/2 - r$, whenever $1 < r < 2$. □

It should perhaps be noted that $2r/3r - 2 \leqq p \leqq 2r/2 - r$, $1 < r < 2$, is equivalent to $2r'/2 + r' \leqq p \leqq 2r'/r' - 2$, $2 < r' < \infty$, where $1/r + 1/r' = 1$.

Theorem 4.7.2. *Let G be a noncompact locally compact Abelian group. If $1 < p < \infty$ then $M(G)\hat{} \subset \mathcal{M}(L_p(G))$ and the inclusion is proper.*

Proof. As we have remarked several times before, we always have $M(G)\hat{} \subset \mathcal{M}(L_p(G))$. If $p = 2$ then $\mathcal{M}(L_2(G)) = L_\infty(\hat{G})$ and the inclusion is obviously proper as G is infinite. If $p \neq 2$ then we may assume from Theorem 4.1.2 that $1 < p < 2$. Let $r = 2p/3p - 2$. Then from the previous theorem we see that $1 < r < 2$, $p = 2r/3r - 2$, and $L_r(\hat{G}) * L_{r'}(\hat{G}) \subset \mathcal{M}(L_p(G))$. Thus to show that the inclusion is proper when $1 < p < 2$ it suffices to show that not every element of $L_r(\hat{G}) * L_{r'}(\hat{G})$ belongs to $M(G)\hat{}$.

Suppose then that $L_r(\hat{G}) * L_{r'}(\hat{G}) \subset M(G)\hat{}$. For $\hat{\mu} \in M(G)\hat{}$ set $\|\hat{\mu}\| = \|\mu\|$. Since $M(G)$ is a semi-simple Banach algebra this defines a complete norm on $M(G)\hat{}$. For each $g \in L_r(\hat{G})$ define $T_g f = g * f$, $f \in L_{r'}(\hat{G})$, Clearly T_g is a linear transformation from $L_{r'}(\hat{G})$ to $M(G)\hat{}$. Moreover, suppose that $\lim_n \|f_n - f\|_{r'} = 0$ and $\lim_n \|T_g f_n - \hat{\mu}\| = 0$. Then

$$\begin{aligned}\|T_g f - \hat{\mu}\|_\infty &\leqq \|T_g f - T_g f_n\|_\infty + \|T_g f_n - \hat{\mu}\|_\infty \leqq \|g * f - g * f_n\|_\infty + \|T_g f_n - \hat{\mu}\| \\ &\leqq \|g\|_r \, \|f - f_n\|_{r'} + \|T_g f_n - \hat{\mu}\|,\end{aligned}$$

which shows that $T_g f = \hat{\mu}$. Thus by the Closed Graph Theorem (Theorem D.6.1) we conclude that each T_g is continuous. In particular, for each $g \in C_c(\hat{G}) \subset L_r(\hat{G})$ there exists a constant $B(g) > 0$ such that

$$\|T_g f\| = \|g * f\| \leqq B(g)\, \|f\|_{r'} \qquad (f \in C_c(\hat{G})).$$

But if $g, f \in C_c(\hat{G})$ then $g * f \in L_2(\hat{G}) * L_2(\hat{G}) = L_1(G)\hat{}$ (F.10a), and $(\hat{g}\hat{f})\check{} = g * f$ where as usual $\check{}$ denotes the inverse Fourier transform. Thus, given $g \in C_c(\hat{G})$ we see that $\hat{g}\hat{f} \in L_1(G)$, $f \in C_c(\hat{G})$, and

$$\|\hat{g}\hat{f}\|_1 = \|(\hat{g}\hat{f})\check{}\| = \|g * f\| \leqq B(g)\, \|f\|_{r'} \qquad (f \in C_c(\hat{G})).$$

Moreover, if $\varphi \in C_0(G)$, $\|\varphi\|_\infty \leqq 1$, then we also have

$$\|\varphi \hat{g} \hat{f}\|_1 \leqq \|\varphi\|_\infty \|\hat{g} \hat{f}\|_1 \leqq B(g) \|f\|_{r'} \qquad (f \in C_c(\hat{G})).$$

However, for each such φ, we deduce immediately from the Plancherel Theorem (Theorem F.8.2) that $(\varphi \hat{g} \hat{f})^\vee = (\varphi \hat{g})^\vee * f \in C_0(\hat{G})$ for each $f \in C_c(\hat{G})$. Hence we have

$$|\langle(\varphi \hat{g})^\vee, f\rangle| = |(\varphi \hat{g})^\vee * f(e)| \leqq \|(\varphi \hat{g} \hat{f})^\vee\|_\infty \leqq \|\varphi \hat{g} \hat{f}\|_1 \leqq B(g) \|f\|_{r'} \qquad (f \in C_c(\hat{G})).$$

Consequently $(\varphi \hat{g})^\vee$ defines a continuous linear functional on $L_{r'}(\hat{G})$. Since $C_c(\hat{G})$ is norm dense in $L_{r'}(\hat{G})$ we conclude that $(\varphi \hat{g})^\vee \in L_r(\hat{G})$, and so $\varphi \hat{g} \in L_r(\hat{G})^\wedge$ for each $\varphi \in C_0(G)$, $\|\varphi\|_\infty \leqq 1$. It is then evident that $\varphi \hat{g} \in L_r(\hat{G})^\wedge$ for each $\varphi \in C_0(G)$ and so by Theorem 4.6.2 we conclude that $\hat{g} = 0$ locally almost everywhere. But this is clearly impossible as g can be any function in $C_c(\hat{G})$.

Therefore $L_r(\hat{G}) * L_{r'}(\hat{G})$ is not contained in $M(G)^\wedge$, and hence $M(G)^\wedge$ is properly contained in $\mathscr{M}(L_p(G))$, $1 < p < 2$. □

4.8. Notes. The subject of multipliers for $L_p(G)$ has been considered, in various forms, by a great number of authors and it has not been possible to discuss all of this work here. We shall try however to indicate some of the origins of the results developed in this chapter, outline some additional results and point out other areas of investigation for the reference of the reader.

The characterization of the elements of $M(L_p(G))$ as bounded functions is generally well known. Proofs of these results for various groups can be found in Brainerd and Edwards [1_{I}], Edwards [14_{II}], 16, Eymard [1], Gaudry [6], V, Hörmander [1] and Zygmund [6], IV.11. Most of the other results of 4.1 can also be found in these references. The description of the translation invariant subspaces of $L_2(G)$ is discussed in Mackey [1].

Pseudomeasures have appeared in many places connected with multipliers or harmonic analysis, of which we mention Brainerd and Edwards [1_{I}], Edwards [5, 12, 13], Eymard [1],[1] Gaudry [1], Kahane [1], Katznelson [2], VI and Larsen [2]. Gaudry [6], VI, contains an extensive discussion of *quasimeasures*, a super set of pseudomeasures. We shall discuss much of this material in the following chapter.

The characterization of $M(L_p(G))$ as pseudomeasures is based on the development of Hörmander [1] and Gaudry [1]. A similar approach is also available in Brainerd and Edwards [1_{I}], Edwards [14_{II}], 16.4, and Gaudry [6], V.

[1] See also Eymard [2].

An alternative description of $M(L_p(G))$ has been given in Figà-Talamanca [1, 3]. Let $1<p<\infty$ and $1/p+1/p'=1$. Denote by $A_p^p(G)$ the set of all functions h on G which are of the form $h=\sum_i f_i * g_i$ where $f_i \in L_p(g), g_i \in L_{p'}(G)$ and $\sum_i \|f_i\|_p \|g_i\|_{p'} < \infty$. Define the norm of $h \in A_p^p(G)$ by

$$|||h||| = \inf\{\sum_i \|f_i\|_p \|g_i\|_{p'} \mid h = \sum_i f_i * g_i\}.$$

Then $A_p^p(G)$ is a Banach space of continuous functions on G with this norm. Figà-Talamanca [3] proves the following result.

Theorem 4.8.1. *Let G be a locally compact Abelian group and $1<p<\infty$. Then $M(L_p(G))$ is isometrically linearly isomorphic to $A_p^p(G)^*$.*

He also shows that if one identifies $M(L_p(G))$ with $A_p^p(G)^*$ then the finite linear combinations of translation operators are weak* dense in $M(L_p(G))$.[1]

A similar characterization of $M(L_p(G), L_q(G))$ has been developed in Figà-Talamanca and Gaudry [1]. We shall examine these results in some detail in the succeeding chapter. Various aspects of the spaces $A_p^p(G)$ are considered in Figà-Talamanca and Gaudry [2, 3], Gaudry [6, 8], Herz [3] and Máté [6, 7]. An elegant characterization of these spaces is given in Rieffel [4].

The development of 4.4 is based on Figà-Talamanca and Gaudry [2]. The theorem of the section was established for the circle group Γ in Katznelson [1]. See also Katznelson [2], IV. 2.5, and Rosenthal [1]. Connections between the spectra of multipliers for l_p and sets of uniqueness are studied in Devinatz and Hirschman [1].

The application of the existence of sets of uniqueness to prove that the inclusions $M(L_p(G)) \subset M(L_r(G)) \subset M(L_2(G))$ are proper when $1<p, r<\infty$ and $|1/p-\frac{1}{2}| > |1/r-\frac{1}{2}| > 0$ is taken from Figà-Talamanca and Gaudry [2]. The proofs of Theorems 4.5.3 and 4.5.4 are also from this paper. Lemma 4.5.2 and Theorem 4.5.1 are essentially to be found in Edwards [10], as is Theorem 4.5.2 for the case of infinite compact Abelian groups. Here and in Gaudry [6], VI, the notion of random Fourier series is utilized. Some other work on this subject which has connections with multipliers is available in Figà-Talamanca and Rider [1, 2] and Helgason [2, 3]. Inclusion results have also been studied in Price [2, 3], 4 and 7.

In Gaudry [6], VI, it is shown when G is an infinite locally compact Abelian group that there exist $\varphi \in C_0(\hat{G})$ such that $\varphi \notin \bigcup_{p \neq 2} \mathscr{M}(L_p(G))$. A proof of this result with the additional assumption that G contain an infinite discrete subgroup is to be found in Gaudry [8].

The material on the derived space for $L_p(G)$ is based primarily on Figà-Talamanca [2] and Figà-Talamanca and Gaudry [2]. In parti-

[1] See also Eymard [3].

cular Theorem 4.6.2 comes from the latter paper. A different proof will be given in 5.5. Proofs of the result for various types of groups can also be found in Figà-Talamanca [2] and Gaudry [2], the proof in the last paper being due to R. E. Edwards. In this regard see also Rosenthal [1]. The other theorems of the section are taken from Figà-Talamanca [2]. Results for Euclidean groups $G=R^n$ similar to some of those in 4.6 are given in Hörmander [1].

Theorem 4.7.1 is taken from Hahn [1] while Theorem 4.7.2 comes essentially from Gaudry [8]. Proofs of the existence of multipliers for $L_p(G)$, $1<p<\infty$, which are not convolution with an element of $M(G)$, can also be found in Brainerd and Edwards [1_{I}], Edwards [14_{II}], 16, Gaudry [6], V, and Hörmander [1].

In particular, we observe that the theorem of M. Riesz [2] provides an example of a multiplier for $L_p(\Gamma)$, $1<p<\infty$, which is not given by convolution with a bounded measure. This theorem asserts that the transformation on $L_p(\Gamma)$, $1<p<\infty$, which assigns to $f\in L_p(\Gamma)$ the function $\tilde{f}\in L_p(\Gamma)$ whose Fourier series is the conjugate series to that of f is a multiplier. However, this multiplier corresponds to multiplication of the Fourier transform of f by the characteristic function χ_{Z_+} of the positive integers, and χ_{Z_+} does not belong to $M(\Gamma)\hat{}$. Indeed, if it did then by the F. and M. Riesz Theorem (Theorem F.12.1) we would have to have $\chi_{Z_+}\in C_0(Z)$, which is clearly not the case. Discussions of the M. Riesz Theorem in this context can be found in Edwards [14_{II}], 16, and Gaudry [6], V. Proofs of the theorem can be found in Edwards [14_{II}], 12.9, Katznelson [2], III.1.8, M. Riesz [2] and Zygmund [6], VII.2.4. We also refer the reader to the following papers which deal with the M. Riesz Theorem, the Hilbert transform and multipliers: Benedek, Calderón and Panzone [1], Calderón [2, 3], Calderón and Zygmund [1–3], J. Schwartz [1], L. Schwartz [1], G. Weiss [1] and Zygmund [3].

The M. Riesz Theorem also supplies an example of a characteristic function which does not determine a multiplier. The question of determining when the characteristic function of a set $E\subset\hat{G}$ determines a multiplier for $L_p(G)$ has been investigated by a number of authors besides M. Riesz [2]. In L. Schwartz [1] it is shown that if $G=R^n$ then the characteristic function of the unit sphere in R^n does not determine a multiplier for $L_p(R^n)$ when $n\geqq 2$ and $p\geqq 2n/n-1$ or $p<2n/n+1$. On the other hand M. Riesz [2] has shown that the characteristic functions of intervals in R and Z determine multipliers for $L_p(R)$ and $L_p(\Gamma)$, $1<p<\infty$, respectively, while L. Schwartz [1] has shown that the characteristc functions of convex polyhedra in R^n determine multipliers for $L_p(R^n)$, $1<p<2$. Other results of this nature are available in de Leeuw [1], Herz [1, 2] and Rudin [5], 5.7.8.

The contents of Hahn [1] are closely related to the work of Hirschman [2]. The general from of the results in both papers is to give sufficient conditions on a function $\varphi\in L_\infty(\hat{G})$ which will insure that $\varphi\in\mathscr{M}(L_p(G))$ for certain values of p. Such results are given for toral groups, Euclidean groups and the group of the integers. The sufficient conditions are in the form of Lipschitz conditions and certain variational restrictions. We will illustrate the type of theorem obtained by stating three of the results from Hirschman's paper.

Let φ be a measurable function on Γ where as usual Γ denotes the group of complex numbers of unit modulus. Then $\varphi\in\operatorname{Lip}\alpha$, $\alpha>0$, if for some constant $K>0$ and all t and h we have

$$|\varphi(e^{i2\pi(t+h)})-\varphi(e^{i2\pi t})|\leqq K|h|^\alpha.$$

For $r\geqq 1$ let

$$V_r(\varphi)=\sup\left\{\sum_{k=0}^{n-1}|\varphi(e^{i2\pi t_{k+1}})-\varphi(e^{i2\pi t_k})|^r\right\}^{1/r}$$

where the supremum is taken over all tuples $t_0, t_1, \ldots, t_n$ of real numbers such that $t_0<t_1<\cdots<t_n=t_0+1$. $V_r(\varphi)$ is called the *r-variation* of φ and it may be finite or infinite.

Theorem 4.8.2. *If* $\varphi\in L_\infty\Gamma)\cap\operatorname{Lip}\alpha$ *then* $\varphi\in\mathscr{M}(L_p(Z))$, $2/1+2\alpha<p<2/1-2\alpha$.

Theorem 4.8.3. *If* $\varphi\in L_\infty(\Gamma)$ *is such that* $V_r(\varphi)<\infty$, $r\geqq 2$, *and* $\varphi\in\operatorname{Lip}\alpha$ *for some* $\alpha>0$ *then* $\varphi\in\mathscr{M}(L_p(Z))$, $2r/r+2<p<2r/r-2$.

Theorem 4.8.4. *If* $\varphi\in L_\infty(\Gamma)$ *and* $V_r(\varphi)<\infty$, $r\geqq 1$ *then* $\varphi\in\mathscr{M}(L_p(Z))$, $2r/r+1<p<2r/r-1$.

In a somewhat similar vein Stein [2] has proved what is perhaps the closest thing to a complete characterization of $\mathscr{M}(L_p(\Gamma))$. Let $V(r)$, $r\geqq 1$, denote the subset of $L_r(\Gamma)$ consisting of those ψ such that $\sup\left\|\sum_k \tau_{t_k}\psi-\tau_{s_k}\psi\right\|_r<\infty$, the supremum being taken over all finite sets of nonoverlapping subintervals $[s_k, t_k]$ of $0\leqq t\leqq 1$. And for $\varphi\in L_\infty(Z)$ set

$$\Phi(e^{i2\pi t})=\sum_{n\neq 0}(in)^{-1}\varphi(n)\,e^{i2\pi nt}.$$

The latter function clearly belongs to $L_2(\Gamma)$.

Theorem 4.8.5. *Let* $2<p'<\infty$, $1/p+1/p'=1$, *and suppose* $\varphi\in L_\infty(Z)$.
(i) *If* $\varphi\in\mathscr{M}(L_r(\Gamma))$, $p\leqq r\leqq p'$, *then* $\Phi\in V(p')$.
(ii) *If* $\Phi\in V(p')$ *then* $\varphi\in\mathscr{M}(L_r(\Gamma))$, $p<r<p'$.

A succinct development of some of the major results of both Hirschman [2] and Stein [2] is available in Edwards [14_{II}], 16.4.7 and 16.5.

In general the problem of determining sufficient or necessary conditions for a function $\varphi \in L_\infty(\hat{G})$ to belong to $\mathcal{M}(L_p(G))$ has been considered in numerous papers. In this connection, besides the papers already cited, we refer the reader to Edwards [8, 10, 15], Figà-Talamanca and Rider [1, 2], Gaudry [6], V and VI, Helgason [3], Hörmander [1], Littlewood and Paley [1], Littman [1], Marcinkiewicz [1], Máté [7], Sunouchi [2] and Zygmund [5, 6].

The results of de Leeuw [1] and Saeki [1] in this regard are particularly interesting, and we shall state several results from these papers. Given a locally compact Abelian group G, and H a closed subgroup of G, we denote by $A(\hat{G}, H)$ the annihilator of H in $\hat{G}$. Then $A(\hat{G}, H)$ is a closed subgroup of $\hat{G}$ and $\hat{H}$ is topologically isomorphic to $\hat{G}/A(\hat{G}, H)$ (B.2). Denote the canonical homomorphism of $\hat{G}$ onto $\hat{G}/A(\hat{G}, H)$ by π.

Theorem 4.8.6. *Let G be a locally compact Abelian group and H a closed subgroup of G. If $1 \leqq p \leqq 2$ then the mapping α: $M(L_p(H)) \to M(L_p(G))$ defined by*

$$\big(\alpha(T)\big)\hat{} = \hat{T} \circ \pi \qquad \big(T \in L_p(H)\big),$$

is an isometric algebra isomorphism of $M(L_p(H))$ onto $M(L_p(G))$.

Thus we see, in this case, that if $\psi \in L_\infty(\hat{G}/A(\hat{G}, H))$ and $\varphi = \psi \circ \pi$ then $\varphi \in \mathcal{M}(L_p(G))$ if and only if $\psi \in \mathcal{M}(L_p(H))$. This type of result can also be extended to certain situations where H is not a closed subgroup provided certain additional restrictions are placed on the functions considered. $\varphi \in L_\infty(\hat{G})$ is said to be *regulated* if there exists a net $\{u_\alpha\} \subset L_1(\hat{G})$ such that:

(i) $\|u_\alpha\|_1 = 1$ for each α,

(ii) $\{u_\alpha * \varphi\}$ converges to φ in the weak* topology on $L_\infty(\hat{G})$ and pointwise on $\hat{G}$.

The Bohr compactification (B.2) of a locally compact Abelian group G is denoted as $\beta(G)$.

Theorem 4.8.7. *Let G be a locally compact Abelian group and $1 \leqq p \leqq 2$.*

(i) *If $\varphi \in C(\hat{G})$ then $\varphi \in \mathcal{M}(L_p(G))$ if and only if $\varphi \in \mathcal{M}(L_p(\beta(G)))$.*

(ii) *If $\varphi \in L_\infty(\hat{G})$ and φ is regulated then $\varphi \in \mathcal{M}(L_p(G))$ if and only if $\varphi \in \mathcal{M}(L_p(\beta(G)))$.*

Furthermore, in either case, if T and T' denote the elements of $M(L_p(G))$ and $M(L_p(\beta(G)))$, respectively, which correspond to φ then $\|T\|_p = \|T'\|_p$.

Proofs of these results are available in Saeki [1]. Theorem 4.8.7 was first established in de Leeuw [1] for $G = R^n$. An exposition of de Leeuw's work can be found in Gaudry [6], VII.

Similar sorts of results have been obtained in Figà-Talamanca and Gaudry [3]. Γ^n and Z^n denote the n-fold product of Γ and Z with itself.

Theorem 4.8.8. *Suppose* $1<p<\infty$ *and* $\varphi\in\mathcal{M}(L_p(\Gamma^n))$. *Then there exists* $\psi\in C(R^n)$ *such that* $\psi\in\mathcal{M}(L_p(R^n))$ *and* φ *is the restriction to* Z^n *of* ψ. *If* $\varphi\in C_0(Z^n)$ *then* ψ *can be chosen in* $C_0(R^n)$.

This theorem is used in Figà-Talamanca and Gaudry [3] to study the spaces m_p which are the operator closures in $\mathcal{M}(L_p(G))$ of $L_1(G)\hat{}$. They show, among other things, when $G=R^n$ or Γ^n that m_p is a proper subspace of $\mathcal{M}(L_p(G))\cap C_0(\hat{G})$, $1<p<\infty$, $p\neq 2$. The result is also valid for $p=1$ and $p=\infty$. The spaces m_p are also discussed in Edwards [14_{II}], 16.6, and Hörmander [1].

We wish to indicate a number of other areas where the multipliers for $L_p(G)$ have been studied or applied. Edwards [8], Gilbert [1] and Máté [6] have used them in connection with certain approximation problems; Edwards [8], Figà-Talamanca and Rider [1] and Máté [6] have utilized multipliers in characterizing Sidon and Helson sets; and the role of multipliers in factorization problems has been examined in Edwards [8], Figà-Talamanca and Gaudry [2] and Gaudry [6, 9]. Hörmander [1] and Igari [2] have studied the functions which operate in $\mathcal{M}(L_p(G))$. Finally, we mention a sampling of miscellaneous references which contain material about the multipliers for $L_p(G)$, including some of the results we have considered in this chapter, and which have not been previously cited. These references are: Edwards [9], Guy [1], Hille and Tamarkin [1], Jones [2], Krée [1, 2], Mihlin [1, 4], Orlicz [1], Peyrière and Spector [1], M. Riesz [1], Rivière and Sagher [1], Salem [1], M. Weiss and Zygmund [1] and Zygmund [6].[1]

Expositions of many of the topics discussed in this chapter and in these notes, as well as other topics not mentioned, can be found in Edwards [14_{II}], 16, Gaudry [6], V, VI and VII, and Hewitt and Ross [2], 35 and 36. The latter reference also contains an excellent historical survey of the problem.

---

[1] See also Littman, McCarthy and Rivière [1, 2].

Chapter 5

The Multipliers for the Pair $(L_p(G), L_q(G))$, $1 \leqq p, q \leqq \infty$

5.0. Introduction. In the previous chapter we discussed multipliers for the pair $(L_p(G), L_q(G))$ when $p=q$. Our attention in this chapter will be focused on the case where $p \neq q$. The problem of describing the multipliers in this situation is equally if not more difficult than in the case $p=q$. In order to obtain a description of the multipliers as convolution operators we shall have to introduce a class of mathematical objects which properly contains the space of pseudomeasures employed previously, namely, the space of quasimeasures. The description of multipliers as multiplication by bounded functions is no longer possible, but an analogous result will be obtained using the Fourier transform of certain quasimeasures. Unfortunately these transforms are again quasimeasures and not in general functions. We shall define a Fourier transform for $L_p(G)$, $p>2$, in terms of quasimeasures and show, in particular, that there exist $f \in L_p(G)$, $p>2$, whose Fourier transforms are not measures. We shall also examine various inclusion relationships for the spaces $M(L_p(G), L_q(G))$ and obtain a characterization of these spaces as dual spaces of certain Banach spaces.

5.1. Quasimeasures. Let G be a locally compact Abelian group and suppose $K \subset G$ is compact. We denote by $C_c^K(G)$ the subspace of $C_c(G)$ consisting of all functions in $C_c(G)$ whose support lies in K. The linear space $D_K(G)$ is defined as follows:

$$D_K(G) = \{h \mid h = \sum_i f_i * g_i,\ f_i,\ g_i \in C_c^K(G),\ \sum_i \|f_i\|_\infty \|g_i\|_\infty < \infty\}.$$

The index i runs over the positive integers. Clearly $D_K(G)$ is a subspace of $C_c^{KK}(G)$. For each $h \in D_K(G)$ we define

$$\|h\|_K = \inf\{\sum_i \|f_i\|_\infty \|g_i\|_\infty,\ h = \sum_i f_i * g_i,\ f_i,\ g_i \in C_c^K(G),\ \sum_i \|f_i\|_\infty \|g_i\|_\infty < \infty\}.$$

It is apparent that $\|\cdot\|_K$ is well defined and that $\|h\|_\infty \leqq \lambda(K) \|h\|_K$. Moreover we have the following theorem.

Theorem 5.1.1. *Let G be a locally compact Abelian group and $K \subset G$ a compact set. Then $\|\cdot\|_K$ is a complete norm for the linear space $D_K(G)$.*

Proof. It is easily seen that $\|\cdot\|_K$ defines a norm on $D_K(G)$. Suppose that $\{h_n\} \subset D_K(G)$ is a Cauchy sequence in the norm $\|\cdot\|_K$. Choose a subsequence $\{k_n\}$ of $\{h_n\}$ such that $\|k_{n+1}-k_n\|_K < 1/2^n$, $n=1, 2, \ldots$. From the definition of $\|\cdot\|_K$, we can find $\{f_{nj}\}$ and $\{g_{nj}\}$ in $C_c^K(G)$ such that

(i) $k_1 = \sum_j f_{1j} * g_{1j}$,

(ii) $\sum_j \|f_{1j}\|_\infty \|g_{1j}\|_\infty < \|k_1\|_K + 1$,

(iii) $k_{n+1} - k_n = \sum_j f_{n+1j} * g_{n+1j}$,

(iv) $\sum_j \|f_{n+1j}\|_\infty \|g_{n+1j}\|_\infty < 1/2^{n-1}$, $n=1, 2, \ldots$.

Set $h = \sum_j f_{1j} * g_{1j} + \sum_n [\sum_j f_{n+1j} * g_{n+1j}]$. The definition makes sense since

$$\begin{aligned}\|h\|_\infty &\leqq \sum_j \|f_{1j} * g_{1j}\|_\infty + \sum_n [\sum_j \|f_{n+1j} * g_{n+1j}\|_\infty] \\ &\leqq \lambda(K)\{\sum_j \|f_{1j}\|_\infty \|g_{1j}\|_\infty + \sum_n [\sum_j \|f_{n+1j}\|_\infty \|g_{n+1j}\|_\infty]\} \\ &< \lambda(K)\{\|k_1\|_K + 1 + \sum_n 1/2^{n-1}\} \\ &= \lambda(K)\{\|k_1\|_K + 3\}.\end{aligned}$$

Furthermore, since the convergence is absolute and uniform it follows immediately that $h \in D_K(G)$.

Let $\varepsilon > 0$ and let $N(\varepsilon)$ be a positive integer such that for $n > N(\varepsilon)$ we have $\sum_{r=n}^{\infty} 1/2^{r-1} < \varepsilon$. Then for $n > N(\varepsilon)$ we see that

$$\begin{aligned}\|h - k_{n+1}\|_\infty &= \left\| h - \left[k_1 + \sum_{r=1}^{n} (k_{r+1} - k_r)\right] \right\|_\infty \\ &\leqq \sum_{r=n+1}^{\infty} [\sum_j \|f_{r+1j} * g_{r+1j}\|_\infty] \\ &< \lambda(K) \sum_{r=n+1}^{\infty} 1/2^{r-1} < \lambda(K)\,\varepsilon.\end{aligned}$$

Consequently, for $n > N(\varepsilon)$ we conclude that $\|h - k_{n+1}\|_K < \lambda(K)\,\varepsilon$. Thus the sequence $\{k_n\}$ converges in $D_K(G)$ to h.

Therefore, since $\{h_n\}$ is Cauchy, it follows that $\{h_n\}$ converges to h in $D_K(G)$, and $D_K(G)$ is complete. □

We now define $D(G)$ as the internal inductive limit of the Banach spaces $D_K(G)$ (D.1). That is, $D(G)$ is the vector space $\bigcup_K D_K(G)$ with the topology which has for a neighborhood base at the origin open sets of the form $U_\varepsilon = \bigcup_K \{f \mid f \in D_K(G), \|f\|_K < \varepsilon\}$. $D(G)$ is then a locally convex topological linear space (D.1). Clearly $D(G) \subset C_c(G)$.

We recall from the preceding chapter that $A(G)$ was the Banach space of all continuous functions on G which are Fourier transforms of elements in $L_1(\hat{G})$ with $\|\hat{f}\|_A = \|f\|_1$. And $A_c(G) = A(G) \cap C_c(G)$.

Theorem 5.1.2. *Let G be a locally compact Abelian group. Then:*

(i) $D(G) \subset A_c(G)$.

(ii) *$D(G)$ is a dense subspace of the Banach space $A(G)$.*

(iii) *$D(G)$ is a dense subspace of the normed linear space $C_c(G)$.*

(iv) *The topology on $D(G)$ is stronger than the topology induced on $D(G)$ as a subspace of $A(G)$.*

Proof. Let $h \in D(G)$. Then there exists a compact $K \subset G$ such that $h \in D_K(G)$. Suppose $h = \sum_i f_i * g_i$ where $f_i, g_i \in C_c^K(G)$ and $\sum_i \|f_i\|_\infty \|g_i\|_\infty < \infty$. Define $h_n = \sum_{i=1}^n f_i * g_i$. Obviously the h_n define elements of $D_K(G)$ and $\lim_n \|h - h_n\|_K = 0$. Since $\|h - h_n\|_\infty \leqq \lambda(K) \|h - h_n\|_K$ it follows that $\lim_n \|h - h_n\|_\infty = 0$.

On the other hand f_i, $g_i \in L_2(G)$ and so $f_i * g_i \in L_1(\hat{G})\hat{}$ (F.10a). Thus $h_n \in A(G) \cap C_c(G) = A_c(G)$. Moreover, appealing to the Plancherel Theorem (Theorem F.8.2) and Hölder's inequality, we see for $n \geqq m$ that

$$\begin{aligned}\|h_n - h_m\|_A &= \left\| \sum_{i=m+1}^n f_i * g_i \right\|_A = \left\| \sum_{i=m+1}^n \hat{f}_i \hat{g}_i \right\|_1 \\ &\leqq \sum_{i=m+1}^n \|\hat{f}_i\|_2 \|\hat{g}_i\|_2 = \sum_{i=m+1}^n \|f_i\|_2 \|g_i\|_2 \\ &\leqq \lambda(K) \sum_{i=m+1}^n \|f_i\|_\infty \|g_i\|_\infty,\end{aligned}$$

where $\hat{f}_i$, $\hat{g}_i$ denote the elements in $L_1(\hat{G}) \cap L_2(\hat{G})$ which are the Fourier-Plancherel transforms of f_i, g_i. Hence for $n \geqq m$, $\|h_n - h_m\|_A \leqq \lambda(K) \|h_n - h_m\|_K$, and $\{h_n\}$ is a Cauchy sequence in $A(G)$. Let $k \in A(G)$ be such that $\lim_n \|h_n - k\|_A = 0$. Then $\lim_n \|h_n - k\|_\infty = 0$ as $\|h_n - k\|_\infty \leqq \|h_n - k\|_A$.

Therefore $h = k \in A(G)$, that is, $D(G) \subset A(G)$.

Actually, since $D(G) \subset C_c(G)$, it is even the case that $D(G) \subset A_c(G)$.

The preceding argument also shows for any $h \in D(G)$ that $\|h\|_A = \lim_n \|h_n\|_A \leqq \lambda(K) \sum_i \|f_i\|_\infty \|g_i\|_\infty$. Thus $\|h\|_A \leqq \lambda(K) \|h\|_K$. From this it is evident that the topology on $D(G)$ is stronger than the one inherited from $A(G)$.

To prove that $D(G)$ is dense in $A(G)$ and $C_c(G)$ it is sufficient to show that $D(G)$ is dense in $A_c(G)$ because this space is dense in $A(G)$ and in $C_c(G)$. Let $\{u_\alpha\} \subset L_1(G)$ be an approximate identity such that $\{u_\alpha\} \subset C_c^{K_0}(G)$ for some fixed compact set K_0 which contains the identity of G and $\|u_\alpha\|_1 = 1$. It is easily seen that $\|\hat{u}_\alpha\|_\infty \leqq 1$ and that $\{\hat{u}_\alpha\}$ converges uniformly to one on each compact subset of $\hat{G}$. Suppose $\hat{f} \in A_c(G)$. Then $\{u_\alpha * \hat{f}\} \subset D(G)$ as $\{u_\alpha\} \subset C_c(G)$ and $\hat{f} \in C_c(G)$. Since $f \in L_1(\hat{G})$, given $\varepsilon > 0$ there exists a compact subset $W \subset \hat{G}$ such that $\int_{\hat{G} \sim W} |f(\gamma)| \, d\eta(\gamma) < \varepsilon/4$. Moreover, there exists an α_0 such that if $\alpha \succ \alpha_0$ then $\sup_{\gamma \in W} |\hat{u}_\alpha(\gamma) - 1| < \varepsilon / \|f\|_1$. Consequently for $\alpha \succ \alpha_0$ we have

$$\begin{aligned} \|\hat{f} - u_\alpha * \hat{f}\|_A &= \int_{\hat{G}} |f(\gamma) - \hat{u}_\alpha(\gamma) f(\gamma)| \, d\eta(\gamma) \\ &= \int_{\hat{G} \sim W} |f(\gamma) - \hat{u}_\alpha(\gamma) f(\gamma)| \, d\eta(\gamma) \\ &\quad + \int_W |f(\gamma) - \hat{u}_\alpha(\gamma) f(\gamma)| \, d\eta(\gamma) \\ &\leqq 2 \int_{\hat{G} \sim W} |f(\gamma)| \, d\eta(\gamma) + \sup_{\gamma \in W} |1 - \hat{u}_\alpha(\gamma)| \int_{\hat{G}} |f(\gamma)| \, d\eta(\gamma) \\ &< \varepsilon. \end{aligned}$$

Therefore $D(G)$ is dense in the space $A_c(G)$. □

It should be noted that the proof of the theorem shows that if $f \in C_c^K(G)$ then there exists a sequence $\{f_n\} \subset D_{K_0}(G)$ for some fixed compact set $K_0 \subset G$ such that $\lim_n \|f - f_n\|_\infty = 0$. Moreover, the proof also shows that $D(G)$ is dense in $C_c(G)$ considered as the internal inductive limit of $C_c^K(G)$.

We are now in a position to define quasimeasures and to give some of their elementary properties.

The space of continuous linear functionals on $D(G)$ is called the space of *quasimeasures*. We denote the space of quasimeasures by $Q(G)$. Clearly $Q(G)$ is a locally convex topological linear space. As usual we shall denote the pairing between $h \in D(G)$ and $\sigma \in Q(G)$ by $\langle h, \sigma \rangle$.

Let $V(G)$ denote the linear space of all complex valued regular Borel measures on G. As is well known the dual space of $C_c(G)$ considered as an internal inductive limit of the spaces $C_c^K(G)$ can be identified with $V(G)$ (Theorem D.9.5) by the formula

$$\langle f, v \rangle = \int_G f(t^{-1}) \, dv(t) \qquad \big(f \in C_c(G),\ v \in V(G)\big).$$

Obviously $M(G)$ is a linear subspace of $V(G)$. Our next result shows that $Q(G)$ contains both $V(G)$ and $P(G)$.

Theorem 5.1.3. *Let G be a locally compact Abelian group. Then $V(G) \subset Q(G)$ and $P(G) \subset Q(G)$.*

Proof. Let $v \in V(G)$. Then for each compact set $K \subset G$ if $h \in D_K(G)$ we have

$$|\langle h, v\rangle| = \Big|\int_G h(t^{-1})\, dv(t)\Big| \leqq \|h\|_\infty\, |v|(K^{-1}) \leqq \lambda(K)\, \|h\|_K\, |v|(K^{-1}).$$

Thus v defines a continuous linear functional on each $D_K(G)$, and hence on all of $D(G)$. Hence $V(G) \subset Q(G)$.

On the other hand, since by Theorem 5.1.2(iv) the topology of $D(G)$ is stronger than the topology induced by $A(G)$ we can conclude that any continuous linear functional on $A(G)$ is a continuous linear functional on $D(G)$ as $D(G) \subset A(G)$.

Therefore $P(G) \subset Q(G)$. □

A description of those quasimeasures which are measures is given in the next result.

Theorem 5.1.4. *Let G be a locally compact Abelian group. Then the following are equivalent:*

(i) $\sigma \in V(G)$.

(ii) *$\sigma \in Q(G)$ and for each compact $K \subset G$ the quasimeasure σ when restricted to $D_K(G)$ with the supremum norm topology is continuous.*

Proof. Obviously (i) implies (ii). Suppose (ii) holds. Since σ is continuous with respect to the supremum norm topology on $D_K(G)$ for each compact K it has a unique continuous extension to the supremum norm closure $D_K(G)^-$ of $D_K(G)$ in $C_c^{KK}(G)$. Furthermore by Theorem 5.1.2(iii) we see that $\bigcup_K D_K(G)^- = C_c(G)$. Let $h \in C_c(G)$ and suppose $\{h_n\} \subset D_{K_1}(G)$ and $\{g_n\} \subset D_{K_2}(G)$ are such that $\lim_n \|h - h_n\|_\infty = 0$ and $\lim_n \|h - g_n\|_\infty = 0$. To show that $\sigma \in V(G)$ it is sufficient to prove that $\lim_n \langle h_n - g_n, \sigma\rangle = 0$. Clearly we may assume that $e \in K_1 \cap K_2$ and hence $\{h_n\} \subset D_{K_1 K_2}(G)$ and $\{g_n\} \subset D_{K_1 K_2}(G)$. Since $K_1 K_2$ is compact it follows immediately that $\lim_n \langle h_n - g_n, \sigma\rangle = 0$ because σ restricted to $D_{K_1 K_2}(G)$ is continuous with respect to the supremum norm topology. □

For our subsequent investigations of quasimeasures and multipliers we shall have need of several lemmas. The main purpose of the lemmas is to enable us to define for $\sigma \in Q(G)$ the convolution of σ and $f \in C_c(G)$, and the product of σ and $\hat{\mu}$ for $\mu \in M(\hat{G})$.

Lemma 5.1.1. *Let G be a locally compact Abelian group and $f \in C_c(G)$. If $h \in D(G)$ then $f * h \in D(G)$ and the mapping $h \to f * h$ is continuous from $D(G)$ to $D(G)$.*

Proof. If $h\in D(G)\subset C_c(G)$ then clearly $f*h\in D(G)$. To show that the mapping $h\to f*h$ is continuous it is sufficient to show that its restriction to each $D_K(G)$ is continuous from $D_K(G)$ to $D(G)$. But suppose $\lim_n \|h_n-h\|_K=0$. Since $\|h_n-h\|_\infty \leqq \lambda(K)\|h_n-h\|_K$ we see that $\lim_n \|h_n-h\|_\infty=0$. Consequently

$$\|f*h_n-f*h\|_{K'}=\|f*(h_n-h)\|_{K'}\leqq \|f\|_\infty \|h_n-h\|_\infty$$

where K' is the product of the support of f and K. Thus $\{f*h_n\}$ converges to $f*h$ in $D(G)$. □

This lemma allows us to define $\sigma*f$ for $\sigma\in Q(G)$ and $f\in C_c(G)$. Indeed, $\sigma*f$ is defined to be that element in $Q(G)$ such that

$$\langle h, \sigma*f\rangle=\langle f*h, \sigma\rangle \qquad (h\in D(G)).$$

The preceding lemma insures that this definition is meaningful.

Lemma 5.1.2. *Let G be a locally compact Abelian group. If $\mu\in M(\hat{G})$ and $h=\sum_i f_i*g_i\in D_K(G)$ then the vector valued integrals*

$$H=\int_{\hat{G}} \gamma^{-1} h\, d\mu(\gamma)=\sum_i \int_{\hat{G}} (\gamma^{-1} f_i)*(\gamma^{-1} g_i)\, d\mu(\gamma)$$

belong to $D_K(G)$, and for each $t\in G$

$$H(t)=\sum_i \hat{\mu}(t)\, f_i*g_i(t)=\hat{\mu}(t)\, h(t).$$

Proof. Evidently $\gamma^{-1} h=\sum_i (\gamma^{-1} f_i)*(\gamma^{-1} g_i)$ belongs to $D_K(G)$ for each $\gamma\in\hat{G}$. Consider the mapping $\gamma\to\gamma^{-1}h$ from $\hat{G}$ to $D_K(G)$. Clearly $\|\gamma^{-1}h\|_K\leqq\|h\|_K$ for each $\gamma\in\hat{G}$. Moreover we claim that the mapping is continuous. Indeed let $\gamma_0\in\hat{G}$ and suppose $\{\gamma_\alpha\}\subset\hat{G}$ converges to γ_0. Then $(\gamma_\alpha^{-1} f_i)*(\gamma_\alpha^{-1} g_i)-(\gamma_0^{-1} f_i)*(\gamma_0^{-1} g_i)$ belongs to $D_K(G)$ and for each i we have

$$\begin{aligned}
&\|(\gamma_\alpha^{-1} f_i)*(\gamma_\alpha^{-1} g_i)-(\gamma_0^{-1} f_i)*(\gamma_0^{-1} g_i)\|_K\\
&\quad\leqq \|(\gamma_\alpha^{-1} f_i)*(\gamma_\alpha^{-1} g_i)-(\gamma_\alpha^{-1} f_i)*(\gamma_0^{-1} g_i)\|_K\\
&\qquad+\|(\gamma_\alpha^{-1} f_i)*(\gamma_0^{-1} g_i)-(\gamma_0^{-1} f_i)*(\gamma_0^{-1} g_i)\|_K\\
&\quad\leqq \|\gamma_\alpha^{-1} f_i\|_\infty \|\gamma_\alpha^{-1} g_i-\gamma_0^{-1} g_i\|_\infty+\|\gamma_\alpha^{-1} f_i-\gamma_0^{-1} f_i\|_\infty \|\gamma_0^{-1} g_i\|_\infty\\
&\quad=\|f_i\|_\infty \|(\gamma_\alpha^{-1}-\gamma_0^{-1})\, g_i\|_\infty+\|g_i\|_\infty \|(\gamma_\alpha^{-1}-\gamma_0^{-1})\, f_i\|_\infty .
\end{aligned}$$

Thus

$$\begin{aligned}
\|\gamma_\alpha^{-1} h-\gamma_0^{-1} h\|_K&\leqq\sum_i [\|f_i\|_\infty \|(\gamma_\alpha^{-1}-\gamma_0^{-1})\, g_i\|_\infty+\|g_i\|_\infty \|(\gamma_\alpha^{-1}-\gamma_0^{-1})\, f_i\|_\infty]\\
&\leqq \sup_K |\gamma_\alpha^{-1}-\gamma_0^{-1}|\,(2\sum_i \|f_i\|_\infty \|g_i\|_\infty).
\end{aligned}$$

But from the definition of the topology of the dual group $\hat{G}$ (B.2) the net $\{\gamma_\alpha\}$ converges to γ_0 in $\hat{G}$ if and only if the net of functions $\{(\cdot, \gamma_\alpha)\}$ converges uniformly to $(\cdot, \gamma_0)$ on compact subsets of G. Consequently if $\{\gamma_\alpha\}$ converges to γ_0 in $\hat{G}$ then $\lim_\alpha \|\gamma_\alpha^{-1} h - \gamma_0^{-1} h\|_K = 0$, that is, the mapping $\gamma \to \gamma^{-1} h$ is continuous.

Since $\mu \in M(\hat{G})$ and $D_K(G)$ is complete by Theorem 5.1.1, it follows that the vector valued integral $\int_{\hat{G}} \gamma^{-1} h \, d\mu(\gamma)$ exists and belongs to $D_K(G)$ (Edwards [11], 8.14.14). Moreover, since $\sum_i (\gamma^{-1} f_i) * (\gamma^{-1} g_i)$ converges to $\gamma^{-1} h$ in $D_K(G)$ uniformly with respect to γ, it is apparent that

$$\int_{\hat{G}} \gamma^{-1} h \, d\mu(\gamma) = \sum_i \int_{\hat{G}} (\gamma^{-1} f_i) * (\gamma^{-1} g_i) \, d\mu(\gamma).$$

The last assertion of the lemma is now obvious. □

Lemma 5.1.3. *Let G be a locally compact Abelian group. If $\mu \in M(\hat{G})$ and $h \in D_K(G)$ then $\hat{\mu} h \in D(G)$ and the mapping $h \to \hat{\mu} h$ from $D_K(G)$ to $D(G)$ is continuous.*

Proof. By the preceding lemma $\hat{\mu} h \in D_K(G)$ and so the mapping is certainly from $D_K(G)$ to $D(G)$. Moreover we have by Lemma 5.1.2 that

$$\|\hat{\mu} h\|_K = \Big\| \int_{\hat{G}} \gamma^{-1} h \, d\mu(\gamma) \Big\|_K \leqq \int_{\hat{G}} \|\gamma^{-1} h\|_K \, d|\mu|(\gamma)$$
$$\leqq \|h\|_K \, \|\mu\|,$$

whence the mapping $h \to \hat{\mu} h$ is continuous. □

Lemma 5.1.3 now allows us to define $\hat{\mu}\sigma$. If $\sigma \in Q(G)$ and $\mu \in M(\hat{G})$ then we define $\hat{\mu}\sigma$ to be that element in $Q(G)$ such that

$$\langle h, \hat{\mu}\sigma \rangle = \langle \hat{\tilde{\mu}} h, \sigma \rangle = \langle \check{\mu} h, \sigma \rangle \qquad (h \in D(G)),$$

where $\check{\mu}(t) = \int_{\hat{G}} (t, \gamma) \, d\mu(\gamma)$. As before $\tilde{\mu}(E) = \mu(E^{-1})$.

Our final concern in this section will be to give a complete characterization of certain linear transformations from $C_c(G)$ to $V(G)$. As noted previously, $V(G)$ can be considered as the space of continuous linear functionals on $C_c(G)$ where the latter is considered as the internal inductive limit of $C_c^K(G)$. The weak topology induced on $V(G)$ by the elements of $C_c(G)$ is called the *vague topology* for $V(G)$. Obviously a net $\{v_\alpha\} \subset V(G)$ converges to $v \in V(G)$ in the vague topology if and only if $\lim_\alpha \langle f, v_\alpha \rangle = \langle f, v \rangle$ for each $f \in C_c(G)$.

Our aim is to characterize all linear transformations $T\colon C_c(G) \to V(G)$ which commute with translations and are continuous from the inductive limit topology on $C_c(G)$ to the vague topology on $V(G)$. That is, we shall characterize the multipliers from $C_c(G)$ to $V(G)$. As usual we shall denote these transformations by $M(C_c(G), V(G))$ where we shall always assume that the topologies are as indicated above.

Before proving the indicated result we need one further lemma.

Lemma 5.1.4. *Let G be a locally compact Abelian group and suppose $T\in M(C_c(G), V(G))$. Then there exists a net of multipliers $\{T_\alpha\}\subset M(C_c(G), V(G))$ with the following properties:*

(i) *For each α there exists a $v_\alpha\in V(G)$ such that $T_\alpha f=v_\alpha * f$ for each $f\in C_c(G)$.*

(ii) *If $K\subset G$ is compact then there exists a constant $c_K>0$ such that for all α, $|T_\alpha f*g(e)|\leqq c_K\|f\|_\infty\|g\|_\infty$ for each $f, g\in C_c^K(G)$.*

(iii) *If $f, g\in C_c(G)$ then $\lim_\alpha T_\alpha f*g(e)=Tf*g(e)$.*

Before beginning the proof of the lemma we remark that if $T\in M(C_c(G), V(G))$ then $Tf*g$ is a continuous function on G for each $f, g\in C_c(G)$ since it is the convolution of a measure Tf and a function $g\in C_c(G)$. Thus we can evaluate $Tf*g$ at the identity of G. It is apparent that in this case we also have $Tf*g(e)=\langle g, Tf\rangle$. Here $\langle\cdot,\cdot\rangle$ denotes the pairing between $C_c(G)$ and its dual $V(G)$.

Proof. Let $\{u_\alpha\}\subset C_c(G)$ be an approximate identity for $L_1(G)$ such that the support of each u_α lies in the compact neighborhood K_0 of the identity in G and $\|u_\alpha\|_1=1$. Define $T_\alpha f=u_\alpha * Tf$ for each $f\in C_c(G)$. Clearly this defines a linear transformation from $C_c(G)$ to $V(G)$ which commutes with translation as $T\in M(C_c(G), V(G))$. By an argument used several times previously we see that the fact that T commutes with translation implies that $Tf*g=f*Tg$ for each $f, g\in C_c(G)$. Thus $T_\alpha f=u_\alpha * Tf=Tu_\alpha * f=v_\alpha * f$ for each $f\in C_c(G)$ where $v_\alpha=Tu_\alpha\in V(G)$. If $f, g\in C_c(G)$ then $\langle g, T_\alpha f\rangle=\langle g, u_\alpha * Tf\rangle=\langle g*u_\alpha, Tf\rangle$. Hence T_α is a continuous linear transformation from $C_c(G)$ to $V(G)$ since T is continuous. Thus $T_\alpha\in M(C_c(G), V(G))$. Thus (i) is established.

Suppose $K\subset G$ is compact and let T_K denote the restriction of T to $C_c^{K\cup K^{-1}}(G)$. For each $g\in C_c^{K\cup K^{-1}}(G)$ define the continuous linear functional $v_g\in[C_c^{K\cup K^{-1}}(G)]^*$ by

$$v_g(f)=\langle g, T_K f\rangle=\langle g, Tf\rangle \qquad (f\in C_c^{K\cup K^{-1}}(G)).$$

Moreover, for each $f\in C_c^{K\cup K^{-1}}(G)$ we have

$$\begin{aligned}|v_g(f)|&=|\langle g, Tf\rangle|=\Big|\int_{K\cup K^{-1}} g(t^{-1})\,dTf(t)\Big|\\ &\leqq\|g\|_\infty\int_{K\cup K^{-1}} d|Tf|(t)=\int_{K\cup K^{-1}} d|Tf|(t)\end{aligned}$$

for all $g \in C_c^{K \cup K^{-1}}(G)$ for which $\|g\|_\infty \leqq 1$. Hence, applying the Uniform Boundedness Theorem (Theorem D.6.4) to the family of functionals $\{v_g | g \in C_c^{K \cup K^{-1}}(G), \|g\|_\infty \leqq 1\}$, we deduce the existence of a constant $b_K > 0$ such that

$$|\langle g, Tf \rangle| = |v_g(f)| \leqq b_K \|f\|_\infty$$

for each $f, g \in C_c^{K \cup K^{-1}}(G)$ where $\|g\|_\infty \leqq 1$. Clearly then we also have

$$|\langle g, Tf \rangle| \leqq b_K \|g\|_\infty \|f\|_\infty$$

for each $f, g \in C_c^{K \cup K^{-1}}(G)$, and so *a fortiori*

$$|Tf * g(e)| = |\langle g, Tf \rangle| \leqq b_K \|g\|_\infty \|f\|_\infty$$

for each $f, g \in C_c^K(G)$.

But then if $K \subset G$ is compact and $f, g \in C_c^K(G)$ we have for each α that

$$\begin{aligned} |T_\alpha f * g(e)| &= |g * u_\alpha * Tf(e)| \leqq b_{KK_0} \|g * u_\alpha\|_\infty \|f\|_\infty \\ &\leqq b_{KK_0} \|u_\alpha\|_1 \|g\|_\infty \|f\|_\infty = b_{KK_0} \|g\|_\infty \|f\|_\infty, \end{aligned}$$

since $g * u_\alpha \in C_c^{KK_0}(G)$ and $f \in C_c^K(G) \subset C_c^{KK_0}(G)$. Setting $c_K = b_{KK_0}$ we see that (ii) of the lemma is proven.

Finally, suppose $f, g \in C_c(G)$ and the supports of f and g are contained in the compact set K. Then

$$\begin{aligned} |Tf * g(e) - T_\alpha f * g(e)| &= |Tf * (g - g * u_\alpha)(e)| \\ &\leqq b_{KK_0} \|g - g * u_\alpha\|_\infty \|f\|_\infty, \end{aligned}$$

since $g * u_\alpha \in C_c^{KK_0}(G)$ and $g \in C_c^K(G) \subset C_c^{KK_0}(G)$. Obviously then $\lim_\alpha T_\alpha f * g(e) = Tf * g(e)$ because $\lim_\alpha \|g - g * u_\alpha\|_\infty = 0$. This proves (iii). □

Theorem 5.1.5. *Let G be a locally compact Abelian group. Then the following are equivalent:*

(i) $T \in M(C_c(G), V(G))$.

(ii) *There exists a unique quasimeasure $\sigma \in Q(G)$ such that $Tf = \sigma * f$ for each $f \in C_c(G)$.*

Moreover the correspondence between T and σ defines a linear isomorphism from $M(C_c(G), V(G))$ onto $Q(G)$.

Proof. Suppose that $T \in M(C_c(G), V(G))$. If $h = \sum_i f_i * g_i \in D_K(G)$ then by the proof of (ii) of Lemma 5.1.4 we have

$$\left|\sum_i Tf_i * g_i(e)\right| \leqq \sum_i |Tf_i * g_i(e)| \leqq b_K \sum_i \|f_i\|_\infty \|g_i\|$$

where $b_K > 0$ depends only on K. Thus $\sum_i Tf_i * g_i(e)$ converges absolutely since $\sum_i \|f_i\|_\infty \|g_i\|_\infty < \infty$. Define $\langle h, \sigma\rangle = \sum_i Tf_i * g_i(e)$. Clearly σ defines a linear functional on $D(G)$ provided it is well defined, that is, provided $\sum_i f_i * g_i = 0$ with $f_i, g_i \in C_c^K(G)$ and $\sum_i \|f_i\|_\infty \|g_i\|_\infty < \infty$ imply that $\sum_i Tf_i * g_i(e) = 0$.

To see that this is indeed the case let us assume that $\sum_i f_i * g_i = 0$ and let $\{T_\alpha\} \subset M(C_c(G), V(G))$ be a net of multipliers which satisfies properties (i)–(iii) of Lemma 5.1.4. For each α part (i) of the lemma shows that

$$\Big|\sum_i T_\alpha f_i * g_i(e)\Big| \leqq \sum_i |T_\alpha f_i * g_i(e)| \leqq c_K \sum_i \|f_i\|_\infty \|g_i\|_\infty$$

where $c_K > 0$ is independent of α. Hence $\sum_i T_\alpha f_i * g_i(e)$ converges uniformly with respect to α. Moreover part (iii) of the lemma asserts that $\lim_\alpha T_\alpha f_i * g_i(e) = Tf_i * g_i(e)$ for each i and hence by the uniformity of the series convergence we conclude that

$$\lim_\alpha \sum_i T_\alpha f_i * g_i(e) = \sum_i Tf_i * g_i(e).$$

Let $v_\alpha \in V(G)$ be the measures such that $T_\alpha f = v_\alpha * f$ for each $f \in C_c(G)$. Then

$$T_\alpha f_i * g_i(e) = v_\alpha * f_i * g_i(e) = \int_{(KK)^{-1}} f_i * g_i(t^{-1})\, d v_\alpha(t)$$

since $f_i, g_i \in C_c^K(G)$. However $\sum_i f_i * g_i$ is a uniformly convergent series of functions in $C_c^{KK}(G)$. Therefore we have

$$\begin{aligned}
\sum_i Tf_i * g_i(e) &= \lim_\alpha \sum_i T_\alpha f_i * g_i(e) \\
&= \lim_\alpha \sum_i \Big[\int_{(KK)^{-1}} f_i * g_i(t^{-1})\, d v_\alpha(t)\Big] \\
&= \lim_\alpha \int_{(KK)^{-1}} \Big[\sum_i f_i * g_i(t^{-1})\Big]\, d v_\alpha(t) \\
&= 0.
\end{aligned}$$

because $\sum_i f_i * g_i = 0$. Consequently σ is a well defined linear functional on $D(G)$.

Furthermore, from the proof of (ii) of Lemma 5.1.4, we see that if $h = \sum_i f_i * g_i \in D_K(G)$ then

$$|\langle h, \sigma\rangle| = \Big|\sum_i Tf_i * g_i(e)\Big| \leqq b_K \sum_i \|f_i\|_\infty \|g_i\|_\infty .$$

Thus $|\langle h, \sigma\rangle| \leqq b_K \|h\|_K$ and σ restricted to $D_K(G)$ is a continuous linear functional on $D_K(G)$.

Therefore σ is a quasimeasure.

Moreover, we claim that $Tf = \sigma * f$ for each $f \in C_c(G)$. Indeed, from the definition of convolution between elements of $Q(G)$ and $C_c(G)$ and the construction of σ we see for each $f \in C_c(G)$ and $h \in D(G)$ that

$$\begin{aligned}\langle h, \sigma * f\rangle &= \langle f * h, \sigma\rangle = Tf * h(e)\\ &= \int_G h(t^{-1})\, dTf(t) = \langle h, Tf\rangle.\end{aligned}$$

Since $D(G)$ is dense in $C_c(G)$ it follows that $Tf = \sigma * f$ for each $f \in C_c(G)$.

The quasimeasure constructed above is unique. As suppose there exists $\sigma, \sigma' \in Q(G)$ such that $Tf = \sigma * f = \sigma' * f$ for each $f \in C_c(G)$. Then we would have for each $f \in C_c(G)$ and $h \in D(G)$ that

$$\langle f * h, \sigma\rangle = \langle h, \sigma * f\rangle = \langle h, \sigma' * f\rangle = \langle f * h, \sigma'\rangle.$$

But, as is easily shown, the set $\{f * h \mid f \in C_c(G), h \in D(G)\} \subset D(G)$ is dense in $D(G)$. Consequently $\sigma = \sigma'$.

Therefore (i) implies (ii).

Suppose now that $Tf = \sigma * f$ for each $f \in C_c(G)$. Clearly $T: C_c(G) \to Q(G)$ is linear and commutes with translations. But it is not entirely apparent that $Tf \in V(G)$ for each $f \in C_c(G)$. From Theorem 5.1.4 in order to show that $Tf \in V(G)$ it is sufficient to prove for each compact $K \subset G$ that $Tf = \sigma * f$ restricted to $D_K(G)$ defines a continuous linear functional on $D_K(G)$ with respect to the topology of $C_c^{KK}(G)$. But if $\{g_\alpha\} \subset D_K(G)$ is such that $\lim_\alpha \|g_\alpha\|_\infty = 0$ then, with the support of f denoted by K_0, we have $\lim_\alpha \|f * g_\alpha\|_{KKK_0} = 0$ since $\|f * g_\alpha\|_{KKK_0} \leqq \|f\|_\infty \|g_\alpha\|_\infty$. Hence $\lim_\alpha \langle g_\alpha, \sigma * f\rangle = \lim_\alpha \langle f * g_\alpha, \sigma\rangle = 0$ and $\sigma * f$ restricted to $D_K(G)$ is continuous.

Thus $Tf = \sigma * f$ defines a linear mapping from $C_c(G)$ to $V(G)$ which commutes with translation.

Let $f, g \in C_c(G)$. From the comment following Theorem 5.1.2 there exists a net $\{g_\alpha\} \subset D_{K_0}(G)$ for some fixed compact set K_0 such that $\lim_\alpha \|g - g_\alpha\|_\infty = 0$. Since $\sigma * f \in V(G)$ we can then conclude that $\lim_\alpha \langle g_\alpha, \sigma * f\rangle = \langle g, \sigma * f\rangle$. Moreover, one may assume that $K_0 K_0$ contains the supports of both g and g_α. It then follows that $\{f * g_\alpha\}$ converges to $f * g$ in $D(G)$. Hence $\lim_\alpha \langle f * g_\alpha, \sigma\rangle = \langle f * g, \sigma\rangle$ as $\sigma \in Q(G)$. But $\langle f * g_\alpha, \sigma\rangle = \langle g_\alpha, \sigma * f\rangle$. Thus $\langle g, \sigma * f\rangle = \langle f * g, \sigma\rangle$ for all $f, g \in C_c(G)$.

Now if $\{f_\alpha\} \in C_c^K(G)$ and $\lim_\alpha \|f_\alpha - f\|_\infty = 0$ where $f \in C_c^K(G)$ then for each $g \in C_c(G)$ we would have

$$\lim_\alpha \langle g, \sigma * f_\alpha\rangle = \lim_\alpha \langle f_\alpha * g, \sigma\rangle = \langle f * g, \sigma\rangle = \langle g, \sigma * f\rangle$$

since $\lim_\alpha \|f_\alpha - f\|_\infty = 0$ implies that $\{f_\alpha * g\}$ converges to $f*g$ in $D(G)$ and $\sigma \in Q(G)$. Hence T restricted to $C_c^K(G)$ is continuous from $C_c^K(G)$ to $V(G)$ with the vague topology.

Therefore T is continuous from $C_c(G)$ to $V(G)$, that is,

$$T \in M(C_c(G), V(G)).$$

The last assertion of the theorem is now obvious. ▯

We shall also have need of the version of the preceding theorem which is valid for the special case of compact groups.

Theorem 5.1.6. *Let G be a compact Abelian group. Then the following are equivalent:*

(i) $T \in M(C(G), M(G))$.

(ii) *There exists a unique pseudomeasure $\sigma \in P(G)$ such that $Tf = \sigma * f$ for each $f \in C(G)$.*

Moreover the correspondence between T and σ defines a linear isomorphism from $M(C(G), M(G))$ onto $P(G)$.

Proof. The previous theorem shows that (ii) implies (i). Suppose on the other hand that $T \in M(C(G), M(G))$. From the proof of Lemma 5.1.4 we know that for each compact $K \subset G$ there is a constant $b_K > 0$ such that

$$\int_{K \cup K^{-1}} d|Tf|(t) \leqq b_K \|f\|_\infty \qquad (f \in C_c^K(G)).$$

In particular then we have

$$\int_G d|Tf|(t) \leqq b_G \|f\|_\infty \qquad (f \in C(G)),$$

since G is compact.

Since T is continuous and commutes with translations, a standard argument utilized in previous sections shows that $T(f*g) = Tf*g$ for $f, g \in C(G)$. Let $\{u_\alpha\} \subset C(G)$ be an approximate identity in $L_1(G)$ for which $\|u_\alpha\|_1 = 1$. Then $\lim_\alpha \|u_\alpha * f - f\|_\infty = 0$ for each $f \in C(G)$ and so

$$\langle g, Tf \rangle = \lim_\alpha \langle g, T(u_\alpha * f) \rangle = \lim_\alpha \langle g, Tu_\alpha * f \rangle = \lim_\alpha \langle g, \mu_\alpha * f \rangle$$

for each $g \in C(G)$ because T is weak* continuous. Of course $Tu_\alpha = \mu_\alpha \in M(G)$. Moreover, for each $f \in C(G)$ we have

$$\|\mu_\alpha * f\|_1 = \int_G d|\mu_\alpha * f|(t) = \int_G d|T(u_\alpha * f)|(t) \leqq b_G \|u_\alpha * f\|_\infty$$
$$\leqq b_G \|f\|_\infty .$$

Hence $\|\hat{\mu}_\alpha \hat{f}\|_\infty \leqq b_G \|f\|_\infty$ for each $f \in C(G)$ and all α. Since G is compact this inequality implies that $\sup_\alpha \|\hat{\mu}_\alpha\|_\infty \leqq b_G$.

Considering $\{\mu_\alpha\}$ as a subset of $P(G)$, the space of pseudomeasures, and recalling that the Fourier transform of pseudomeasures is an isometry, we see that $\{\mu_\alpha\}$ forms a norm bounded set of pseudomeasures. Thus, by Alaoglu's Theorem (Theorem D.4.3), there exists a subset $\{\mu_\beta\}$ of $\{\mu_\alpha\}$ which is weak* convergent to a pseudomeasure σ, that is,

$$\langle g, \sigma \rangle = \lim_\beta \langle g, \mu_\beta \rangle \qquad (g \in A(G)).$$

But then for each $f \in C(G)$ we have

$$\langle g, \sigma * f \rangle = \langle g * f, \sigma \rangle = \lim_\beta \langle g * f, \mu_\beta \rangle = \lim_\beta \langle g, \mu_\beta * f \rangle = \langle g, Tf \rangle \quad (g \in A(G)),$$

because $A(G)$ is an ideal in $C(G)$ with respect to convolution. Since $A(G)$ is dense in $C(G)$ (F.7c), we conclude that $Tf = \sigma * f$ for each $f \in C(G)$.

The remainder of the theorem is proved as before. □

The preceding theorem also provides a characterization of $M(C(G), L_1(G))$ when G is compact. Indeed, in this case, $M(C(G), L_1(G))$ and $M(C(G), M(G))$ are identical. As clearly if $T \in M(C(G), L_1(G))$ then $T \in M(C(G), M(G))$. Conversely, if $T \in M(C(G), M(G))$ then $Tf \in M(G)$ for each $f \in C(G)$. However, since T commutes with translation, the estimate

$$\|\tau_s Tf - Tf\| = \|T(\tau_s f - f)\| \leqq \|T\| \, \|\tau_s f - f\|_\infty \quad (f \in C(G), \ s \in G)$$

reveals that the mapping $s \to \tau_s Tf$ is continuous from G to $M(G)$ for each $f \in C(G)$. Hence $Tf \in L_1(G)$ (F.7a) for each $f \in C(G)$, that is, $T \in M(C(G), L_1(G))$.

5.2. The Multipliers for the Pair $(L_p(G), L_q(G))$, $1 \leqq p, q \leqq \infty$. Now let us turn our attention to a discussion of the spaces $M(L_p(G), L_q(G))$. As indicated earlier the spaces $M(L_p(G), L_q(G))$ are Banach spaces of continuous linear transformations from $L_p(G)$ to $L_q(G)$. The norm of an element $T \in M(L_p(G), L_q(G))$ will be denoted by $\|T\|_{p,q}$.

Our first result shows that certain of these spaces may be identified with each other.

Theorem 5.2.1. *Let G be a locally compact Abelian group and suppose $1 \leqq p < \infty$, $1 < q \leqq \infty$, $1/p + 1/p' = 1$, $1/q + 1/q' = 1$. Then there exists an isometric linear isomorphism of $M(L_p(G), L_q(G))$ onto $M(L_{q'}(G), L_{p'}(G))$.*

Proof. Let $T \in M(L_p(G), L_q(G))$. If $1 < p, q < \infty$ we definie $T^*: L_{q'}(G) \to L_{p'}(G)$ to be the operator adjoint to T, that is, the linear operator determined by the equation

$$\langle f, T^* g \rangle = \langle Tf, g \rangle \qquad (f \in L_p(G), \ g \in L_{q'}(G)).$$

Clearly T^* is continuous. Moreover, $T^* \tau_s = \tau_s T^*$ for each $s \in G$ since as usual we have for each $f \in L_p(G)$ and $g \in L_{q'}(G)$ that

$$\begin{aligned}\langle f, T^* \tau_s g\rangle &= \langle Tf, \tau_s g\rangle = \langle \tau_s Tf, g\rangle = \langle T \tau_s f, g\rangle = \langle \tau_s f, T^* g\rangle \\ &= \langle f, \tau_s T^* g\rangle.\end{aligned}$$

Thus $T^* \in M(L_{q'}(G), L_{p'}(G))$ and $\|T\|_{p,q} = \|T^*\|_{q',p'}$. The reflexivity of $L_p(G)$ and $L_q(G)$ shows immediately that the mapping $T \to T^*$ is surjective. Hence this mapping defines an isometric linear isomorphism from $M(L_p(G), L_q(G))$ onto $M(L_{q'}(G), L_{p'}(G))$ when $1 < p, q < \infty$.

The assertion of the theorem for the cases $p=1$, $1 < q \leqq \infty$ and $1 \leqq p < \infty$, $q = \infty$, follows immediately from Theorems 3.1.1 and 3.3.1. ∎

Theorem 5.2.5 shows that the preceding result is also trivially valid when $1 \leqq p < \infty$, $q = 1$ and $p = \infty$, $1 < q < \infty$, provided G is noncompact. The situation when $p = q = 1$ or $p = q = \infty$ is discussed in Theorems 3.1.1 and 3.4.3.

The most general theorem describing the elements of $M(L_p(G), L_q(G))$ is the following one. Its proof is an application of the characterization of $M(C_c(G), V(G))$.

Theorem 5.2.2. *Let G be a locally compact Abelian group and suppose $1 \leqq p, q \leqq \infty$. If $T \in M(L_p(G), L_q(G))$ then there exists a unique quasi-measure $\sigma \in Q(G)$ such that $Tf = \sigma * f$ for each $f \in C_c(G)$.*

Proof. If $f \in C_c(G)$ then Tf defines an element of $V(G)$ because for each compact $K \subset G$ we have by Hölder's inequality that

$$\left|\int_G \chi_K(t)\, Tf(t)\, d\lambda(t)\right| \leqq \|\chi_K\|_{q'} \|Tf\|_q$$

where $1/q + 1/q' = 1$ and χ_K is the characteristic function of K. Thus T restricted to $C_c(G)$ defines a linear mapping from $C_c(G)$ into $V(G)$. Clearly T commutes with translations. Moreover suppose $\{f_\alpha\} \subset C_c^K(G)$, $f \in C_c^K(G)$ and $\lim_\alpha \|f_\alpha - f\|_\infty = 0$. Then $\lim_\alpha \|f_\alpha - f\|_p = 0$ and so, by the continuity of T, we have that $\lim_\alpha \|Tf_\alpha - Tf\|_q = 0$. But then

$$\begin{aligned}|\langle g, Tf_\alpha - Tf\rangle| &= \left|\int_G g(t)\,[Tf_\alpha(t^{-1}) - Tf(t^{-1})]\, d\lambda(t)\right| \\ &\leqq \|g\|_{q'} \|Tf_\alpha - Tf\|_q\end{aligned}$$

for each $g \in C_c(G)$. Consequently $\{Tf_\alpha\}$ converges to Tf in the vague topology on $V(G)$. Therefore T restricted to $C_c(G)$ defines an element of $M(C_c(G), V(G))$.

The conclusion of the theorem now follows by applying Theorem 5.1.5. ∎

Corollary 5.2.1. *Let G be a locally compact Abelian group and $1 \leqq p$, $q \leqq \infty$, $p \neq \infty$. Then $M(L_p(G), L_q(G))$ is isomorphic to a linear subspace of $Q(G)$.*

The restriction $p \neq \infty$ is to insure that the mapping from $M(L_p(G), L_q(G))$ into $Q(G)$ is injective. This is necessary as $C_c(G)$ is not norm dense in $L_\infty(G)$ when G is infinite.

The argument given in proof of Theorem 5.2.2 coupled with Theorem 5.1.6 immediately establishes the counterpart of the preceding theorem and corollary for compact groups.

Theorem 5.2.3. *Let G be a compact Abelian group and $1 \leqq p$, $q \leqq \infty$. If $T \in M(L_p(G), L_q(G))$ then there exists a unique pseudomeasure $\sigma \in P(G)$ such that $Tf = \sigma * f$ for each $f \in C(G)$.*

Corollary 5.2.2. *Let G be a compact Abelian group and $1 \leqq p$, $q < \infty$, $p \neq \infty$. Then $M(L_p(G), L_q(G))$ is isomorphic to a linear subspace of $P(G)$.*

For certain values of p and q it is not difficult to see that this isomorphism is a surjective isometry.

Theorem 5.2.4. *Let G be a compact Abelian group. If $2 \leqq p < \infty$ and $1 < q \leqq 2$ then the following are equivalent:*

(i) $T \in M(L_p(G), L_q(G))$.

(ii) *There exists a unique pseudomeasure $\sigma \in P(G)$ such that $Tf = \sigma * f$ for each $f \in C(G)$.*

Moreover the correspondence between T and σ defines an isometric isomorphism of $M(L_p(G), L_q(G))$ onto $P(G)$.

Proof. We already know that (i) implies (ii) for arbitrary p, q. Suppose then that $\sigma \in P(G)$ and set $Tf = \sigma * f$ for each $f \in C(G)$. For each $f \in C(G)$ we can consider $\sigma * f$ as a pseudomeasure. Its Fourier transform is $(\sigma * f)\hat{} = \hat{\sigma}\hat{f}$ where $\hat{\sigma} \in L_\infty(\hat{G})$. Since $f \in L_2(G)$ the Plancherel Theorem (Theorem F.8.2) shows that $(\sigma * f)\hat{} \in L_2(\hat{G})$ and so $\sigma * f \in L_2(G)$. Hence applying Hölder's inequality and the Plancherel Theorem we have

$$\|\sigma * f\|_q \leqq \|\sigma * f\|_2 = \|\hat{\sigma}\hat{f}\|_2 \leqq \|\hat{\sigma}\|_\infty \|\hat{f}\|_2 = \|\sigma\|_P \|f\|_2 \leqq \|\sigma\|_P \|f\|_p$$

because $1 < q \leqq 2 \leqq p$ and G is compact. Consequently $\|Tf\|_q \leqq \|\sigma\|_P \|f\|_p$ for each $f \in C(G)$, and T obviously commutes with translations when applied to functions in $C(G)$. Since $C(G)$ is norm dense in $L_p(G)$, it follows that T can be uniquely extended to a continuous linear transformation from $L_p(G)$ to $L_q(G)$ which commutes with translations.

Therefore $T \in M(L_p(G), L_q(G))$.

The density of $C(G)$ in $L_p(G)$ also shows that $\|T\|_{p,q} \leqq \|\sigma\|_P$. On the other hand, since $1<q\leqq 2$, for each $\gamma\in\hat{G}$ we have

$$|\hat{\sigma}(\gamma)| = \|\hat{\sigma}\,\chi_{\{\gamma\}}\|_{q'} = \|(\sigma*(\cdot,\gamma))\hat{}\,\|_{q'} \leqq \|\sigma*(\cdot,\gamma)\|_q \leqq \sup_{\substack{f\in C(G)\\ \|f\|_p=1}} \|\sigma*f\|_q = \|T\|_{p,q},$$

where $1/q+1/q'=1$. The first inequality is valid because of the Hausdorff-Young Theorem (Theorem F.8.4). Thus $\|\sigma\|_P = \|\hat{\sigma}\|_\infty \leqq \|T\|_{p,q}$.

Hence the correspondence between $M(L_p(G), L_q(G))$ and $P(G)$ is a surjective isometry. ▯

Corollary 5.2.3. *Let G be a compact Abelian group. If $2\leqq p_1, p_2<\infty$ and $1<q_1, q_2\leqq 2$ then there is an isometric linear isomorphism of $M(L_{p_1}(G), L_{q_1}(G))$ onto $M(L_{p_2}(G), L_{q_2}(G))$.*

In general for compact G it should be noted that $M(L_p(G), L_q(G))$ always contains $L_q(G)$. Indeed, if $g\in L_q(G)$ then for each $f\in L_p(G)\subset L_1(G)$ we have $g*f\in L_q(G)$ and

$$\|g*f\|_q \leqq \|g\|_q\,\|f\|_1 \leqq \|g\|_q\,\|f\|_p.$$

Thus $Tf=g*f$ defines a bounded linear transformation from $L_p(G)$ to $L_q(G)$ which commutes with translation.

This is in marked contrast with the situation for noncompact groups. For such groups if $1<q<p<\infty$ then $M(L_p(G), L_q(G))=\{0\}$. In this case the conclusions of Theorem 5.2.2 and Corollary 5.2.1 are only trivially valid.

Theorem 5.2.5. *Let G be a noncompact locally compact Abelian group. If $1<q<p<\infty$ then $M(L_p(G), L_q(G))=\{0\}$.*

Proof. Suppose $T\in M(L_p(G), L_q(G))$ where $1<q<p<\infty$ and $T\neq 0$. Then for each $f\in L_p(G)$ and $s\in G$ we have

$$\|Tf+\tau_s Tf\|_q = \|T(f+\tau_s f)\|_q \leqq \|T\|_{p,q}\,\|f+\tau_s f\|_p.$$

Applying Lemma 3.5.1 (i) we conclude that

$$\|Tf\|_q \leqq 2^{1/p-1/q}\,\|T\|_{p,q}\,\|f\|_p$$

for each $f\in L_p(G)$. But this is a contradiction since $p>q$ implies that $2^{1/p-1/q}\,\|T\|_{p,q} < \|T\|_{p,q}$.

Therefore $M(L_p(G), L_q(G))=\{0\}$. ▯

Thus for noncompact groups only the spaces $M(L_p(G), L_q(G))$, $1<p\leqq q$, are of any interest. We have already discussed the case of $p=q$ in the preceding chapter. Among other things the results of the next section show that $M(L_p(G), L_q(G))\neq\{0\}$ for $1<p<q$.

5.3. Inclusion Results. We now wish to discuss some results concerning the relationship between the spaces $M(L_p(G), L_q(G))$ for various values of p and q. Our first results are simple applications of the Riesz-Thorin Convexity Theorem (Theorem D.11.2) and Theorem 5.2.1.

Theorem 5.3.1. *Let G be a locally compact Abelian group and $1 \leqq p_i$, $q_i \leqq \infty$, $i=1,2$. If $T \in M(L_{p_i}(G), L_{q_i}(G))$, $i=1,2$, then T defines a unique element of $M(L_p(G), L_q(G))$ for each pair (p, q) for which there exists an α, $0<\alpha<1$, such that $1/p=\alpha/p_1+(1-\alpha)/p_2$, $1/q=\alpha/q_1+(1-\alpha)/q_2$.*

Proof. Since the space $\mathscr{S}(G)$ of integrable simple functions is contained in $L_{p_i}(G)$, $i=1,2$, we see that T restricted to $\mathscr{S}(G)$ defines a linear transformation which commutes with translation from this space into the space of all measurable functions on G. Hence we may apply the Riesz-Thorin Convexity Theorem (Theorem D.11.2) to conclude that $\log \|T\|_{1/a, 1/b}$ is a convex function on $0 \leqq a, b \leqq 1$, where $\|T\|_{1/a, 1/b}$ denotes the norm of the extension of T restricted to $\mathscr{S}(G)$ to a continuous linear transformation from $L_{1/a}(G)$ to $L_{1/b}(G)$ if such an extension exists, and $\|T\|_{1/a, 1/b}=\infty$ otherwise. In particular, one has

$$\|T\|_{p, q} \leqq (\|T\|_{p_1, q_1})^{\alpha} (\|T\|_{p_2, q_2})^{1-\alpha},$$

from which it follows that T defines an element of $M(L_p(G), L_q(G))$. The multiplier so defined is unique as $\mathscr{S}(G)$ is norm dense in $L_p(G)$. □

Theorem 5.3.2. *Let G be a locally compact Abelian group and $1 \leqq p < \infty$, $1 < q \leqq \infty$. If $1 \leqq r$, $s \leqq \infty$ is such that the point $(1/r, 1/s)$ in the rectangle $0 \leqq a$, $b \leqq 1$ is symmetric with respect to the line $a+b=1$ to the point $(1/p, 1/q)$ then there exists an isometric linear isomorphism of $M(L_p(G), L_q(G))$ onto $M(L_r(G), L_s(G))$.*

Proof. Since the points $(1/p, 1/q)$ and $(1/r, 1/s)$ are symmetric with respect to $a+b=1$ we see that $1/p+1/s=1$ and $1/q+1/r=1$. That is, $1/s=1/p'$ and $1/r=1/q'$. The conclusion then follows immediately from Theorem 5.2.1. □

Before we can continue the development of this section we need to prove a rather technical lemma. The lemma and some of its immediate consequences will be used in both this and the succeeding section.

Given an open set $U \subset \hat{G}$ with compact closure we denote by $C_c^U(\hat{G})$ the subspace of $C_c(\hat{G})$ consisting of those functions whose support is contained in U.

Lemma 5.3.1. *Let G be a noncompact locally compact Abelian group and let $U \subset \hat{G}$ be an open set with compact closure. If $\varphi_0 \in C_c^U(\hat{G})$ is not* *identically zero and* $\hat{\varphi}_0 \in L_1(G)$ *then given* $\delta, 0<\delta<1$, *there exists a sequence* $\{\varphi_n\} \subset C_c^U(\hat{G})$ *such that for each nonnegative integer* n:

(i) $\hat{\varphi}_n \in L_1(G)$.
(ii) $\|\varphi_n\|_\infty \leqq 2^{(n+1)/2} \|\varphi_0\|_\infty$.
(iii) $\|\hat{\varphi}_n\|_\infty \leqq (1+\delta)^n \|\hat{\varphi}_0\|_\infty$.
(iv) $\|\hat{\varphi}_n\|_2 \geqq (2-\delta)^{n/2} \|\varphi_0\|_2$.

Proof. We shall actually define two sequences of functions which satisfy the conclusions of the lemma. Set $\psi_0 = \varphi_0$. And for each positive integer n we define

$$\begin{aligned}\varphi_n(\gamma) &= \varphi_{n-1}(\gamma) + (t_{n-1}, \gamma)\, \psi_{n-1}(\gamma)\\ \psi_n(\gamma) &= \varphi_{n-1}(\gamma) - (t_{n-1}, \gamma)\, \psi_{n-1}(\gamma) \qquad (\gamma \in \hat{G}),\end{aligned}$$

where $t_{n-1} \in G$ is so chosen that

$$\begin{aligned}\|\hat{\varphi}_n\|_\infty &\leqq (1+\delta) \max\{\|\hat{\varphi}_{n-1}\|_\infty, \|\hat{\psi}_{n-1}\|_\infty\}\\ \|\hat{\psi}_n\|_\infty &\leqq (1+\delta) \max\{\|\hat{\varphi}_{n-1}\|_\infty, \|\hat{\psi}_{n-1}\|_\infty\},\end{aligned}$$

and

$$\begin{aligned}(\|\hat{\varphi}_n\|_2)^2 &\geqq (2-\delta) \min\{(\|\hat{\varphi}_{n-1}\|_2)^2, (\|\hat{\psi}_{n-1}\|_2)^2\}\\ (\|\hat{\psi}_n\|_2)^2 &\geqq (2-\delta) \min\{(\|\hat{\varphi}_{n-1}\|_2)^2, (\|\hat{\psi}_{n-1}\|_2)^2\}.\end{aligned}$$

To see how to choose such a $t_{n-1} \in G$, we note first that from the recursive nature of the definition given any $t_{n-1} \in G$ we would have $\varphi_n, \psi_n \in C_c^U(\hat{G})$ and $\hat{\varphi}_{n-1}, \hat{\psi}_{n-1} \in L_1(G) \cap L_2(G)$. Let

$$0 < \varepsilon_{n-1} = \min \begin{cases} \delta \max\{\|\hat{\varphi}_{n-1}\|_\infty, \|\hat{\psi}_{n-1}\|_\infty\} \\ \dfrac{\delta \min\{\|\hat{\varphi}_{n-1}\|_2)^2, (\|\hat{\psi}_{n-1}\|_2)^2\}}{2(\|\hat{\varphi}_{n-1}\|_1 + \|\hat{\psi}_{n-1}\|_1)} \end{cases}.$$

Then since $\hat{\varphi}_{n-1}, \hat{\psi}_{n-1} \in C_0(G)$ there exists a compact set $K_{n-1} \subset G$ such that if $t \notin K_{n-1}$ then

$$\begin{aligned}|\hat{\varphi}_{n-1}(t)| &\leqq \varepsilon_{n-1}\\ |\hat{\psi}_{n-1}(t)| &\leqq \varepsilon_{n-1}.\end{aligned}$$

Let $t_{n-1} \in G \sim K_{n-1} K_{n-1}^{-1}$. Such a choice of t_{n-1} is possible as G is not compact. Since $K_{n-1} K_{n-1}^{-1}$ is a symmetric set, it follows easily that if $t \in K_{n-1}$ then $t\, t_{n-1}^{-1} \notin K_{n-1}$.

Having chosen t_{n-1} in this manner we see that for $t \in K_{n-1}$,

$$\begin{aligned}|\hat{\varphi}_n(t)| = |\hat{\varphi}_{n-1}(t) + \hat{\psi}_{n-1}(t\, t_{n-1}^{-1})| &\leqq |\hat{\varphi}_{n-1}(t)| + \varepsilon_{n-1}\\ &\leqq \|\hat{\varphi}_{n-1}\|_\infty + \delta \max\{\|\hat{\varphi}_{n-1}\|_\infty, \|\hat{\psi}_{n-1}\|_\infty\}\\ &\leqq (1+\delta) \max\{\|\hat{\varphi}_{n-1}\|_\infty, \|\hat{\psi}_{n-1}\|_\infty\}.\end{aligned}$$

And if $t \notin K_{n-1}$, then

$$\begin{aligned}|\hat{\varphi}_n(t)| &\leqq \varepsilon_{n-1} + |\hat{\psi}_{n-1}(t\,t_{n-1}^{-1})| \\ &\leqq \delta \max\{\|\hat{\varphi}_{n-1}\|_\infty, \|\hat{\psi}_{n-1}\|_\infty\} + \|\hat{\psi}_{n-1}\|_\infty \\ &\leqq (1+\delta) \max\{\|\hat{\varphi}_{n-1}\|_\infty, \|\hat{\psi}_{n-1}\|_\infty\}.\end{aligned}$$

Hence we have

$$\|\hat{\varphi}_n\|_\infty \leqq (1+\delta) \max\{\|\hat{\varphi}_{n-1}\|_\infty, \|\hat{\psi}_{n-1}\|_\infty\}.$$

Essentially the same argument establishes the analogous estimate for $\|\hat{\psi}_n\|_\infty$.

On the other hand, if $t \in K_{n-1}$ we have

$$\begin{aligned}|\hat{\varphi}_n(t)|^2 &= |\hat{\varphi}_{n-1}(t) + \hat{\psi}_{n-1}(t\,t_{n-1}^{-1})|^2 \\ &\geqq \big||\hat{\varphi}_{n-1}(t)| - |\hat{\psi}_{n-1}(t\,t_{n-1}^{-1})|\big|^2 \\ &= |\hat{\varphi}_{n-1}(t)|^2 + |\hat{\psi}_{n-1}(t\,t_{n-1}^{-1})|^2 - 2|\hat{\varphi}_{n-1}(t)|\,|\hat{\psi}_{n-1}(t\,t_{n-1}^{-1})| \\ &\geqq |\hat{\varphi}_{n-1}(t)|^2 + |\hat{\psi}_{n-1}(t\,t_{n-1}^{-1})|^2 - 2\varepsilon_{n-1}|\hat{\varphi}_{n-1}(t)|.\end{aligned}$$

While for $t \notin K_{n-1}$ we have

$$|\hat{\varphi}_n(t)|^2 \geqq |\hat{\varphi}_{n-1}(t)|^2 + |\hat{\psi}_{n-1}(t\,t_{n-1}^{-1})|^2 - 2\varepsilon_{n-1}|\hat{\psi}_{n-1}(t\,t_{n-1}^{-1})|.$$

Consequently,

$$\begin{aligned}(\|\hat{\varphi}_n\|_2)^2 &= \int_{K_{n-1}} |\hat{\varphi}_n(t)|^2\, d\lambda(t) + \int_{G\sim K_{n-1}} |\hat{\varphi}_n(t)|^2\, d\lambda(t) \\ &\geqq \int_{K_{n-1}} (|\hat{\varphi}_{n-1}(t)|^2 + |\hat{\psi}_{n-1}(t\,t_{n-1}^{-1})|^2\, d\lambda(t) \\ &\quad + \int_{G\sim K_{n-1}} (|\hat{\varphi}_{n-1}(t)|^2 + |\hat{\psi}_{n-1}(t\,t_{n-1}^{-1})|^2)\, d\lambda(t) \\ &\quad - 2\varepsilon_{n-1} \int_{K_{n-1}} |\hat{\varphi}_{n-1}(t)|\, d\lambda(t) \\ &\quad - 2\varepsilon_{n-1} \int_{G\sim K_{n-1}} |\hat{\psi}_{n-1}(t\,t_{n-1}^{-1})|\, d\lambda(t) \\ &\geqq (\|\hat{\varphi}_{n-1}\|_2)^2 + (\|\hat{\psi}_{n-1}\|_2)^2 \\ &\quad - \frac{\delta \min\{(\|\hat{\varphi}_{n-1}\|_2)^2, (\|\hat{\psi}_{n-1}\|_2)^2\}}{\|\hat{\varphi}_{n-1}\|_1 + \|\hat{\psi}_{n-1}\|_1}\, \|\hat{\varphi}_{n-1}\|_1 + \|\hat{\psi}_{n-1}\|_1 \\ &> (2-\delta) \min\{(\|\hat{\varphi}_{n-1}\|_2), (\|\hat{\psi}_{n-1}\|_2)^2\}.\end{aligned}$$

Once again similar reasoning proves the analogous estimate for $\|\hat{\psi}_n\|_2$.

Now, given the two sequences $\{\varphi_n\}$ and $\{\psi_n\}$ in $C_c^U(\hat{G})$ constructed as above, we see by means of an elementary computation that for each n

$$|\varphi_n(\gamma)|^2+|\psi_n(\gamma)|^2=2\big(|\varphi_{n-1}(\gamma)|^2+|\psi_{n-1}(\gamma)|^2\big) \qquad (\gamma\in\hat{G}).$$

This identity immediately yields the fact that for each n

$$|\varphi_n(\gamma)|^2+|\psi_n(\gamma)|^2=2^{n+1}\,|\varphi_0(\gamma)|^2 \qquad (\gamma\in\hat{G})$$

from which we deduce at once that for each n

$$\|\varphi_n\|_\infty \leqq 2^{(n+1)/2}\,\|\varphi_0\|_\infty$$
$$\|\psi_n\|_\infty \leqq 2^{(n+1)/2}\,\|\varphi_0\|_\infty .$$

Clearly $\|\hat{\varphi}_0\|_\infty \leqq (1+\delta)^0\,\|\hat{\varphi}_0\|_\infty$ and $\|\hat{\psi}_0\|_\infty \leqq (1+\delta)^0\,\|\hat{\varphi}_0\|_\infty$. Suppose then that $\|\hat{\varphi}_{n-1}\|_\infty \leqq (1+\delta)^{n-1}\,\|\hat{\varphi}_0\|_\infty$ and $\|\hat{\psi}_{n-1}\|_\infty \leqq (1+\delta)^{n-1}\,\|\hat{\varphi}_0\|_\infty$. Then by the definition of φ_n we have

$$\|\hat{\varphi}_n\|_\infty \leqq (1+\delta)\max\{\|\hat{\varphi}_{n-1}\|_\infty,\|\hat{\psi}_{n-1}\|_\infty\}$$
$$\leqq (1+\delta)^n\,\|\hat{\varphi}_0\|_\infty .$$

Similarly we obtain

$$\|\hat{\psi}_n\|_\infty \leqq (1+\delta)^n\,\|\hat{\varphi}_0\|_\infty .$$

Thus by induction we conclude that for all n we have

$$\|\hat{\varphi}_n\|_\infty \leqq (1+\delta)^n\,\|\hat{\varphi}_0\|_\infty$$
$$|\hat{\psi}_n\|_\infty \leqq (1+\delta)^n\,\|\hat{\varphi}_0\|_\infty .$$

An analogous induction argument establishes the validity of (iv) for both the sequences $\{\varphi_n\}$ and $\{\psi_n\}$.

Therefore both of the sequences $\{\varphi_n\}$ and $\{\psi_n\}$ satisfy (i)–(iv), and the lemma is proved. □

If G is an infinite discrete group then an examination of the proof of the preceding lemma reveals that the following corollary is valid.

Corollary 5.3.1. *Let G be an infinite discrete Abelian group. Then there exists a sequence of trigonometric polynomials $\{\varphi_n\}\subset C(\hat{G})$ such that for each n:*

(i) $\hat{\varphi}_n\in L_1(G)$.

(ii) $\hat{\varphi}_n$ *takes only the values* 0 *and* ± 1 *and has precisely* 2^n *distinct points of support.*

(iii) $\|\varphi_n\|_\infty \leqq 2^{(n+1)/2}$.

(iv) $\|\hat{\varphi}_n\|_\infty = 1$.

(v) $\|\hat{\varphi}_n\|_p = 2^{n/p}$, $1\leqq p<\infty$.

Proof. Apply the construction of Lemma 5.3.1 to $\varphi_0 \equiv 1$. $\square$

In the previous section we noted for compact groups G that $M(L_p(G), L_q(G))$ contained $L_q(G)$ in the sense that $L_q(G)$ could be continuously embedded in $M(L_p(G), L_q(G))$. Again, as in 4.5, we shall slightly abuse the terminology by writing $A \subset B$ whenever there is a continuous linear isomorphism of the space A into B. We now wish to establish a result for arbitrary infinite groups which is similar to the one for compact groups just mentioned.

Theorem 5.3.3. *Let G be an infinite locally compact Abelian group and suppose $1 < p < q < \infty$. If $r > 1$ is such that $1/p - 1/q = 1 - 1/r$ then $L_r(G) \subset M(L_p(G), L_q(G))$ and the inclusion is proper.*

Proof. From Theorems 3.1.1 and 3.3.1 we know that $L_r(G)$ can be identified with $M(L_1(G), L_r(G))$ and with $M(L_{r'}(G), L_\infty(G))$, $1/r + 1/r' = 1$, by the mapping $f \to T_f$ where $T_f g = f * g$. And that $\|T_f\|_{1,r} = \|T_f\|_{r',\infty} = \|f\|_r$. Moreover, it is easy to verify that, since $1/p - 1/q = 1 - 1/r$, we have $1/p = \alpha/r' + (1-\alpha)$ and $1/q = (1-\alpha)/r$ where $0 < \alpha = 1 - r/q < 1$. Thus by the Riesz-Thorin Convexity Theorem (Theorem D.11.2) we see that T_f is a linear transformation from $L_p(G)$ to $L_q(G)$ which commutes with translation and

$$\|T_f\|_{p,q} \leqq (\|T_f\|_{r',\infty})^\alpha (\|T_f\|_{1,r})^{1-\alpha} = \|f\|_r.$$

Therefore $f \to T_f$ defines a continuous linear isomorphism from $L_r(G)$ into $M(L_p(G), L_q(G))$, that is $L_r(G) \subset M(L_p(G), L_q(G))$.

To show the inclusion is proper we first note that if $L_r(G) = M(L_p(G), L_q(G))$ as Banach spaces, then, since $\|T_f\|_{p,q} \leqq \|f\|_r$, $f \in L_r(G)$, we can deduce from the Two Norm Theorem (Theorem D.6.3) that there exists a constant $B > 0$ such that $\|f\|_r \leqq B\|T\|_{p,q}$, $f \in L_r(G)$. Moreover, we claim that we may, without loss of generality, assume that the point $(1/p, 1/q)$ lies in the triangle bounded by $a = 1$, $a + b = 1$ and $a = b$. Indeed, since $1 < p < q < \infty$, it is evident that $(1/p, 1/q)$ lies in the triangle bounded by $a = 1, b = 0, a = b$. It then follows immediately from Theorem 5.3.2 that we may also assume that $(1/p, 1/q)$ lies in the triangle bounded by $a = 1$, $a + b = 1$ and $a = b$. Consider the straight line through the points $(1/2, 1/2)$ and $(1/p, 1/q)$. Denote its intersection with $a = 1$ by $(1, 1/s)$. Let α, $0 < \alpha < 1$ be such that

$$\frac{1}{p} = \frac{\alpha}{2} + (1-\alpha)$$

$$\frac{1}{q} = \frac{\alpha}{2} + \frac{(1-\alpha)}{s}.$$

Now if $f \in C_c(G)$ then $f \in L_2(G) \cap L_s(G)$ and so by Theorems 4.1.1 and 3.1.1 we have $T_f \in M(L_2(G))$ and $T_f \in M(L_1(G), L_s(G))$. Hence by Theorem

5.3.1 we see that

$$\|f\|_r \leqq B\|T_f\|_{p,q} \leqq B(\|T_f\|_2)^\alpha (\|T_f\|_{1,s})^{1-\alpha}$$
$$= B(\|\hat{f}\|_\infty)^\alpha (\|f\|_s)^{1-\alpha} \qquad (f \in C_c(G)).$$

If G is discrete then by Corollary 5.3.1 there exists a sequence of trigonometric polynomials $\{\varphi_n\} \subset C(\hat{G})$ such that $\{\hat{\varphi}_n\} \subset C_c(G)$, $\|\varphi_n\|_\infty \leqq 2^{(n+1)/2}$, $\|\hat{\varphi}_n\|_r = 2^{n/r}$ and $\|\hat{\varphi}_n\|_s = 2^{n/s}$. Applying the previous estimate to these polynomials we see for each positive integer n that

$$2^{n/r} \leqq B\, 2^{\alpha(n+1)/2}\, 2^{(1-\alpha)n/s},$$

and hence,

$$2^{n(1/r - 1/q)} \leqq B\, 2^{\alpha/2}.$$

But such an inequality cannot hold for all n since $1/p - 1/q = 1 - 1/r$ implies that $1/r - 1/q = 1 - 1/p > 0$.

Thus when G is discrete we see that $L_r(G)$ is properly contained in $M(L_p(G), L_q(G))$.

Now assume that G is nondiscrete and let $U \subset G$ be an open set with compact closure. Since $\hat{G}$ is noncompact there exists a sequence $\{\varphi_n\} \subset C_c^U(G)$ such that

(i) $\|\varphi_n\|_\infty \leqq 2^{(n+1)/2}$,
(ii) $\|\hat{\varphi}_n\|_\infty \leqq C(1+1/n)^n$,
(iii) $\|\hat{\varphi}_n\|_2 \geqq D(2-1/n)^{n/2}$,

where C and D are constants independent of n. We construct such a sequence by applying Lemma 5.3.1 repeatedly with any φ_0 such that $\|\varphi_0\|_\infty \leqq 1$ and $\delta_n = 1/n$. Define $\psi_n = \varphi_n / 2^{(n+1)/2}$. Then $\{\psi_n\} \subset C_c^U(G)$ and

(i) $\|\psi_n\|_\infty \leqq 1$,
(ii) $\|\hat{\psi}_n\|_\infty \leqq C(1+1/n)^n / 2^{(n+1)/2}$,
(iii) $\|\hat{\psi}_n\|_2 \geqq D(2-1/n)^{n/2} / 2^{(n+1)/2}$.

It is immediately apparent that $\lim_n \|\hat{\psi}_n\|_\infty = 0$ and that $\liminf_n \|\psi_n\|_2 = \liminf_n \|\hat{\psi}_n\|_2 \neq 0$.

Further we claim that $\liminf_n \|\psi_n\|_r \neq 0$, $1 \leqq r \leqq \infty$. As if $1 \leqq r < 2$ then

$$\int_G |\psi_n(t)|^2\, d\lambda(t) = \int_G |\psi_n(t)|^{2-r} |\psi_n(t)|^r\, d\lambda(t)$$
$$\leqq (\|\psi_n\|_\infty)^{2-r} \int_G |\psi_n(t)|^r\, d\lambda(t)$$
$$\leqq \int_G |\psi_n(t)|^r\, d\lambda(t),$$

from which it follows that $\liminf_n \|\psi_n\|_r \neq 0$ as $\liminf_n \|\psi_n\|_2 \neq 0$.

While if $2 \leqq r \leqq \infty$ then

$$\int_G |\psi_n(t)|^2 \, d\lambda(t) = \int_G |\psi_n(t)|^2 \chi_U(t) \, d\lambda(t)$$

$$\leqq \Big(\int_G |\psi_n(t)|^r \, d\lambda(t)\Big)^{2/r} \Big(\int_G \chi_U(t) \, d\lambda(t)\Big)^{1-2/r}$$

$$= (\|\psi_n\|_r)^2 (\lambda(U))^{1-2/r}$$

from which we again deduce that $\liminf_n \|\psi_n\|_r \neq 0$.

However, for each positive integer n, we also have that

$$\|\psi_n\|_r \leqq B(\|\hat{\psi}_n\|_\infty)^\alpha (\|\psi_n\|_s)^{1-\alpha}$$

where r is such that $1/p - 1/q = 1 - 1/r$. But $\|\psi_n\|_s$ is bounded in n as

$$\|\psi_n\|_s = \Big(\int_G |\psi_n(t)|^s \, d\lambda(t)\Big)^{1/s} \leqq \|\psi_n\|_\infty \Big(\int_G \chi_U(t) \, d\lambda(t)\Big)^{1/s} \leqq \lambda(U)^{1/s}.$$

Thus, since $\lim_n \|\hat{\psi}_n\|_\infty = 0$, we conclude that $\liminf_n \|\psi_n\|_r = 0$, $1/p - 1/q = 1 - 1/r$, contradicting the fact that $\liminf_n \|\psi_n\|_r \neq 0$, $1 \leqq r \leqq \infty$.

Therefore we also have that $L_r(G)$ is properly included in $M(L_p(G), L_q(G))$ when G is nondiscrete, and the proof is complete. □

Next we wish to examine the relationship between the spaces $M(L_p(G), L_q(G))$ for various choices of p and q such that $1/p - 1/q = 1 - 1/r$ for a fixed r. The results we shall prove are valid for any infinite locally compact Abelian group. However, we shall only give the proof for the case of noncompact groups. The proof for infinite compact groups requires somewhat different technical machinery than we have developed in this section, and for this reason we choose to omit the proof.

In order to reduce the length of the proof of Theorem 5.3.4 we shall first establish a lemma.

Lemma 5.3.2. *Let G be a noncompact locally compact Abelian group and let $\varphi_0, f \in C_c(G)$. Then there exists a sequence $\{\varphi_n\} \subset C_c(G)$ such that for each nonnegative integer n:*

(i) $\|\varphi_n\|_\infty = \|\varphi_0\|_\infty$.
(ii) $\|\hat{\varphi}_n\|_\infty \leqq 2^{(n+1)/2} \|\hat{\varphi}_0\|_\infty$.
(iii) $\|\varphi_n * f\|_p = 2^{n/p} \|\varphi_0 * f\|_p$, $1 \leqq p \leqq \infty$.

Proof. The proof is, in many respects, similar to that of Lemma 5.3.1 so we shall not give as many details as previously. We shall construct two sequences $\{\varphi_n\}$ and $\{\psi_n\}$ in $C_c(G)$ by defining $\psi_0 = \varphi_0$ and

$$\varphi_n = \varphi_{n-1} + \tau_{t_{n-1}} \psi_{n-1}$$

$$\psi_n = \varphi_{n-1} - \tau_{t_{n-1}} \psi_{n-1},$$

where $t_{n-1} \in G$ is chosen in such a way that φ_{n-1} and $\tau_{t_{n-1}} \psi_{n-1}$ have disjoint supports, and that $\varphi_{n-1} * f$ and $\tau_{t_{n-1}}(\psi_{n-1} * f)$ have disjoint supports. To see that this can always be done we denote the compact supports of $\varphi_{n-1}, \psi_{n-1}$ and f as K_{n-1}, L_{n-1} and K respectively. Then the compact supports of $\varphi_{n-1} * f$ and $\psi_{n-1} * f$ are contained in $K_{n-1} K$ and $L_{n-1} K$, respectively. Choose $t_{n-1} \in G \sim K_{n-1} L_{n-1}^{-1} \cup K_{n-1} K L_{n-1}^{-1} K^{-1}$. Such a t_{n-1} exists since G is noncompact. It is easily seen with such a choice of t_{n-1} that the supports of $\varphi_{n-1}, \tau_{t_{n-1}} \psi_{n-1}, \varphi_{n-1} * f$ and $\tau_{t_{n-1}}(\psi_{n-1} * f)$ have the desired properties.

Since the supports of φ_{n-1} and $\tau_{t_{n-1}} \psi_{n-1}$ are disjoint, it is apparent that $\|\varphi_n\|_\infty = \|\psi_n\|_\infty = \|\varphi_0\|_\infty$. From the equations

$$\begin{aligned} \hat{\varphi}_n(\gamma) &= \hat{\varphi}_{n-1}(\gamma) + (t_{n-1}^{-1}, \gamma)\, \hat{\psi}_{n-1}(\gamma) \\ \hat{\psi}_n(\gamma) &= \hat{\varphi}_{n-1}(\gamma) - (t_{n-1}^{-1}, \gamma)\, \hat{\psi}_{n-1}(\gamma) \qquad (\gamma \in \hat{G}), \end{aligned}$$

we deduce *mutatis mutandis* as in the proof of Lemma 5.3.1 that

$$\begin{aligned} \|\hat{\varphi}_n\|_\infty &\leqq 2^{(n+1)/2} \|\hat{\varphi}_0\|_\infty \\ \|\hat{\psi}_n\|_\infty &\leqq 2^{(n+1)/2} \|\hat{\varphi}_0\|_\infty . \end{aligned}$$

Finally, given $1 \leqq p < \infty$ if (iii) holds for $\|\varphi_{n-1} * f\|_p$ and $\|\psi_{n-1} * f\|_p$ then, since the supports of $\varphi_{n-1} * f$ and $\tau_{t_{n-1}}(\psi_{n-1} * f)$ are disjoint, we have

$$\begin{aligned} \|\varphi_n * f\|_p &= \|\varphi_{n-1} * f + \tau_{t_{n-1}}(\psi_{n-1} * f)\|_p \\ &= [(\|\varphi_{n-1} * f\|_p)^p + (\|\psi_{n-1} * f\|_p)^p]^{1/p} \\ &\leqq [2^{n-1}(\|\varphi_0 * f\|_p)^p + 2^{n-1}(\|\varphi_0 * f\|_p)^p]^{1/p} \\ &= 2^{n/p} \|\varphi_0 * f\|_p . \end{aligned}$$

A similar estimate is clearly valid for $\|\psi_n * f\|_p$.

The validity of the estimate for $p = \infty$ is apparent. □

Theorem 5.3.4. *Let G be a noncompact locally compact Abelian group, $1 < p_0 < 2 < q_0 = p_0' < \infty$ and set $1/p_0 - 1/q_0 = 1 - 1/r$. If $1 < p_2 < p_1 < p_0$ and q_i are such that $1/p_i - 1/q_i = 1 - 1/r$, $i = 1, 2$, then $M(L_{p_2}(G), L_{q_2}(G)) \subset M(L_{p_1}(G), L_{q_1}(G)) \subset M(L_{p_0}(G), L_{q_0}(G))$ and the inclusions are proper.*

Proof. First we note that $0 < 1/q_0 < 1/q_1 < 1/q_2 < \frac{1}{2} < 1/p_0 < 1/p_1 < 1/p_2 < 1$ and $1/p_0 + 1/q_0 = 1/p_0 + 1/p_0' = 1$. Moreover $1/q_i = 1/p_i - (1 - 1/r) = 1/p_i - 1/r'$, $i = 1, 2, 3$. Thus the points $(1/p_i, 1/q_i)$, $i = 1, 2, 3$, lie in the square bounded by $a = 1$, $b = 0$, $a = \frac{1}{2}$, $b = \frac{1}{2}$, and they lie on a straight line with slope one. Clearly $(1/p_0, 1/q_0)$ lies on the line $a + b = 1$, and by Theorem 5.3.2 we may assume, without loss of generality, that $(1/p_i, 1/q_i)$, $i = 1, 2$, lie in the triangle bounded by $a = 1$, $b = \frac{1}{2}$ and $a + b = 1$. Furthermore the points $(1/q_i', 1/p_i')$, $i = 1, 2$, lie on the same straight line as

$(1/p_i, 1/q_i)$, $i=1,2$, by Theorem 5.3.2. Hence we deduce from the Riesz-Thorin Convexity Theorem (Theorem D.11.2) and Theorem 5.2.1 that for $i=1,2$

$$\|T\|_{p_0,q_0} \leqq (\|T\|_{p_i,q_i})^\alpha (\|T\|_{q_i',p_i'})^{1-\alpha} = \|T\|_{p_i,q_i} \qquad (T \in M(L_{p_i}(G), L_{q_i}(G)),$$

and

$$\|T\|_{p_1,q_1} \leqq (\|T\|_{p_2,q_2})^\alpha (\|T\|_{p_2',q_2'})^{1-\alpha} = \|T\|_{p_q,q_2} \qquad (T \in M(L_{p_2}(G), L_{q_2}(G)),$$

where α, $0<\alpha<1$, is suitably chosen in each case.

Therefore $M(L_{p_2}(G), L_{q_2}(G)) \subset M(L_{p_1}(G), L_{q_1}(G)) \subset M(L_{p_0}(G), L_{q_0}(G))$.

Suppose that $M(L_{p_1}(G), L_{q_1}(G)) = M(L_{p_0}(G), L_{q_0}(G))$. Then the Two Norm Theorem (Theorem D.6.3) yields the existence of a constant $B>0$ such that

$$\|T\|_{p_1,q_1} \leqq B\,\|T\|_{p_0,q_0} \qquad (T \in M(L_{p_1}(G), L_{q_1}(G)).$$

In particular, if $\varphi \in C_c(G)$ then $\varphi \in L_r(G)$ and $T_\varphi \in M(L_{p_1}(G), L_{q_1}(G))$ by Theorem 5.3.3. As before $T_\varphi f = \varphi * f$, $f \in L_{p_1}(G)$. Thus we have

$$\|T_\varphi\|_{p_1,q_1} \leqq B\,\|T_\varphi\|_{p_0,q_0} \qquad (\varphi \in C_c(G)).$$

Furthermore by Theorems 4.1.1 and 3.1.1 it is evident that $T_\varphi \in M(L_2(G))$ and $T_\varphi \in M(L_1(G), L_\infty(G))$, $\varphi \in C_c(G)$. Since $1/p_0 + 1/q_0 = 1$ there exists an α, $0<\alpha<1$, for which $1/p_0 = \alpha/2 + (1-\alpha)$ and $1/q_0 = \alpha/2$. Another application of the Riesz-Thorin Convexity Theorem then shows that

$$\begin{aligned}\|T_\varphi\|_{p_1,q_1} &\leqq B\,\|T_\varphi\|_{p_0,q_0} \leqq B(\|T_\varphi\|_2)^\alpha (\|T_\varphi\|_{1,\infty})^{1-\alpha} \\ &= B(\|\hat\varphi\|_\infty)^\alpha (\|\varphi\|_\infty)^{1-\alpha} \qquad (\varphi \in C_c(G)).\end{aligned}$$

Consequently for each φ, $f \in C_c(G)$ we have

$$\|\varphi * f\|_{q_1} = \|T_\varphi f\|_{q_1} \leqq \|T_\varphi\|_{p_1,q_1} \|f\|_{p_1} \leqq B(\|\hat\varphi\|_\infty)^\alpha (\|\varphi\|_\infty)^{1-\alpha} \|f\|_{p_1}.$$

For given φ_0, $f \in C_c(G)$, let $\{\varphi_n\} \subset C_c(G)$ be a sequence which satisfies the conclusions of Lemma 5.3.2. Then we deduce from the preceding estimate that for each positive integer n

$$\begin{aligned}2^{n/q_1} \|\varphi_0 * f\|_{q_1} = \|\varphi_n * f\|_{q_1} &\leqq B(\|\hat\varphi_n\|_\infty)^\alpha (\|\varphi_n\|_\infty)^{1-\alpha} \|f\|_{p_1} \\ &\leqq B\,2^{\alpha(n+1)/2} (\|\hat\varphi_0\|_\infty)^\alpha (\|\varphi_0\|_\infty)^{1-\alpha} \|f\|_{p_1}.\end{aligned}$$

Hence

$$2^{n(1/q_1 - \alpha/2)} \|\varphi_0 * f\|_{q_1} \leqq B\,2^{\alpha/2} (\|\hat\varphi_0\|_\infty)^\alpha (\|\varphi_0\|_\infty)^{1-\alpha} \|f\|_{p_1}$$

for each positive integer n. But this is impossible for all choices of φ_0, and $f \in C_c(G)$ since $1/q_1 - \alpha/2 = 1/q_1 - 1/q_0 > 0$.

Thus $M(L_{p_1}(G), L_{q_1}(G))$ is properly contained in $M(L_{p_0}(G), L_{q_0}(G))$.

Now suppose that $M(L_{p_2}(G), L_{q_2}(G))=M(L_{p_1}(G), L_{q_1}(G))$. As before this implies the existence of some constant $B>0$ such that

$$\|T\|_{p_2,q_2}\leqq B\,\|T\|_{p_1,q_1}\quad (T\in M(L_{p_2}(G),\ L_{q_2}(G))).$$

However $(1/p_1, 1/q_1)$ lies between $(1/p_0, 1/q_0)$ and $(1/p_2, 1/q_2)$ on the straight line passing through these three points. Hence there exists an α, $0<\alpha<1$, for which $1/p_1=\alpha/p_0+(1-\alpha)/p_2$ and $1/q_1=\alpha/q_0+(1-\alpha)/q_2$. The Riesz-Thorin Convexity Theorem then yields that

$$\|T\|_{p_1,q_1}\leqq(\|T\|_{p_0,q_0})^{\alpha}(\|T\|_{p_2,q_2})^{1-\alpha}\quad (T\in M(L_{p_2}(G),\ L_{q_2}(G))).$$

From the argument of the preceding paragraphs we know that $M(L_{p_2}(G), L_{q_2}(G))$ is properly contained in $M(L_{p_0}(G), L_{q_0}(G))$, and so the topologies induced on $M(L_{p_2}(G), L_{q_2}(G))$ by the norms $\|\cdot\|_{p_2,q_2}$ and $\|\cdot\|_{p_0,q_0}$ can not be equivalent. Hence there exists a sequence $\{T_n\}\subset M(L_{p_2}(G), L_{q_2}(G))$ such that $\|T_n\|_{p_2,q_2}=1$ and $\lim_n \|T_n\|_{p_0,q_0}=0$. From the estimate

$$\|T_n\|_{p_1,q_1}\leqq(\|T_n\|_{p_0,q_p})^{\alpha}(\|T_n\|_{p_2,q_2})^{1-\alpha}$$

we conclude that $\lim_n \|T_n\|_{p_1,q_1}=0$. But this leads to a contradiction since

$$1=\|T_n\|_{p_2,q_2}\leqq B\,\|T_n\|_{p_1 q_1}.$$

Therefore $M(L_{p_2}(G), L_{q_2}(G))$ is properly contained in $M(L_{p_1}(G), L_{q_1}(G))$. □

We note that the hypothesis of noncompactness for G was used only to establish the properness of the inclusions. Thus the proof of the theorem proves the inclusion relations for arbitrary groups.

5.4. The Fourier Transform for $L_p(G)$, $1<p\leqq\infty$, and $M(L_p(G), L_q(G))$, $1\leqq p$, $q\leqq\infty$. As a tool in the discussion of some additional results concerning $M(L_p(G), L_q(G))$, we wish to discuss the concept of the Fourier transform for elements of $L_p(G)$ and for the elements of $M(L_p(G), L_q(G))$. If G is compact or $1<p\leqq 2$ there is, of course, no difficulty in discussing the Fourier transform for $L_p(G)$. However for noncompact G and $p>2$ the problem is more delicate. We shall restrict our attention mainly to noncompact groups. Consequently, in view of Theorem 5.2.5 the Fourier transform for $M(L_p(G), L_q(G))$ only needs to be defined when $1<p\leqq q$ since in the other cases $M(L_p(G), L_q(G))=\{0\}$.

First we shall define the Fourier transform for $L_p(G)$, $1<p\leqq\infty$. The Fourier transform of $f\in L_p(G)$ will be seen to be a quasimeasure in $Q(\hat{G})$. In particular, if $f\in L_p(G)$ then $\hat{f}$ is that quasimeasure for which

$$\langle h,\hat{f}\rangle=\langle\hat{h},f\rangle=\int_G \hat{h}(t)f(t^{-1})\,d\lambda(t)\qquad (h\in D(\hat{G}))$$

where $\hat{h}$ denotes the usual L_1-Fourier transform of h, that is, $\hat{h}(t) = \int_{\hat{G}} (t^{-1}, \gamma)\, h(\gamma)\, d\eta(\gamma)$. A little work is involved in showing that this definition makes sense.

By Theorem 5.1.2 the space $D(\hat{G})$ is contained in $A_c(\hat{G})$. Thus, if $h \in D(\hat{G})$ then there exists a $g \in L_1(G)$ such that $\check{g} = h$, where $\check{g}(\gamma) = \int_G (t, \gamma)\, g(t)\, d\lambda(t)$ is the usual inverse L_1-Fourier transform. Moreover, since $h \in C_c(\hat{G}) \subset L_1(\hat{G})$ the Fourier Inversion Theorem (Theorem F.8.1) applies to show that $\hat{h} = g \in L_1(G)$. Hence $\hat{h} \in L_1(G) \cap L_\infty(G) \subset L_r(G)$, $r > 1$, for each $h \in D(\hat{G})$. Consequently if $f \in L_p(G)$ then $\langle h, \hat{f} \rangle = \langle \hat{h}, f \rangle$ for each $h \in D(\hat{G})$ defines a linear functional on $D(\hat{G})$ as $\hat{h} \in L_{p'}(G)$, $1/p + 1/p' = 1$.

To see that this functional is continuous we note first that for each compact $K \subset \hat{G}$ the topology on $D_K(\hat{G})$ is stronger than the topology induced on $D_K(\hat{G})$ as a subspace of $L_1(\hat{G})$. Indeed, if $h = \sum_i f_i * g_i$ where $f_i, g_i \in C_c^K(\hat{G})$ and $\sum_i \|f_i\|_\infty \|g_i\|_\infty < \infty$ then

$$\|h\|_1 \leqq \sum_i \|f_i * g_i\|_1 \leqq \sum_i \|f_i\|_1 \|g_i\|_1 \leqq \lambda(K)^2 \sum_i \|f_i\|_\infty \|g_i\|_\infty .$$

Hence $\|h\|_1 \leqq \lambda(K)^2 \|h\|_K$. Now suppose $\{g_\alpha\} \subset D_K(\hat{G})$ and $\lim_\alpha \|g_\alpha\|_K = 0$. From the previous observation we see that $\lim_\alpha \|g_\alpha\|_1 = 0$, and hence $\lim_\alpha \|\hat{g}_\alpha\|_\infty = 0$. Furthermore, Theorem 5.1.2 (iv) says that the topology on $D_K(\hat{G})$ is stronger than that induced by $A(\hat{G})$ and so $\lim_\alpha \|g_\alpha\|_A = \lim_\alpha \|\hat{g}_\alpha\|_1 = 0$. Thus $\lim_\alpha \langle g_\alpha, \hat{f} \rangle = \lim_\alpha \langle \hat{g}_\alpha, f \rangle = 0$ since

$$|\langle \hat{g}_\alpha, f \rangle| \leqq \|\hat{g}_\alpha\|_{p'} \|f\|_p \leqq (\|\hat{g}_\alpha\|_\infty)^{1/p} (\|\hat{g}_\alpha\|_1)^{1/p'} \|f\|_p .$$

Consequently $\hat{f}$ defines a continuous linear functional on each $D_K(\hat{G})$ and hence on $D(\hat{G})$. Therefore $\hat{f}$ is a quasimeasure.

Thus we have defined a transformation from $L_p(G)$ to $Q(\hat{G})$ which we call the Fourier transform. The transformation is clearly linear. Moreover, it is injective. As suppose $f \in L_p(G)$ and $\langle h, \hat{f} \rangle = \langle \hat{h}, f \rangle = 0$ for all $h \in D(\hat{G})$. To show that $f = 0$ it is sufficient to prove that $D(\hat{G})^\wedge = \{\hat{g} \mid g \in D(\hat{G})\}$ is norm dense in $L_{p'}(G)$. Since $p' \neq \infty$, the set of all $k \in L_1(G) \cap L_{p'}(G)$ such that $\hat{k} \in C_c(\hat{G})$ is norm dense in $L_{p'}(G)$ (F.7 d), the Fourier transform being the usual L_1-transform. If k is such an element of $L_1(G) \cap L_{p'}(G)$ and $\{k_\alpha\} \subset C_c(\hat{G})$ is an approximate identity for $L_1(\hat{G})$ such that $\|k_\alpha\|_1 = 1$ then $\hat{k}_\alpha k \in L_1(G) \cap L_{p'}(G)$ and $\{\hat{k}_\alpha\}$ converges to one uniformly on compact subsets of G. But then essentially the same argument as given in the last portion of the proof of Theorem 5.1.2 reveals that $\lim_\alpha \|\hat{k}_\alpha k - k\|_{p'} = 0$.

However, $\hat{k}_\alpha k=(k_\alpha * \check{k})^\wedge=(k_\alpha*\hat{\tilde{k}})^\wedge\in D(\hat{G})^\wedge$ since k_α and $\hat{\tilde{k}}$ belong to $C_c(\hat{G})$. Thus $D(\hat{G})^\wedge$ is norm dense in $L_{p'}(G)$ and $f=0$.

Furthermore, the transformation is continuous from $L_p(G)$ to $Q(\hat{G})$ provided we consider $Q(\hat{G})$ in the weak* topology induced on it by $D(\hat{G})$. Indeed, let $\lim_n \|f_n-f\|_p=0$ and suppose $h\in D(\hat{G})$. Then

$$\lim_n \langle h,\hat{f}_n\rangle=\lim_n\langle \hat{h}, f_n\rangle=\langle\hat{h},f\rangle=\langle h,\hat{f}\rangle$$

since $\hat{h}\in L_r(G)$, $r\geqq 1$.

Finally, we note that if $f\in L_1(G)\cap L_p(G)$ then the usual Fourier transform of f agrees with the transform of f as an element of $L_p(G)$. This is evident from the relations

$$\begin{aligned}\langle h,\hat{f}\rangle=\langle\hat{h},f\rangle&=\int_G \hat{h}(t)f(t^{-1})\,d\lambda(t)\\ &=\int_G f(t^{-1})\Big[\int_G (t,\gamma^{-1})\,h(\gamma)\,d\eta(\gamma)\Big]\,d\lambda(t)\\ &=\int_{\hat{G}} h(\gamma)\Big[\int_G (t^{-1},\gamma^{-1})f(t)\,d\lambda(t)\Big]\,d\eta(\gamma)\\ &=\int_{\hat{G}} h(\gamma)\hat{f}(\gamma^{-1})\,d\eta(\gamma),\end{aligned}$$

which are valid for each $h\in D(\hat{G})$ and $f\in L_1(G)\cap L_p(G)$. Moreover, for each $f\in L_p(G)$ and $g\in L_1(G)$ we have by the definition following Lemma 5.1.3, that

$$\begin{aligned}\langle h,\hat{f}\hat{g}\rangle&=\langle \check{g}\,h,\hat{f}\rangle=\langle(\check{g}\,h)^\wedge,f\rangle=\langle g*\hat{h},f\rangle\\ &=\int_G g*\hat{h}(t)f(t^{-1})\,d\lambda(t)\\ &=\int_G\Big[\int_G g(t\,s^{-1})\,\hat{h}(s)\,d\lambda(s)\Big]f(t^{-1})\,d\lambda(t)\\ &=\int_G \hat{h}(s)\Big[\int_G g(s^{-1}t^{-1})f(t)\,d\lambda(t)\Big]\,d\lambda(s)\\ &=\int_G \hat{h}(s)f*g(s^{-1})\,d\lambda(s)=\langle\hat{h},f*g\rangle=\langle h,(f*g)^\wedge\rangle\end{aligned}$$

for each $h\in D(G)$. That is, $(f*g)^\wedge=\hat{f}\hat{g}$ for each $f\in L_p(G)$, $g\in L_1(G)$.

We collect the preceding discussion in a theorem.

Theorem 5.4.1. *Let G be a locally compact Abelian group and $1<p\leqq\infty$. The equation*

$$\langle h,\hat{f}\rangle=\langle\hat{h},f\rangle\qquad (h\in D(\hat{G}),\ f\in L_p(G))$$

defines a continuous injective linear transformation $f\to\hat{f}$ from $L_p(G)$ into $Q(\hat{G})$ when $Q(\hat{G})$ is given the weak topology induced by $D(\hat{G})$. If*

$g \in L_1(G) \cap L_p(G)$ then $\hat{g}$ is the usual L_1-Fourier transform of g and $(f*g)^\wedge = \hat{f}\hat{g}$ for each $f \in L_p(G)$ and $g \in L_1(G)$.

$\hat{f}$ is called the *Fourier transform* of f.

We note that if $1 < p \leqq 2$ then the Fourier transform as just defined is equivalent to the usual definition of the Hausdorff-Young and Fourier-Plancherel transformations. In particular, $\hat{f} \in L_{p'}(G)$ when $1 < p \leqq 2$ and $\|\hat{f}\|_{p'} \leqq \|f\|_p$ when $1 < p < 2$ and $\|\hat{f}\|_2 = \|f\|_2$. These assertions are immediate consequences of the inequalities

$$|\langle h, \hat{f} \rangle| = |\langle \hat{h}, f \rangle| \leqq \|\hat{h}\|_{p'} \|f\|_p \leqq \|h\|_p \|f\|_p \qquad (h \in D(\hat{G}),\ f \in L_p(G)),$$

which are valid because of the Hausdorff-Young Theorem (Theorem F.8.4) and the density of $D(\hat{G})$ in $L_{p'}(\hat{G})$. When $p = \infty$ the preceding inequality becomes

$$|\langle h, \hat{f} \rangle| = |\langle \hat{h}, f \rangle| \leqq \|\hat{h}\|_1 \|f\|_\infty = \|h\|_A \|f\|_\infty \qquad (h \in D(\hat{G}),\ f \in L_\infty(G)).$$

But $D(\hat{G})$ is dense in $A(\hat{G})$ by Theorem 5.1.2 (ii). Thus $\hat{f}$ defines a continuous linear functional on $A(\hat{G})$, that is, $\hat{f}$ is a pseudomeasure. Hence $L_\infty(G)^\wedge \subset P(\hat{G})$.

Next we wish to define the Fourier transform of elements in $M(L_p(G), L_q(G))$. We remark that the previous discussion supplies us with a suitable definition for $M(L_1(G), L_q(G))$ and $M(L_{q'}(G), L_\infty(G))$ since by Theorems 3.1.1 and 3.3.1 these spaces are isometrically isomorphic to $L_q(G)$, $1 < q \leqq \infty$. Furthermore, we have seen from Theorems 3.1.1 and 3.4.3 that $M(L_1(G))$ and $M(L_\infty(G))$ can be identified with $M(G)$ so that the usual Fourier-Stieltjes transform provides us with a Fourier transform for these multipliers. When $p = q$ or G is compact Theorems 4.3.1, 4.3.2, 5.2.3 and 5.2.4 show that $M(L_p(G), L_q(G))$ can be identified with subspaces of $P(G)$ for which we also have an adequate Fourier transform. Consequently, we shall now restrict our attention completely to noncompact G and indices $1 < p < q < \infty$.

Before giving the definition of the transform for $M(L_p(G), L_q(G))$ in these cases we need some new ideas and preliminary results.

A family of open sets $\{U_\alpha\}$ is called a *locally finite cover* for G if $G = \bigcup_\alpha U_\alpha$ and each point $t \in G$ has a neighborhood U_t which intersects only finitely many of the U_α. Every locally compact Abelian group possesses a locally finite cover consisting of open sets with compact closures (B.4). A family of continuous functions $\{f_\alpha\}$ on G is called *locally finite* if for each point t in G there exists a neighborhood U_t with compact closure such that all but a finite number of the f_α vanish identically on the closure $\overline{U}_t$ of U_t. Clearly, if $\{f_\alpha\}$ is a locally finite family and U is any open set with compact closure then all but a finite number of the elements of $\{f_\alpha\}$ vanish identically on $\overline{U}$.

We shall need the following lemma.

Lemma 5.4.1. *Let G be a locally compact Abelian group. If $\{U_\alpha\}$ is a locally finite cover of G by open sets with compact closures then there exists a locally finite family of functions $\{f_\alpha\}\subset A_c(G)$ such that:*

(i) *The support of f_α is contained in U_α.*
(ii) $0\leqq f_\alpha(t)\leqq 1$ *for each $t\in G$ and all α.*
(iii) $\sum_\alpha f_\alpha(t)=1$ *for each $t\in G$.*

Proof. Let $\{V_\alpha\}$ and $\{W_\alpha\}$ be coverings of G by open sets with compact closures such that for each α we have $W_\alpha\subset\overline{W}_\alpha\subset V_\alpha\subset\overline{V}_\alpha\subset U_\alpha$ (Theorem A.1.1). Then for each α let $g_\alpha\in A(G)$ be such that $g_\alpha\equiv 1$ on $\overline{W}_\alpha$, $g_\alpha\equiv 0$ off of V_α and $g_\alpha\geqq 0$. Such a choice of g_α can always be made (F.7e). Since $\{U_\alpha\}$ is locally finite it is evident that the family $\{g_\alpha\}$ is locally finite. Hence $\sum_\alpha g_\alpha(t)$ converges for each t in G as the sum is finite, and $\sum_\alpha g_\alpha(t)>0$ since $\{W_\alpha\}$ forms an open cover for G. Set $f_\alpha=g_\alpha/\sum_\beta g_\beta$. Clearly $\{f_\alpha\}$ is a locally finite family of continuous functions, each f_α has support contained in U_α, $0\leqq f_\alpha(t)\leqq 1$ for each t in G and all α, and $\sum_\alpha f_\alpha(t)=1$ for each $t\in G$.

To complete the proof we have only to show that $f_\alpha\in A_c(G)$. Since $g_\alpha\in A_c(G)$ it is sufficient to show that $1/\sum_\beta g_\beta$ agrees with some element of $A(G)$ on $\overline{V}_\alpha$. Furthermore, because $\overline{V}_\alpha$ is compact and $\{g_\alpha\}$ is locally finite, there exist only a finite number of the g_α which do not vanish identically on $\overline{V}_\alpha$. Call them $g_{\alpha_1},\ldots,g_{\alpha_n}$. Clearly then $\sum_{i=1}^n g_{\alpha_i}=g\in A(G)$ and $g>0$ on $\overline{V}_\alpha$. Thus it is apparent that we need only show that if $K\subset G$ is compact and $g\in A(G)$ is such that $g>0$ on K then there exists $h\in A(G)$ such that $1/g=h$ on K. Let I be the closed ideal in $L_1(\hat{G})$ consisting of all functions in $L_1(\hat{G})$ whose Fourier transforms vanish on K. Set $A=L_1(\hat{G})/I$. Then A is isomorphic to the algebra B_K, under pointwise operations, of the restrictions to K of the members of $A(G)$, and, since $A(G)$ separates points, the maximal ideal space of B_K is K (E.4). Clearly B_K possesses an identity, and so g restricted to K considered as an element of B_K is invertible. Therefore there exists an h in $A(G)$ such that $1/g=h$ on K.

This completes the proof. □

One further preliminary concept is required. Let $\sigma\in Q(G)$ and suppose $U\subset G$ is an open set. Then we define the *restriction of σ to U*, denoted by σ_U, as the continuous linear functional on the internal inductive limit $\bigcup_{K\subset U} D_K(G)$ determined by the equation

$$\langle h,\sigma_U\rangle=\langle h,\sigma\rangle \qquad (h\in D_K(G),\ K\subset U).$$

Consider now the spaces $M(L_p(G), L_q(G))$, $1<p<q<\infty$, where G is noncompact. From Theorem 5.2.2 and Corollary 5.2.1 we may identify $M(L_p(G), L_q(G))$ with a linear subspace of $Q(G)$. In what follows we shall assume such an identification has been made. Ultimately we shall define the Fourier transform $\hat{\sigma}$ of a quasimeasure σ in $M(L_p(G), L_q(G))$ as a quasimeasure on $\hat{G}$. We shall do this by first defining the local behavior of $\hat{\sigma}$ and then piecing together the local definitions to obtain a unique quasimeasure in $Q(\hat{G})$.

Let $\sigma \in M(L_p(G), L_q(G))$. If $U \subset \hat{G}$ is any open set with compact closure then choose $k \in C_c(G)$ such that $\check{k}$ does not vanish on $\bar{U}$, and then choose $g \in L_1(G)$ such that $\check{k}\,\check{g} \equiv 1$ on U. Clearly such a k always exists as $C_c(G)$ is norm dense in $L_1(G)$ and there is an $f \in L_1(G)$ such that $\check{f} \equiv 1$ on $\bar{U}$ (F.7e). The argument at the end of Lemma 5.4.1 shows that one can always choose $g \in L_1(G)$ with the desired property. For each $h \in \bigcup_{K \subset U} D_K(G)$ define

$$\langle h, \hat{\omega}_U \rangle = \langle h, \hat{g}(\sigma * k)\hat{}_U \rangle .$$

We note first that $\hat{g}(\sigma * k)\hat{} \in Q(\hat{G})$. Indeed, since $\sigma \in M(L_p(G), L_q(G))$ we have $\sigma * k \in L_q(G)$, and so, by Theorem 5.4.1, the Fourier transform $(\sigma * k)\hat{}$ exists and belongs to $Q(\hat{G})$. The definition following Lemma 5.1.3 shows that $\hat{g}(\sigma * k)\hat{} \in Q(\hat{G})$ as $\hat{g} \in M(G)\hat{}$. Moreover, $\hat{\omega}_U$ is well defined as the definition is independent of the choice of k and g. To see this let $\{u_\alpha\} \subset C_c(G)$ be an approximate identity for $L_1(G)$ with $\|u_\alpha\|_1 = 1$. Then, since $1<q<\infty$, we have $\{u_\alpha\}$ is an approximate identity for $L_q(G)$ (F.7d) and so $\lim_\alpha \|\sigma * k - \sigma * k * u_\alpha\|_q = 0$. But then if $h \in D_K(\hat{G})$ where $K \subset U$ we have

$$\begin{aligned}
\langle h, \hat{\omega}_U \rangle &= \langle h, \hat{g}(\sigma * k)\hat{}_U \rangle = \langle h, \hat{g}(\sigma * k)\hat{}\, \rangle = \langle \check{g}\, h, (\sigma * k)\hat{}\, \rangle \\
&= \langle (\check{g}\, h)\hat{}\,, \sigma * k \rangle = \langle g * \hat{h}, \sigma * k \rangle = \lim_\alpha \langle g * \hat{h}, \sigma * k * u_\alpha \rangle \\
&= \lim_\alpha \langle k * g * \hat{h}, \sigma * u_\alpha \rangle = \lim_\alpha \langle (\check{k}\, \check{g}\, h)\hat{}\,, \sigma * u_\alpha \rangle \\
&= \lim_\alpha \langle \hat{h}, \sigma * u_\alpha \rangle .
\end{aligned}$$

The last limit is obviously independent of the choice of k and g, and hence $\hat{\omega}_U$ is well defined. Clearly $\hat{\omega}_U$ defines a continuous linear functional on $\bigcup_{K \subset U} D_K(\hat{G})$. Moreover, if U_1, U_2 are two open sets with compact closure such that $U_1 \cap U_2 \neq \emptyset$ and $h \in D_K(\hat{G})$ where $K \subset U_1 \cap U_2$ then it is easily seen from the preceding arguments that

$$\langle h, \hat{\omega}_{U_1} \rangle = \langle h, \hat{\omega}_{U_1 \cap U_2} \rangle = \langle h, \hat{\omega}_{U_2} \rangle .$$

Thus $\hat{\omega}_{U_1} = \hat{\omega}_{U_2}$ on $U_1 \cap U_2$.

Now let $\{U_\alpha\}$ be a locally finite cover of $\hat{G}$ by open sets with compact closure, and let $\{f_\alpha\}\subset A_c(\hat{G})$ be a locally finite family of functions satisfying the conclusions of Lemma 5.4.1. If $h\in D(\hat{G})$ then define

$$\langle h, \hat{\sigma}\rangle = \sum_\alpha \langle f_\alpha h, \hat{\omega}_{U_\alpha}\rangle.$$

Since all but a finite number of the f_α vanish identically on the support of h, the sum is finite. Moreover, by Lemma 5.1.3 each $f_\alpha h\in D(\hat{G})$ as $f_\alpha\in A_c(\hat{G})\subset M(G)\hat{}$. Clearly then the support of $f_\alpha h$ is contained in U_α and so $\langle f_\alpha h, \hat{\omega}_{U_\alpha}\rangle$ is defined. Obviously $\hat{\sigma}$ defines a linear functional on $D(\hat{G})$. If $K\subset\hat{G}$ is compact and $\{h_\beta\}\subset D_K(\hat{G})$ is such that $\lim_\beta \|h_\beta\|_\infty=0$ then apparently $\lim_\beta \|f_\alpha h_\beta\|_\infty=0$ for each α since $0\leqq f_\alpha\leqq 1$. Thus $\{f_\alpha h_\beta\}$ converges to zero in $D_{U_\alpha\cap K}(\hat{G})$ which implies $\lim_\beta\langle f_\alpha h_\beta, \hat{\omega}_{U_\alpha}\rangle=0$ for each α. Consequently $\lim_\beta\langle h_\beta, \hat{\sigma}\rangle=0$, that is, $\hat{\sigma}$ is a continuous linear functional on $D_K(\hat{G})$. Therefore $\hat{\sigma}$ is a quasimeasure.

It is this quasimeasure which we call the *Fourier transform* of $\sigma\in M(L_p(G), L_q(G))$.

We note further that if $U\subset\hat{G}$ is an open set with compact closure then $\hat{\sigma}$ restricted to U is equal to $\hat{\omega}_U$. Indeed if $h\in D_K(\hat{G})$ where $K\subset U$ then

$$\begin{aligned}\langle h, \hat{\sigma}_U\rangle &= \langle h, \hat{\sigma}\rangle = \sum_\alpha\langle f_\alpha h, \hat{\omega}_{U_\alpha}\rangle = \sum_\alpha\langle f_\alpha h, \hat{\omega}_{U_\alpha\cap U}\rangle\\ &= \sum_\alpha\langle f_\alpha h, \hat{\omega}_U\rangle = \langle h, \hat{\omega}_U\rangle,\end{aligned}$$

as the support of $f_\alpha h$ is contained in $U_\alpha\cap U$ and $\sum_\alpha f_\alpha=1$.

The Fourier transform as defined above obviously establishes a linear mapping from $M(L_p(G), L_q(G))$ to $Q(\hat{G})$. Furthermore, if $\sigma\in M(L_p(G), L_q(G))$ and $f\in C_c(G)$ then $(\sigma*f)\hat{}=\hat{\sigma}\hat{f}$, where $(\sigma*f)\hat{}$ is the Fourier transform of $\sigma*f\in L_q(G)$ as defined in Theorem 5.4.1. To see this let $h\in D_K(\hat{G})$ and let U be an open neighborhood of K with compact closure. Choose k and g as before so that $\check{k}\check{g}=1$ on U. Then we have

$$\begin{aligned}\langle h, \hat{\sigma}\hat{f}\rangle &= \langle\check{f}h, \hat{\sigma}\rangle = \langle\check{f}h, \hat{g}(\sigma*k)\hat{}_U\rangle = \langle\check{f}h, \hat{g}(\sigma*k)\hat{}\rangle\\ &= \langle\check{g}\check{f}h, (\sigma*k)\hat{}\rangle = \langle g*f*\hat{h}, \sigma*k\rangle\\ &= \langle k*g*\hat{h}, \sigma*f\rangle = \langle(\check{k}\check{g}h)\hat{}, \sigma*f\rangle = \langle\hat{h}, (\sigma*f)\rangle\\ &= \langle h, (\sigma*f)\hat{}\rangle.\end{aligned}$$

Since this holds for every $h\in D(\hat{G})$ we conclude that $(\sigma*f)\hat{}=\hat{\sigma}\hat{f}$ for each $f\in C_c(G)$.

Moreover, the mapping defined by the Fourier transform is injective. As let $\sigma\in M(L_p(G), L_q(G))$ and suppose $\hat{\sigma}=0$. Then by the preceding result if $f \in C_c(G)$ and $h \in D(G)$ then

$$\langle h, (\sigma * f)^\wedge \rangle = \langle h, \hat{\sigma} \hat{f} \rangle = \langle \check{\hat{f}} h, \hat{\sigma} \rangle = 0.$$

However by Theorem 5.4.1 the Fourier transform for $L_q(G)$ is one-to-one, hence $\sigma * f = 0$ for each $f \in C_c(G)$. Therefore we conclude from Theorem 5.1.5 that $\sigma = 0$.

Finally, from the definition of the Fourier transform for $M(L_p(G), L_q(G))$ and the continuity properties of the Fourier transform for $L_q(G)$, it is easily seen that the Fourier transform of $M(L_p(G), L_q(G))$ is a continuous mapping into $Q(\hat{G})$ provided this space is considered with the weak* topology induced by $D(\hat{G})$.

The previous development can be summarized in the following theorem.

Theorem 5.4.2. *Let G be a noncompact locally compact Abelian group and suppose $1 < p < q < \infty$. Then the Fourier transform $\sigma \to \hat{\sigma}$ defines a continuous injective linear transformation from $M(L_p(G), L_q(G))$ into $Q(\hat{G})$ when $Q(\hat{G})$ is given the weak* topology induced by $D(\hat{G})$.*

We saw above that if $\sigma \in M(L_p(G), L_q(G))$ and $f \in C_c(G)$ then $(\sigma * f)^\wedge = \hat{\sigma} \hat{f}$. This result combined with Theorems 3.1.1, 5.2.2 and 5.4.1 immediately establishes the next theorem.

Theorem 5.4.3. *Let G be a noncompact locally compact Abelian group and suppose $1 < p, q < \infty$, $p < q$. If $T \in M(L_p(G), L_q(G))$ then there exists a unique quasimeasure $\hat{\sigma} \in Q(\hat{G})$ such that $(Tf)^\wedge = \hat{\sigma} \hat{f}$ for each $f \in C_c(G)$.*

Of course similar results are valid in general for $M(L_p(G), L_q(G))$, but these have essentially been discussed in previous sections.

As usual the space of Fourier transforms of $M(L_p(G), L_q(G))$ will be denoted by $M(L_p(G), L_q(G))^\wedge$.

5.5. Some Results Concerning $L_p(G)^\wedge$ and $M(L_p(G), L_q(G))^\wedge$. The main purpose of this section is to show for noncompact G and $1 \leqq p < 2 < q \leqq \infty$ that there exist $T \in M(L_p(G), L_q(G))$ such that $\hat{T} \notin V(\hat{G})$. That is there exist $T \in M(L_p(G), L_q(G))$ such that $\hat{T}$ is not a measure. An immediate consequence of this result will be the fact that for $p > 2$ there exist $f \in L_p(G)$ such that $\hat{f}$ is not a measure.

Theorem 5.5.1. *Let G be a noncompact locally compact Abelian group. If $1 \leqq p < 2 < q \leqq \infty$ then there exists a $T \in M(L_p(G), L_q(G))$ such that $\hat{T} \notin V(\hat{G})$.*

Proof. Suppose $M(L_p(G), L_q(G))^\wedge \subset V(\hat{G})$. Then if $U \subset \hat{G}$ is an open set with compact closure it is evident that the mapping $T \to \hat{T}|_U$ is a linear transformation from $M(L_p(G), L_q(G))$ to $M(U)$, the Banach space of bounded regular complex valued Borel measures on U. $\hat{T}|_U$ of course denotes the restriction to U of the measure $\hat{T}$. If $\lim_n \|T_n - T\|_{p,q} = 0$ and $\lim_n \|\hat{T}_n|_U - \mu\| = 0$ for some $\mu \in M(U)$ then by Theorem 5.4.2 we have

$$\lim_n \langle h, \hat{T}_n\rangle = \langle h, \hat{T}\rangle \qquad (h \in D(\hat{G})).$$

Since $D(\hat{G})$ is norm dense in $C_c(\hat{G})$ by Theorem 5.1.2, we conclude at once that $\hat{T}|_U = \mu$. Consequently, the transformation $T \to \hat{T}|_U$ is closed, and so, by the Closed Graph Theorem (Theorem D.6.1), there exists a constant $B > 0$ such that

$$\|\hat{T}|_U\| = \int_U d|\hat{T}|_U|(\gamma) \leqq B \|T\|_{p,q} \qquad (T \in M(L_p(G), L_q(G))).$$

By Theorem 5.3.2 we may assume, without loss of generality, that the point $(1/p, 1/q)$ lies in the triangle bounded by $a = 1$, $b = \frac{1}{2}$ and $a + b = 1$. Consider the straight line through $(\frac{1}{2}, \frac{1}{2})$ and $(1/p, 1/q)$ and let its intersection with $a = 1$ be denoted by $(1, 1/s)$. Clearly $s > 2$ as $q > 2$. Let α, $0 < \alpha < 1$, be such that $1/p = \alpha/2 + (1-\alpha)$ and $1/q = \alpha/2 + (1-\alpha)/s$.

Let $f \in L_1(G)$ be such that $\hat{f} \in C_c(\hat{G})$. Then by the Fourier Inversion Theorem (Theorem F.8.1) we see that $f \in L_1(G) \cap L_\infty(G)$ and hence, from the convexity of $\log \|f\|_{1/a}$, $0 \leqq a \leqq 1$ (Theorem D.11.3), we deduce that $f \in L_r(G)$, $1 \leqq r \leqq \infty$. In particular, from Theorems 3.1.1, 4.1.1 and 5.3.3 we see that T_f belongs to $M(L_1(G), L_s(G))$, $M(L_2(G))$ and $M(L_p(G), L_q(G))$, where as usual $T_f g = f * g$. By an application of the Riesz-Thorin Convexity Theorem (Theorem D.11.2), we conclude that

$$\begin{aligned}\int_U |\hat{f}(\gamma)|\, d\eta(\gamma) &\leqq B \|T_f\|_{p,q} \leqq B(\|T_f\|_2)^\alpha (\|T_f\|_{1,s})^{1-\alpha} \\ &= B(\|\hat{f}\|_\infty)^\alpha (\|f\|_s)^{1-\alpha}\end{aligned}$$

for each $f \in L_1(G)$ such that $\hat{f} \in C_c(\hat{G})$.

Using Lemma 5.3.1 as in the last portion of the proof of Theorem 5.3.3, we construct a sequence $\{\psi_n\} \subset C_c^U(\hat{G})$ such that $\{\hat{\psi}_n\} \subset L_1(G)$, $\|\psi_n\|_\infty \leqq 1$, $\lim_n \|\hat{\psi}_n\|_\infty = 0$ and $\liminf_n \|\psi_n\|_2 \neq 0$. The latter fact implies, as before, that $\liminf_n \|\psi_n\|_1 \neq 0$.

The previous estimate then shows that for each positive integer n we have

$$\|\psi_n\|_1 = \int_U |\psi_n(\gamma)|\, d\eta(\gamma) \leqq B(\|\psi_n\|_\infty)^\alpha (\|\hat{\psi}_n\|_s)^{1-\alpha} \leqq B(\|\hat{\psi}_n\|_s)^{1-\alpha}$$

since $\tilde{\hat{\psi}}_n = \check{\psi}_n \in L_1(G)$ and $\psi_n \in C_c^U(\hat{G})$. However, from the convexity of $\log \|\cdot\|_{1/a}$, $0 \leqq a \leqq 1$, (Theorem D.11.3), the Plancherel Theorem

(Theorem F.8.2) and the fact that $2<s\leqq\infty$, we deduce that

$$\begin{aligned}\|\hat{\psi}_n\|_s &\leqq (\|\hat{\psi}_n\|_2)^{2/s}(\|\hat{\psi}_n\|_\infty)^{1-2/s} = (\|\psi_n\|_2)^{2/s}(\|\hat{\psi}_n\|_\infty)^{1-2/s}\\ &\leqq \Big[\|\psi_n\|_\infty\Big(\int_{\hat{G}}\chi_U(\gamma)\,d\eta(\gamma)\Big)^{\frac{1}{2}}\Big]^{2/s}(\|\hat{\psi}_n\|_\infty)^{1-2/s}\\ &< \eta(U)^{1/s}(\|\hat{\psi}_n\|_\infty)^{1-2/s}.\end{aligned}$$

Consequently we see for each positive integer n that

$$\|\psi_n\|_1 \leqq B[\eta(U)^{1/s}(\|\hat{\psi}_n\|_\infty)^{1-2/s}]^{1-\alpha}.$$

But this leads to a contradiction of the fact that $\liminf_n \|\psi_n\|_1 \neq 0$ as $\lim_n \|\hat{\psi}_n\|_\infty = 0$.

Therefore there exists some $T\in M(L_p(G), L_q(G))$ such that $\hat{T}\notin V(\hat{G})$. □

By Theorem 3.1.1 we know that $M(L_1(G), L_p(G))$, $1<p\leqq\infty$, can be identified with $L_p(G)$. This fact combined with the preceding theorem immediately establishes the following corollary.

Corollary 5.5.1. *Let G be a noncompact locally compact Abelian group. If $p>2$ then there exists an $f\in L_p(G)$ such that $\hat{f}\notin V(\hat{G})$.*

An argument similar to the one used in the proof of Theorem 5.5.1 allows us to establish a slight generalization of this corollary.

Theorem 5.5.2. *Let G be a noncompact locally compact Abelian group. Then there exists an $f\in \bigcap_{2<p\leqq\infty} L_p(G)$ such that $\hat{f}\notin V(\hat{G})$.*

Proof. As usual suppose the theorem is not true, that is $\big(\bigcap_{2<p\leqq\infty} L_p(G)\big)\hat{}\subset V(\hat{G})$. Choose a sequence $\{r_n\}$ of real numbers such that $r_n>2$ and $\lim_n r_n = 2$, and let $X=\bigcap_{2<p\leqq\infty} L_p(G)$. For each $f\in X$ define

$$|||f|||_n = \sup\{\|f\|_p \mid r_n \leqq p \leqq \infty\}.$$

We first note that for each $f\in X$ we have

$$|||f|||_n \leqq \sup\{(\|f\|_\infty)^{1-r_n/p}(\|f\|_{r_n})^{r_n/p} \mid r_n\leqq p\leqq\infty\}<\infty$$

since $r_n\leqq p$. Furthermore, it is easily checked that each $|||\cdot|||_n$ is a norm on X. Define a metric ρ on X by the formula

$$\rho(f,g)=\sum_{n=1}^{\infty} |||f-g|||_n/2^n(1+|||f-g|||_n) \qquad (f,g\in X).$$

It is evident that ρ is a complete invariant metric on X as each $L_p(G)$, $2<p\leqq\infty$, is complete in the L_p-norm. Hence X is a F-space (D.1).

Let $U \subset \hat{G}$ be an open set with compact closure. Then the mapping $f \to \chi_U \hat{f}$ defines a linear transformation from X to $M(U)$. Essentially the same argument as used in the preceding theorem shows that the transformation is closed, and hence, since X and $M(U)$ are both F-spaces, we conclude via the Closed Graph Theorem (Theorem D.6.1) that the transformation is continuous. In particular for each n there is a constant $B_n > 0$ such that

$$\int_U d|\hat{f}|(\gamma) \leqq B_n |||f|||_n \qquad (f \in X).$$

Consider some fixed r_n. For simplicity of notation we write $r = r_n > 2$, $B = B_n$ and $|||\cdot||| = |||\cdot|||_n$.

Let $\{\psi_n\} \subset C_c^U(\hat{G})$ be chosen as in the proofs of Theorems 5.5.1 or 5.3.3, that is, $\{\hat{\psi}_n\} \subset L_1(G) \cap L_\infty(G)$, $\|\psi_n\|_\infty \leqq 1$, $\lim_n \|\hat{\psi}_n\|_\infty = 0$ and $\liminf_n \|\psi_n\|_2 \neq 0$. Then, as before, $\{\hat{\psi}_n\} \subset \bigcap_{1 \leqq p \leqq \infty} L_p(G)$ and $\liminf_n \|\psi_n\|_1 \neq 0$.

However the previous estimate applied to the sequence $\{\tilde{\hat{\psi}}_n\} = \{\check{\psi}_n\}$ yields

$$\|\psi_n\|_1 = \int_U |\psi_n(\gamma)|\, d\eta(\gamma) \leqq B\, |||\hat{\psi}_n|||$$
$$\leqq B \sup\{(\|\hat{\psi}_n\|_\infty)^{1-r/p} (\|\hat{\psi}_n\|_r)^{r/p} \,|\, r \leqq p \leqq \infty\}.$$

Since $2 < r < \infty$ the convexity of $\log \|\cdot\|_{1/a}$, $0 \leqq a \leqq 1$, (Theorem D.11.3) implies that $\|\hat{\psi}_n\|_r \leqq (\|\hat{\psi}_n\|_2)^{2/r} (\|\hat{\psi}_n\|_\infty)^{1-2/r}$. Moreover, it is apparent that $\|\hat{\psi}_n\|_2 = \|\psi_n\|_2 \leqq (\eta(U))^{\frac{1}{2}}$. Combining these estimates with the preceding one we obtain for each positive integer n that

$$\|\psi_n\|_1 \leqq B \sup\{(\eta(U))^{1/p} (\|\hat{\psi}_n\|_\infty)^{1-2/p} \,|\, r \leqq p \leqq \infty\}.$$

Choose n_0 such that for $n \geqq n_0$ we have $\|\hat{\psi}_n\|_\infty < 1$. This is possible because $\lim_n \|\hat{\psi}_n\|_\infty = 0$. Then for $n \geqq n_0$ we see that $(\|\hat{\psi}_n\|_\infty)^{1-2/p} \leqq (\|\hat{\psi}_n\|_\infty)^{1-2/r}$, $r \leqq p \leqq \infty$, since the exponential function a^x, $0 < a < 1$, is monotone decreasing. Consequently we have for $n \geqq n_0$ that

$$\|\psi_n\|_1 \leqq B \sup\{(\eta(U))^{1/p} \,|\, r \leqq p \leqq \infty\} (\|\hat{\psi}_n\|_\infty)^{1-2/r},$$

from which it follows at once that $\liminf_n \|\psi_n\|_1 = 0$. This however contradicts the fact that $\liminf_n \|\psi_n\|_1 \neq 0$.

Therefore there is some $f \in \bigcap_{2 < p \leqq \infty} L_p(G)$ such that $\hat{f} \notin V(\hat{G})$. □

The next result of this section exhibits a number of assertions which are equivalent to the conclusion of Corollary 5.5.1. We shall say that a function $\psi \in L_\infty(\hat{G})$ determines a multiplier for $(L_p(G), L_q(G))$, $1 \leqq p, q < \infty$, if there exists a $T_\psi \in M(L_p(G), L_q(G))$ for which $(T_\psi f)\hat{} = \psi \hat{f}$ for each $f \in C_c(G)$.

Theorem 5.5.3. *Let G be a locally compact Abelian group and suppose $1 \leqq p, q < \infty$, $q \neq 1$. Then the following are equivalent:*

(i) *If $p > 2$ then there exists an $f \in L_p(G)$ such that $\hat{f} \notin V(\hat{G})$.*

(ii) *If $p > 2$ then there exists a sequence $\{f_n\} \subset C_c(G)$ and a compact set $K \subset \hat{G}$ such that $\sup_n \|f_n\|_p < \infty$ and $\sup_n \|\chi_K \hat{f}_n\|_1 = \infty$.*

(iii) *If $p > 2$ then there exists a compact set $K \subset \hat{G}$ and a function $\psi \in L_\infty(\hat{G})$ such that ψ vanishes almost everywhere off of K and $\psi \notin L_{p'}(G)^\wedge$, $1/p + 1/p' = 1$.*

(iv) *If there exists a function $\varphi \in L_\infty(\hat{G})$ such that $\varphi \geqq 0$, $\varphi > 0$ almost everywhere on a set of positive measure, and which has the property that $\psi \in L_\infty(\hat{G})$ determines a multiplier for $(L_p(G), L_q(G))$ whenever $|\psi(\gamma)| \leqq \varphi(\gamma)$, for almost all $\gamma \in \hat{G}$, then $p \leqq 2 \leqq q$.*

Proof. Let $p > 2$ and $f \in L_p(G)$ be such that $\hat{f} \notin V(\hat{G})$. Then there exists an open set $U \subset \hat{G}$ with compact closure such that $\hat{f}$ restricted to U does not define a measure in $V(\hat{G})$. Since $p \neq \infty$ there exists a sequence $\{f_n\} \subset C_c(G)$ such that $\lim_n \|f_n - f\|_p = 0$. Clearly $\sup_n \|f_n\|_p < \infty$. If $h \in D(\hat{G})$ then $\hat{h} \in L_1(G) \cap C_0(G)$. Hence $\hat{h} \in L_{p'}(G)$, $1/p + 1/p' = 1$. Thus for each $h \in D(\hat{G})$ we have

$$\langle h, \hat{f} \rangle = \langle \hat{h}, f \rangle = \lim_n \langle \hat{h}, f_n \rangle = \lim_n \langle h, \hat{f}_n \rangle .$$

Let $K = \overline{U}$ and suppose $\sup_n \|\chi_K \hat{f}_n\|_1 < \infty$. Then $\{\chi_K \hat{f}_n\}$ is a norm bounded sequence in $M(K)$ and hence, by Alaoglu's Theorem (Theorem D.4.3), there exists a measure $\mu \in M(K)$ and a subnet $\{\chi_K \hat{f}_{n_k}\}$ such that

$$\lim_k \langle h, \hat{f}_{n_k} \rangle = \langle h, \mu \rangle \qquad (h \in C(K)).$$

Since the space of all $h \in D(\hat{G})$ whose support lies in K is norm dense in $C(K)$, it follows immediately that $\chi_K \hat{f} = \mu$, contradicting the choice of f.

Thus $\sup_k \|\chi_K \hat{f}_n\|_1 = \infty$ and (i) implies (ii).

On the other hand, suppose $p > 2$ and $(L_p(G))^\wedge \subset V(\hat{G})$. Let $K \subset \hat{G}$ be compact. Then for each $f \in L_p(G)$ the restriction to K of $\hat{f}$ defines an element of $M(K)$. Clearly the mapping $f \to \chi_K \hat{f}$ is linear. If $\lim_n \|f_n - f\|_p = 0$ and $\lim_n \|\chi_K \hat{f}_n - \mu\| = 0$ then by Theorem 5.4.1 we have

$$\lim_n \langle h, \hat{f}_n \rangle = \langle h, \hat{f} \rangle \qquad (h \in D(\hat{G})).$$

It then follows easily that $\chi_K \hat{f} = \mu$, and so, by the Closed Graph Theorem (Theorem D.6.1), the mapping $f \to \chi_K \hat{f}$ is continuous. Thus there exists a constant $B > 0$ such that $\|\chi_K \hat{f}\| \leqq B \|f\|_p$, $f \in L_p(G)$. But if $f \in C_c(G)$ then $\hat{f}$ is continuous so that $\|\chi_K \hat{f}\| = \|\chi_K \hat{f}\|_1 \leqq B \|f\|_p$. Hence (ii) cannot hold.

Thus (ii) implies (i).

Now suppose (iii) holds and let φ be as in (iv). For each $\psi \in L_\infty(\hat{G})$ we have that $|\varphi\psi(\gamma)|/\|\psi\|_\infty \leqq \varphi(\gamma)$ for almost all $\gamma \in \hat{G}$. Thus, by the hypotheses on φ, we conclude that $\varphi\psi/\|\psi\|_\infty$, and so $\varphi\psi$, determines a multiplier for $(L_p(G), L_q(G))$, as $M(L_p(G), L_q(G))$ is a linear space. That is, there is a multiplier $T_\psi \in M(L_p(G), L_q(G))$ such that $(T_\psi f)^\wedge = \varphi\psi\hat{f}$ for each $f \in C_c(G)$. Clearly this defines a linear mapping from $L_\infty(\hat{G})$ to $M(L_p(G), L_q(G))$. Moreover, suppose $\{\psi_n\} \subset L_\infty(\hat{G})$, $\psi \in L_\infty(\hat{G})$ and $T \in M(L_p(G), L_q(G))$ are such that $\lim_n \|\psi_n - \psi\|_\infty = 0$ and $\lim_n \|T_{\psi_n} - T\|_{p,q} = 0$. Let $\sigma \in Q(G)$ be the unique quasimeasure such that $Tf = \sigma * f$, $f \in C_c(G)$. Then if $f \in C_c(G)$, for each $h \in D(\hat{G})$ we have

$$\begin{aligned}\langle h, (Tf)^\wedge\rangle &= \langle \hat{h}, Tf\rangle = \lim_n \langle \hat{h}, T_{\psi_n} f\rangle = \lim_n \langle h, (T_{\psi_n} f)^\wedge\rangle \\ &= \lim_n \langle h, \varphi\psi_n \hat{f}\rangle = \langle h, \varphi\psi\hat{f}\rangle = \langle h, (T_\psi f)^\wedge\rangle.\end{aligned}$$

Thus $T = T_\psi$ and the mapping from $L_\infty(\hat{G})$ to $M(L_p(G), L_q(G))$ is closed. Therefore, by the Closed Graph Theorem (Theorem D.6.1), there exists a constant $B > 0$ such that $\|T_\psi\|_{p,q} \leqq B\|\psi\|_\infty$.

Now if $f, g \in C_c(G)$ then $T_\psi f$ and g belong to $L_2(G)$, and so by Parseval's Formula (Theorem F.8.3) we have

$$\begin{aligned}\Big|\int_{\hat{G}} \varphi(\gamma)\psi(\gamma)\hat{f}(\gamma)\hat{g}(\gamma)\,d\eta(\gamma)\Big| &= \Big|\int_G T_\psi f(t)\, g(t^{-1})\,d\lambda(t)\Big| \\ &\leqq \|T_\psi f\|_q \|g\|_{q'} \\ &\leqq \|T_\psi\|_{p,q} \|f\|_p \|g\|_{q'} \\ &\leqq B\|\psi\|_\infty \|f\|_p \|g\|_{q'}\end{aligned}$$

where $1/q + 1/q' = 1$. Since for each $f, g \in C_c(G)$ this holds for all $\psi \in L_\infty(\hat{G})$, it is then valid, in particular, for the function $\psi(\gamma) = \exp(-i \arg \hat{f}(\gamma)\hat{g}(\gamma))$. Hence we have

$$\int_{\hat{G}} \varphi(\gamma)|\hat{f}(\gamma)\hat{g}(\gamma)|\,d\eta(\gamma) \leqq B\|f\|_p \|g\|_{q'} \qquad (f, g \in C_c(G)).$$

Since $(f(\cdot, \delta^{-1}))^\wedge(\gamma) = \hat{f}(\gamma\delta)$ for each $\delta \in \hat{G}$ we also have for each $\delta \in \hat{G}$ that

$$\int_G \varphi(\gamma\delta^{-1})|\hat{f}(\gamma)\hat{g}(\gamma)|\,d\eta(\gamma) \leqq B\|f\|_p \|g\|_{q'} \quad (f, g \in C_c(G)).$$

Let V denote the set of positive measure on which φ is positive almost everywhere. Given any compact set $K \subset \hat{G}$, choose a $k \in L_1(\hat{G})$ such that k is zero off of a σ-compact set W and positive on W, k is continuous, $\|k\|_1 = 1$

and $K \subset VW$. Then

$$\begin{aligned}\int_{\hat{G}} \varphi * k(\gamma)\,|\hat{f}(\gamma)\,\hat{g}(\gamma)|\,d\eta(\gamma) &= \int_{\hat{G}} k(\delta)\Big[\int_{\hat{G}} \varphi(\gamma\,\delta^{-1})\,|\hat{f}(\gamma)\,\hat{g}(\gamma)|\,d\eta(\gamma)\Big]\,d\eta(\delta)\\ &\leqq B\,\|f\|_p\,\|g\|_{q'} \qquad (f, g \in C_c(G)).\end{aligned}$$

Moreover, since $\varphi * k$ is continuous and positive on K, we have $\inf\limits_{\gamma\in K} \varphi * k(\gamma) = c_K > 0$. Let $h_K \in C_c(G)$ be such that $\hat{h}_K \geqq 1$ on K and set $g = h_K/c_K$. Then the previous inequality shows that

$$\int_K |\hat{f}(\gamma)|\,d\eta(\gamma) \leqq (B/c_K)\,\|h_K\|_{q'}\,\|f\|_p = b_K\,\|f\|_p \qquad (f \in C_c(G)),$$

since $\varphi * k(\gamma)\,\hat{g}(\gamma) \geqq \varphi * k(\gamma)/c_K \geqq 1$ for $\gamma \in K$.

Consequently for any compact $K \subset \hat{G}$ if $\psi \in L_\infty(\hat{G})$ vanishes almost everywhere off of K then

$$\int_{\hat{G}} |\psi(\gamma)\,\hat{f}(\gamma)|\,d\eta(\gamma) \leqq b_K\,\|\psi\|_\infty\,\|f\|_p \qquad (f \in C_c(G)).$$

Furthermore, a similar estimate is valid for each $f \in L_1(G) \cap C_0(G)$. Indeed, let $f \in L_1(G) \cap C_0(G)$ and $f \geqq 0$. Let $\{f_n\} \subset C_c(G)$ be a monotone increasing sequence of nonnegative functions which converges almost everywhere to f. Then appealing to the Monotone Convergence Theorem (Theorem C.5.1), we see that $\lim\limits_n \|f_n - f\|_1 = 0$ and $\lim\limits_n \|f_n\|_p = \|f\|_p$. Thus we deduce from the inequalities

$$\begin{aligned}\int_{\hat{G}} |\psi(\gamma)\,\hat{f}(\gamma)|\,d\eta(\gamma) &\leqq \int_K |\psi(\gamma)|\,|\hat{f}(\gamma) - \hat{f}_n(\gamma)|\,d\eta(\gamma) + \int_{\hat{G}} |\psi(\gamma)\,\hat{f}_n(\gamma)|\,d\eta(\gamma)\\ &\leqq \|\hat{f} - \hat{f}_n\|_\infty\,\|\psi\|_\infty\,\eta(K) + b_K\,\|\psi\|_\infty\,\|f_n\|_p\end{aligned}$$

that

$$\int_{\hat{G}} |\psi(\gamma)\,\hat{f}(\gamma)|\,d\eta(\gamma) \leqq b_K\,\|\psi\|_\infty\,\|f\|_p \qquad (f \in L_1^+(G) \cap C_0^+(G)).$$

Hence it follows immediately that for any compact $K \subset \hat{G}$, if $\psi \in L_\infty(\hat{G})$ vanishes almost everywhere off of K then

$$\int_{\hat{G}} |\psi(\gamma)\,\hat{f}(\gamma)|\,d\eta(\gamma) \leqq 4 b_K\,\|\psi\|_\infty\,\|f\|_p \qquad (f \in L_1(G) \cap C_0(G)).$$

Clearly then this inequality allows us to conclude that the mapping $T\colon L_1(G) \cap C_0(G) \to L_\infty(G)$, defined by $Tf = (\hat{f}\psi)\check{}$ for each $f \in L_1(G) \cap C_0(G)$, is well defined and continuous when $L_1(G) \cap C_0(G)$ is considered as a subspace of $L_p(G)$. Moreover, the equations

$$(\psi(\tau_s f)\hat{}\,)\check{} = ((s^{-1}, \cdot)\,\psi\,\hat{f})\check{} = \tau_s(\psi\,\hat{f})\check{} \qquad (s \in G)$$

show that T commutes with translations. Since $L_1(G) \cap C_0(G)$ is norm dense in $L_p(G)$, it follows that T can be uniquely extended to all of $L_p(G)$, and that this extension, also denoted by T, belongs to $M(L_p(G), L_\infty(G))$.

Thus, by Theorem 3.3.1, there exists a unique $g \in L_{p'}(G)$ such that $Tf = f * g$ for each $f \in L_p(G)$. In particular, if $h \in D(\hat{G})$ and $f \in L_1(G) \cap C_0(G)$ is such that $\check{f} = h$, then

$$\begin{aligned}\langle h, \psi\rangle &= \langle \check{f}, \psi\rangle = \int_{\hat{G}} \psi(\gamma)\check{f}(\gamma^{-1})\, d\eta(\gamma) = \int_{\hat{G}} \psi(\gamma)\hat{f}(\gamma)\, d\eta(\gamma)\\ &= (\psi \hat{f})\check{}\,(e) = f * g(e) = \langle f, g\rangle = \langle \hat{h}, g\rangle = \langle h, \hat{g}\rangle .\end{aligned}$$

Consequently, by the definition of the Fourier transform for $L_{p'}(G)$, we see that ψ defines the same quasimeasure as $\hat{g}$. Thus, for any compact $K \subset \hat{G}$, if $\psi \in L_\infty(\hat{G})$ vanishes almost everywhere off of K then ψ is the Fourier transform of an element of $L_{p'}(G)$, and so by (iii) we must have $1 < p \leqq 2$.

By Theorems 3.1.1, 3.3.1 and 5.2.1 we know that the spaces $M(L_p(G), L_q(G))$ and $M(L_{q'}(G), L_{p'}(G))$ are isometrically isomorphic. Thus, if we can show that any ψ which satisfies the conditions of (iv) determines a multiplier for $(L_{q'}(G), L_{p'}(G))$, then a repetition of the previous argument reveals that $1 < q' \leqq 2$, that is, $q \geqq 2$.

Recalling the preceding development it is evident that it is sufficient to show for each $\psi \in L_\infty(\hat{G})$ that there exists $S_\psi \in M(L_{q'}(G), L_{p'}(G))$ such that $(S_\psi f)\hat{} = \varphi \psi \hat{f}$ for each $f \in C_c(G)$. Let $T_\psi \in M(L_p(G), L_q(G))$ be defined as before, and define $T_{\tilde{\psi}} \in M(L_p(G), L_q(G))$ by $T_{\tilde{\psi}} f = (T_\psi \tilde{f})\tilde{}$ for each $f \in L_p(G)$. It is easily seen that $(T_{\tilde{\psi}} f)\hat{} = \tilde{\varphi}\tilde{\psi}\hat{f}$ for each $f \in C_c(G)$. Define $S_\psi \in M(L_{q'}(G), L_{p'}(G))$ by $S_\psi f = (T^*_{\tilde{\psi}} f)\tilde{}$ for each $f \in L_{q'}(G)$. Then, if $f \in D(G)$, $h \in D(\hat{G})$ we have by Theorem 5.4.1 that

$$\begin{aligned}\langle \hat{h}, S_\psi f\rangle &= \langle \tilde{\hat{h}}, T^*_{\tilde{\psi}} f\rangle = \langle T_{\tilde{\psi}} \check{h}, f\rangle = \langle (T_{\tilde{\psi}} \check{h})\hat{}, \hat{f}\rangle\\ &= \langle \tilde{\varphi}\tilde{\psi} h, \hat{f}\rangle = \langle h, \varphi\psi\hat{f}\rangle .\end{aligned}$$

Moreover if $f \in C_c(G)$ then $f \in C_c^K(G)$ for some compact set $K \subset G$. By the remark following Theorem 5.1.2 there exists a compact set $K_0 \subset G$ and a sequence $\{f_n\} \subset D_{K_0}(G)$ such that $\lim_n \|f_n - f\|_\infty = 0$, and so, by the Lebesgue Dominated Convergence Theorem (Theorem C.5.2), we also have $\lim_n \|f_n - f\|_1 = 0$ and $\lim_n \|f_n - f\|_{q'} = 0$. Consequently, since $T^*_{\tilde{\psi}}$ is continuous, we have that $\lim_n \|\hat{f}_n - \hat{f}\|_\infty = 0$ and $\lim_n \|T^*_{\tilde{\psi}} f_n - T^*_{\tilde{\psi}} f\|_{p'} = 0$. Hence for each $h \in D(\hat{G})$,

$$\begin{aligned}\langle \hat{h}, S_\psi f\rangle &= \langle \check{h}, T^*_{\tilde{\psi}} f\rangle = \lim_n \langle \check{h}, T^*_{\tilde{\psi}} f_n\rangle = \lim_n \langle \hat{h}, S_\psi f_n\rangle\\ &= \lim_n \langle h, \varphi\psi\hat{f}_n\rangle = \langle h, \varphi\psi\hat{f}\rangle .\end{aligned}$$

Hence for each $h \in D(\hat{G})$ and $f \in C_c(G)$ we have, by Theorem 5.4.1, that

$$\langle h, (S_\psi f)^\wedge \rangle = \langle \hat{h}, S_\psi f \rangle = \langle h, \varphi \psi \hat{f} \rangle.$$

That is, $(S_\psi f)^\wedge = \varphi \psi \hat{f}$, as quasimeasures, which is what we set out to prove.

Therefore $p \leqq 2 \leqq q$, and (iii) implies (iv).

Conversely, suppose that (iv) is valid and $p > 2$. Let $K \subset \hat{G}$ be a compact set such that $\eta(K) > 0$, and let $\varphi \in L_\infty(\hat{G})$ be such that φ vanishes almost everywhere off of K, $\varphi \geqq 0$ and $\varphi > 0$ almost everywhere on a set of positive measure. Then by (iv) there exists a $\psi \in L_\infty(\hat{G})$ such that $|\psi(\gamma)| \leqq \varphi(\gamma)$ for almost all $\gamma \in \hat{G}$ and ψ does not determine a multiplier for $(L_1(G), L_{p'}(G))$, as $1 < p' < 2$. But by Theorem 3.1.1, the space $M(L_1(G), L_{p'}(G))$ can be identified with $L_{p'}(G)$.

Thus $\psi \notin L_{p'}(G)^\wedge$ and (iv) implies (iii).

Next suppose (ii) is valid, and that there exists a φ satisfying the hypotheses of (iv). As we saw in the proof of the implication (iii) implies (iv), this latter assumption entails that for any compact $K \subset \hat{G}$ we have

$$\|\chi_K \hat{f}\|_1 = \int_K |\hat{f}(\gamma)| \, d\eta(\gamma) \leqq b_K \|f\|_p \qquad (f \in C_c(G)).$$

The assumption of the validity of (ii) then requires that $p \leqq 2$. Since $M(L_p(G), L_q(G))$ can be identified with $M(L_{q'}(G), L_{p'}(G))$ by Theorem 5.2.1, the same argument shows that $q' \leqq 2$, that is, $q \geqq 2$.

Hence $p \leqq 2 \leqq q$ and (ii) implies (iv).

Finally, let $p > 2$ and suppose (iii) holds. Then there exists a compact set $K \subset \hat{G}$ and a function $\psi \in L_\infty(\hat{G})$ which vanishes almost everywhere off of K such that $\psi \notin L_{p'}(G)^\wedge$, $1/p + 1/p' = 1$. In analogy with the first portion of the proof of the implication (iii) implies (iv), we define $Th = (\psi \hat{h})^\vee$ for each $h \in D(\hat{G})^\vee$. It is easily seen that T is a linear transformation from $D(\hat{G})^\vee \subset L_1(G) \cap C_0(G) \subset L_p(G)$ to $L_\infty(G)$ which commutes with translation. Moreover, we note that $D(\hat{G})^\vee$ is norm dense in $L_p(G)$ as can be seen by essentially the same argument as that given in the discussion immediately preceding Theorem 5.4.1.

We claim that T is not continuous from $D(\hat{G})^\vee$ considered as a subspace of $L_p(G)$ to $L_\infty(G)$. As if T were continuous then it could be uniquely extended to an element of $M(L_p(G), L_\infty(G))$, since $D(\hat{G})^\vee$ is norm dense in $L_p(G)$. Hence, by Theorem 3.3.1, there would exist a unique $g \in L_{p'}(G)$ such that $(\psi \hat{h})^\vee = Th = h * g$ for each $h \in D(\hat{G})^\vee$. But then, since $(\psi \hat{h})^\vee$ and $h * g$ are continuous for each $h \in D(\hat{G})^\vee$, if $k \in D(\hat{G})$ we have $h = \check{k} = \check{\check{k}} \in D(\hat{G})^\vee$, and

$$\langle k, \hat{g} \rangle = \langle \check{k}, g \rangle = \langle h, g \rangle = h * g(e) = (\psi \hat{h})^\vee(e) = \langle \hat{h}, \psi \rangle = \langle k, \psi \rangle.$$

Consequently $\hat{g}$ and ψ define the same quasimeasure, and hence $\psi \in L_{p'}(G)^\wedge$, contrary to the choice of ψ. Therefore T is not continuous.

Thus there exists a sequence $\{h_n\} \subset D(\hat{G})^\vee \subset L_1(G) \cap C_0(G)$ such that $\sup_n \|h_n\|_p < \infty$ and $\sup_n \|(\psi \hat{h}_n)^\vee\|_\infty = \infty$. The latter assertion clearly implies that $\sup_n \|\psi \hat{h}_n\|_1 = \infty$. It is then easily seen by means of the denseness of $C_c(G)$ in $L_1(G) \cap C_0(G)$ that there exists a sequence $\{f_n\} \subset C_c(G)$ for which $\sup_n \|f_n\|_p < \infty$ and $\sup_n \|\psi \hat{f}_n\|_1 = \infty$. We deduce from this that also $\sup_n \|\chi_K \hat{f}_n\|_1 = \infty$. As if the sequence $\{\|\chi_K \hat{f}_n\|_1\}$ were bounded then so would be $\{\|\psi \hat{f}_n\|_1\}$ because $\psi \in L_\infty(\hat{G})$ and ψ vanishes almost everywhere off of K. Therefore $\sup_n \|f_n\|_p < \infty$ and $\sup_n \|\chi_K \hat{f}_n\|_1 = \infty$, that is, (iii) implies (ii).

Therefore the equivalence of (i)–(iv) is established. ∎

Recalling Corollary 5.5.1 we see that all of (i)–(iv) are valid whenever G is a noncompact locally compact Abelian group.

The techniques of the proofs of Theorems 5.5.1 and 5.5.2 allow us to give an alternative proof of Theorem 4.6.2 which does not utilize the results on the existence of sets of uniqueness for $L_p(G)$. We shall prove a slightly more general theorem than previously from which Theorem 4.6.2 can be immediately deduced.

Theorem 5.5.4. *Let G be a noncompact locally compact Abelian group. If f is a function on $\hat{G}$ such that $\varphi f \in \bigcup_{1 \leqq p < 2} L_p(G)^\wedge$ for each $\varphi \in C_0(\hat{G})$ then $f = 0$ locally almost everywhere.*

Proof. Suppose f is not zero locally almost everywhere. As in the proof of Theorem 4.6.2 we may assume, without loss of generality, that f vanishes off of some compact set K with positive measure.

If $\varphi \in C_0(\hat{G})$ is such that $\varphi \equiv 1$ on K then there exists a p, $1 \leqq p < 2$, for which $\varphi f = f \in L_p(G)^\wedge$. By the Hausdorff-Young Theorem (Theorem F.8.4) we conclude that $f \in L_{p'}(\hat{G})$ for some $p' > 2$. An application of Hölder's inequality then shows that $f \in L_2(\hat{G})$. Thus for any $\varphi \in C_0(\hat{G})$, we have, on the one hand, that there exists $g \in L_p(G)$ for some p, $1 \leqq p < 2$, such that $\varphi f = \hat{g}$, while on the other hand, since $f \in L_2(\hat{G})$ we see that there exists some $g' \in L_2(G)$ such that $\varphi f = \hat{g}'$. Clearly $g = g'$ and so $g \in L_2(G)$. Consequently, by the convexity of $\log \|\cdot\|_{1/a}$, $0 \leqq a \leqq 1$ (Theorem D.11.3), we see that $g \in L_q(G)$, $p \leqq q \leqq 2$. Further, we note that $f \in L_2(\hat{G})$ implies that f is finite almost everywhere.

Let $\{p_n\}$ be a sequence of real numbers such that $1 < p_n < 2$ and $\lim_n p_n = 2$. The preceding observations show that $\varphi f \in \bigcup_{n=1}^\infty L_{p_n}(G)^\wedge$ for each $\varphi \in C_0(\hat{G})$. Define for each pair of positive integers n and k the set

$$C(n, k) = \{\varphi \mid \varphi \in C_0(\hat{G}),\ \varphi f = \hat{g} \text{ for some } g \in L_{p_n}(G),\ \|g\|_{p_n} \leqq k\},$$

where the equality sign here means equality almost everywhere. Then each $C(n,k)$ is a closed subset of $C_0(\hat{G})$. Indeed, suppose $\{\varphi_m\}\subset C(n,k)$ is a sequence which converges uniformly to $\varphi\in C_0(\hat{G})$. Let $\{g_m\}\in L_{p_n}(G)$ be such that $\varphi_m f=\hat{g}_m$, $\|g_m\|_{p_n}\leqq k$, and let $g\in L_p(G)$ for some p, $1\leqq p<2$, be such that $\varphi f=\hat{g}$. From the preceding discussion we note that $\{g_m\}\subset L_2(G)$ and $g\in L_2(G)$. Since $f\in L_2(\hat{G})$ and $\{\varphi_m\}$ converges uniformly to φ, we conclude, by the Lebesgue Dominated Convergence Theorem (Theorem C.5.2), that

$$\lim_m \|\varphi_m f-\varphi f\|_2=\lim_m \|\hat{g}_m-\hat{g}\|_2=0.$$

Thus, by the Plancherel Theorem and Parseval's Formula (Theorems F.8.2 and F.8.3), we see for each $h\in D(G)$ that

$$\begin{aligned}|\langle h,g\rangle| &= |\langle \check{h},\hat{g}\rangle|=\lim_m |\langle \check{h},\hat{g}_m\rangle|=\lim_m |\langle h,g_m\rangle|\\ &\leqq \limsup_m \|h\|_{p_n'}\|g_m\|_{p_n}\leqq \|h\|_{p_n'}\,k\end{aligned}$$

since $D(G)\subset L_{p_n'}(G)$, $1/p_n+1/p_n'=1$, and $\varphi_m\in C(n,k)$. Hence g defines a continuous linear functional on the norm dense subspace $D(G)$ of $L_{p_n'}(G)$ with norm no greater than k. It is then apparent that $g\in L_{p_n}(G)$ and $\|g\|_{p_n}\leqq k$, that is, $C(n,k)$ is a closed subset of $C_0(\hat{G})$. Furthermore it is clear that $C_0(\hat{G})=\bigcup_{n,k=1}^{\infty} C(n,k)$.

From the Baire Category Theorem (Theorem D.2.1), we conclude that there exists a pair of integers n_0, k_0 such that $C(n_0,k_0)$ contains a neighborhood of the zero element in $C_0(\hat{G})$. Set $p_0=p_{n_0}$. Then $1<p_0<2$ and $\varphi f\in L_{p_0}(G)\hat{}$ for each $\varphi\in C_0(\hat{G})$.

Define the mapping $S\colon C_0(\hat{G})\to L_{p_0}(G)$ determined by the formula $(S\varphi)\hat{}=\varphi f$ for each $\varphi\in C_0(\hat{G})$. Clearly S is linear and well-defined. Moreover, the usual argument appealing to the Closed Graph Theorem shows that S is continuous.

Furthermore, we note that we may assume that f is continuous. For if $\delta\in\hat{G}$ then it is evident that for almost all $\gamma\in\hat{G}$ and each $\varphi\in C_0(\hat{G})$ we have $\tau_\delta\varphi(\gamma)\,\tau_\delta f(\gamma)=[(\cdot,\delta)\,S\varphi(\cdot)]\hat{}(\gamma)$ and $\varphi(\gamma)\,\tau_\delta f(\gamma)=[(\cdot,\delta)\,S\tau_{\delta^{-1}}\cdot\varphi(\cdot)]\hat{}(\gamma)$. Consequently, if $h\in C_c(\hat{G})$ then

$$\begin{aligned}\varphi(\gamma)\,f*h(\gamma) &= \varphi(\gamma)\int_{\hat{G}}\tau_\delta f(\gamma)\,h(\delta)\,d\eta(\delta)\\ &= \int_{\hat{G}}[(\cdot,\delta)\,S\tau_{\delta^{-1}}\varphi(\cdot)]\hat{}(\gamma)\,h(\delta)\,d\eta(\delta)\\ &= \Big[\int_{\hat{G}}(\cdot,\delta)\,S\tau_{\delta^{-1}}\varphi(\cdot)\,h(\delta)\,d\eta(\delta)\Big]\hat{}(\gamma).\end{aligned}$$

The function $\int_{\hat{G}} (\cdot, \delta) S\tau_{\delta^{-1}} \varphi(\cdot) h(\delta) d\eta(\delta)$ belongs to $L_{p_0}(G)$, and the last equality is valid because h has compact support and the mapping $\delta \to S\tau_{\delta^{-1}} \varphi$ from $\hat{G}$ to $L_{p_0}(G)$ is continuous (Theorems D.12.1 and D.12.2). Thus we see that for each $h \in C_c(\hat{G})$ we have $\varphi(f*h) \in L_{p_0}(G)^\wedge$. Since f is not zero locally almost everywhere and vanishes off of a compact set, it follows that $f*h \in C_c(\hat{G})$ for each $h \in C_c(\hat{G})$. Moreover, it is apparent that we can choose $h \in C_c(\hat{G})$ such that $f*h$ is not the zero function in $C_c(\hat{G})$. Hence we need only prove the theorem for $f \in C_c(\hat{G})$.

Assuming $f \in C_c(\hat{G})$, we consider the adjoint transformation $S^*: L_{p_0'}(G) \to M(\hat{G})$ of S. If $\varphi \in C_0(\hat{G})$ then $\varphi f \in L_2(\hat{G})$ and so $S\varphi \in L_2(G)$ by the Plancherel Theorem (Theorem F.8.2). Consequently, if $h \in L_2(G) \cap L_{p_0'}(G)$ then Parseval's Formula (Theorem F.8.3) shows that

$$\begin{aligned}\langle S\varphi, h\rangle &= \int_G S\varphi(t)\, h(t^{-1})\, d\lambda(t) = \int_{\hat{G}} (S\varphi)^\wedge(\gamma)\, \hat{h}(\gamma)\, d\eta(\gamma) \\ &= \int_{\hat{G}} \varphi f(\gamma)\, \hat{h}(\gamma)\, d\eta(\gamma) = \langle \varphi, \tilde{f}\check{h}\rangle = \langle \varphi, S^* h\rangle .\end{aligned}$$

Hence $S^* h = \tilde{f}\check{h}$ for each $h \in L_2(G) \cap L_{p_0'}(G)$. Since $f \in C_c(\hat{G})$ and $\check{h} \in L_2(\hat{G})$, we deduce from the continuity of S^* that $\|\tilde{f}\check{h}\|_1 = \|S^* h\| \leqq \|S^*\| \, \|h\|_{p_0'}$ for each $h \in L_2(G) \cap L_{p_0'}(G)$. If $h \in L_{p_0'}(G)$ then there exists a sequence of functions $\{h_n\} \subset D(G) \subset L_2(G) \cap L_{p_0'}(G)$ such that $\lim_n \|h_n - h\|_{p_0'} = 0$. Clearly the sequence $\{\|h_n\|_{p_0}\}$ is bounded and so the sequence of measures $\{\tilde{f}\check{h}_n\}$ is bounded in $M(\hat{G})$.

Now, since f is not identically zero on $\hat{G}$, there exists an open subset $U \subset \hat{G}$ with compact closure W such that $\tilde{f}$ does not vanish on W and $|\tilde{f}(\gamma)| \geqq \varepsilon > 0$ for $\gamma \in W$. Then the sequence of measures $\{\chi_W \check{h}_n\}$ is bounded in $M(W)$, as otherwise we would have

$$\|\tilde{f}\check{h}_n\|_1 \geqq \int_W |\tilde{f}\check{h}_n(\gamma)|\, d\eta(\gamma) \geqq \varepsilon \int_W |\check{h}_n(\gamma)|\, d\eta(\gamma),$$

contradicting the boundedness of the sequence $\{\tilde{f}\check{h}_n\}$ in $M(\hat{G})$. Thus, by Alaoglu's Theorem (Theorem D.4.3), there exists a measure $\mu \in M(W)$ and a subnet $\{\chi_W \check{h}_{n_k}\}$ of $\{\chi_W \check{h}_n\}$ which converges in the weak* topology to μ. Moreover, for each $g \in D_{W^{-1}}(\hat{G})$ we have

$$\begin{aligned}\langle \chi_W \check{h}, g\rangle &= \langle \check{h}, \chi_{W^{-1}}\, g\rangle = \langle \check{h}, g\rangle = \langle h, \check{g}\rangle = \lim_k \langle h_{n_k}, \check{g}\rangle \\ &= \lim_k \langle \check{h}_{n_k}, g\rangle = \lim_k \langle \chi_W \check{h}_{n_k}, g\rangle = \langle \mu, g\rangle .\end{aligned}$$

Thus we conclude that $\chi_W \check{h} = \mu$.

Consequently we see that for each $h \in L_{p_0'}(G)$ we have $\chi_W \check{\check{h}} = \chi_W \hat{h} \in M(W)$. However, an argument essentially like the ones given in the proofs of Theorems 5.5.1 or 5.5.2, employing Lemma 5.3.1, shows, since $p_0' > 2$, that there exists at least one $h \in L_{p'_0}(G)$ such that $\chi_W \hat{h} \notin M(W)$. But this contradicts the preceding assertion.

Therefore f is zero locally almost everywhere. □

We note that the situation for compact groups is markedly different from that for noncompact groups. Clearly if G is a compact Abelian group and $p>2$ then $L_p(G)^\wedge \subset V(\hat{G})$, so that Corollary 5.5.1 and Theorem 5.5.2 fail. Theorem 5.5.4 is also no longer valid for compact Abelian groups as is shown, for example, by Lemma 4.5.2.

When G is locally compact Abelian and $p=q$, or compact Abelian and $1 \leqq p, q \leqq \infty$, we have seen that to each $T \in M(L_p(G), L_q(G))$ there exists a unique pseudomeasure σ such that $Tf = \sigma * f$, $f \in C_c(G)$. This is the content of Theorems 4.3.1, 4.3.2 and 5.2.3. If G is noncompact and $1 \leqq p < q \leqq \infty$ then this is no longer the case, that is, not every $T \in M(L_p(G), L_q(G))$ is obtained by convolution with a pseudomeasure.

Theorem 5.5.5. *Let G be a noncompact locally compact Abelian group and $1 \leqq p < q \leqq \infty$. Then there exists some $T \in M(L_p(G), L_q(G))$ such that $Tf = \sigma * f$, $f \in C_c(G)$, holds for no pseudomeasure σ in $P(G)$.*

Proof. Suppose the contrary is the case, that is, to each $T \in M(L_p(G), L_q(G))$ there exists some $\sigma \in P(G)$ for which $Tf = \sigma * f$, $f \in C_c(G)$. By Theorem 5.2.2 this pseudomeasure must be unique, and by Corollary 5.2.1 the correspondence so determined between T and σ defines a linear injective mapping from $M(L_p(G), L_q(G))$ into $P(G)$. By Theorem 5.4.3 we see that $(Tf)^\wedge = \hat{\sigma} \hat{f}$ for each $f \in C_c(G)$, where $\hat{\sigma}$ denotes the quasimeasure in $Q(\hat{G})$ which is the Fourier transform of T. Since $\sigma \in P(G)$, we know by Theorem 4.2.2 that σ also has a Fourier transform as a pseudomeasure which belongs to $L_\infty(\hat{G})$. For the moment let us denote this transform as $\hat{\sigma}_P$. We wish to show that the two Fourier transforms are identical, that is, they define the same quasimeasure. To this end let $h \in D(\hat{G})$ and suppose the compact support of h is K. Let U be an open neighborhood of K with compact closure and choose $k \in C_c(G)$, $g \in L_1(G)$ as in the development preceding Theorem 5.4.2 such that $\check{k} \check{g} = 1$ on U. Then appealing to Theorem 4.3.1 and the remarks following Theorems 4.3.2 and 5.4.1, we see that

$$\begin{aligned}\langle h, \hat{\sigma} \rangle &= \langle h \check{k} \check{g}, \hat{\sigma} \rangle = \langle h \check{g}, \hat{\sigma} \hat{k} \rangle = \langle h \check{g}, (\sigma * k)^\wedge \rangle \\ &= \langle h \check{g}, \hat{\sigma}_P \hat{k} \rangle = \langle h \check{k} \check{g}, \hat{\sigma}_P \rangle = \langle h, \hat{\sigma}_P \rangle .\end{aligned}$$

It follows at once that $\hat{\sigma}_P$ defines the same quasimeasure as $\hat{\sigma}$. Hence we have that $(Tf)^\wedge = \hat{\sigma} \hat{f}$ for each $f \in C_c(G)$ where $\hat{\sigma} \in L_\infty(\hat{G})$ is the Fourier transform of the pseudomeasure σ.

Moreover, suppose that $\lim_n \| T_n - T \|_{p,q} = 0$ and $\lim_n \| \sigma_n - \sigma \|_P = 0$ where $T_n f = \sigma_n * f$ for each $f \in C_c(G)$. Then, since by Theorems 5.4.1 and

5.4.2 the Fourier transform is continuous from $M(L_p(G), L_q(G))$ to $Q(\hat{G})$ with the weak* topology, we see at once that

$$\lim_n \langle h, \hat{\sigma}_n\rangle = \lim_n \langle h, \hat{T}_n\rangle = \langle h, \hat{T}\rangle \qquad (h\in D(\hat{G})).$$

While on the other hand,

$$\lim_n \langle h, \hat{\sigma}_n\rangle = \lim_n \int_{\hat{G}} h(\gamma)\, \hat{\sigma}_n(\gamma^{-1})\, d\eta(\gamma)$$
$$= \int_{\hat{G}} h(\gamma)\, \hat{\sigma}(\gamma^{-1})\, d\eta(\gamma) = \langle h, \hat{\sigma}\rangle \qquad (h\in D(\hat{G})),$$

since $\lim_n \|\hat{\sigma}_n - \hat{\sigma}\|_\infty = \lim_n \|\sigma_n - \sigma\|_P = 0$ and $D(\hat{G})\subset C_c(\hat{G})$. Consequently, $\langle h, \hat{T}\rangle = \langle h, \hat{\sigma}\rangle$ for each $h\in D(\hat{G})$ and so $\hat{T}=\hat{\sigma}$. Since the Fourier transform on $M(L_p(G), L_q(G))$ is injective, we conclude, via the Closed Graph Theorem (Theorem D.6.1), that the correspondence between $M(L_p(G), L_q(G))$ and $P(G)$ is continuous. Thus there exists some constant $B>0$ such that

$$\|\sigma\|_P \leqq B\, \|T\|_{p,q}$$

where $Tf=\sigma * f$ for each $f\in C_c(G)$.

But now let U be a compact subset of G with a nonempty interior which contains the identity e of G. Since G is noncompact, it is evident that there exists a sequence $\{t_n\}$ of distinct points in G such that $t_{n+1}\notin \bigcup_{i=1}^{n} t_i U$. Let W be a symmetric neighborhood of e such that $W^2\subset U$ (B.1). Then the sets $\{t_n W\}$ are pairwise disjoint. Indeed if $j>k$ and $t_j W\cap t_k W\neq\emptyset$ then there exist $w_1, w_2\in W$ such that $t_j w_1 = t_k w_2$, from which it follows that $t_j = t_k w_2 w_1^{-1}\in t_k W^2\subset \bigcup_{i=1}^{j-1} t_i U$, contrary to the choice of $\{t_n\}$.

Now for each positive integer n let $U_n = \bigcup_{i=1}^{k(n)} t_i W$, where $k(n)$ is chosen so that

$$|\lambda(U_n) - e^n| \leqq \lambda(W).$$

Since the $\{t_n W\}$ are pairwise disjoint, it is easy to verify that it suffices to choose $k(n)$ such that

$$e^n/\lambda(W) - 1 < k(n) < e^n/\lambda(W) + 1.$$

Define $g_n = n e^{-n}\chi_{U_n}$. Clearly $\{g_n\}\subset L_r(G)$, $1/p - 1/q = 1 - 1/r$. Thus, by Theorem 5.3.3, we see that the mappings $T_n\colon L_p(G)\to L_q(G)$ defined by $T_n f = g_n * f$, $f\in L_p(G)$, belong to $M(L_p(G), L_q(G))$. Furthermore, since $\{g_n\}\subset L_1(G)\subset P(G)$ we see that, considering each g_n as a pseudomeasure, we obtain by Theorem 4.2.2 that

$$\|g_n\|_P = \|\hat{g}_n\|_\infty \geqq |\hat{g}_n(e)| = \left|\int_G g_n(t)\, d\lambda(t)\right| = n\, e^{-n} \lambda(U_n)$$
$$\geqq n\, e^{-n}(e^n - \lambda(W)) = n(1 - e^{-n} \lambda(W)).$$

Hence $\limsup_n \|g_n\|_P = \infty$.

However, since $1 \leqq p < q \leqq \infty$ we see that $(1/p, 1/q)$ lies in the triangle bounded by $a=1$, $b=0$ and $a=b$. By Theorems 5.2.1 and 5.3.2 we may assume, without loss of generality, that $(1/p, 1/q)$ lies in the triangle bounded by $a=1$, $a=b$ and $a+b=1$. Denote the intersection of the straight line through $(\frac{1}{2}, \frac{1}{2})$ and $(1/p, 1/q)$ with $a=1$ by $(1, 1/s)$. Then appealing to the Riesz-Thorin Convexity Theorem (Theorem D.11.2) once again, we conclude that for each n

$$\|T_n\|_{p,q} \leqq (\|T_n\|_{2,2})^\alpha (\|T_n\|_{1,s})^{1-\alpha}$$

where $1/p = \alpha/2 + (1-\alpha)$, $1/q = \alpha/2 + (1-\alpha)/s$. Using Theorems 3.1.1 and 4.1.1, we deduce that for each n such that $e^n > \lambda(W)$ we have

$$\|T_n\|_{p,q} \leqq (\|\hat{g}_n\|_\infty)^\alpha (\|g_n\|_s)^{1-\alpha} \leqq (\|g_n\|_1)^\alpha (\|g_n\|_s)^{1-\alpha}$$
$$= [n\, e^{-n} \lambda(U_n)]^\alpha [n\, e^{-n}]^{1-\alpha} [\lambda(U_n)]^{(1-\alpha)/s}$$
$$\leqq n\, e^{-n} [e^n + \lambda(W)]^{\alpha + (1-\alpha)/s} \leqq n\, e^{-n} [2 e^n]^{\alpha + (1-\alpha)/s}$$
$$= 2^{\alpha + (1-\alpha)/s}\, n\, e^{n[-1 + \alpha + (1-\alpha)/s]}.$$

Since $p < q$, we have $0 \leqq \alpha < 1$ and so $-1 + \alpha + (1-\alpha)/s < 0$. Thus $\lim_n \|T_n\|_{p,q} = 0$.

But this result combined with the previous one that $\limsup_n \|g_n\|_P = \infty$, contradicts the fact that for each n

$$\|g_n\|_P \leqq B\, \|T_n\|_{p,q}.$$

This contradiction proves the theorem. □

5.6. $M(L_p(G), L_q(G))$ as a Dual Space, $1 \leqq p, q < \infty$. In this section we shall show that $M(L_p(G), L_q(G))$ is isometrically isomorphic with the space of continuous linear functionals for certain Banach spaces. Before we can begin this development we need to establish a result concerning the approximation of elements of $M(L_p(G), L_q(G))$ in the strong operator topology by multipliers determined by convolution with an element of $C_c(G)$. This theorem is of considerable interest in its own right.

We recall that if G is a noncompact locally compact Abelian group and $1 < q < p < \infty$ then $M(L_p(G), L_q(G)) = \{0\}$ by Theorem 5.2.5. Thus it is natural to restrict our attention to either the case where G is an arbitrary locally compact Abelian group and $1 \leqq p \leqq q < \infty$ or G is a compact Abelian group and $1 \leqq q < p < \infty$. We begin with the following elementary lemma.

Lemma 5.6.1. *Let G be either an arbitrary locally compact Abelian group and $1 \leqq p \leqq q < \infty$ or a compact Abelian group and $1 \leqq q < p < \infty$. If $g \in C_c(G)$ and $Tf = g * f$, $f \in L_p(G)$, then $T \in M(L_p(G), L_q(G))$.*

Proof. If G is arbitrary and $1 \leqq p \leqq q < \infty$ then it is apparent that $Tf = g * f \in L_\infty(G) \cap L_p(G)$, $f \in L_p(G)$, T is linear and commutes with translation. Moreover, since $\log \|\cdot\|_{1/a}$ is a convex function of a, $0 \leqq a \leqq 1$ (Theorem D.11.3), we conclude for each $f \in L_p(G)$ that

$$\|Tf\|_q = \|g * f\|_q \leqq (\|g * f\|_p)^{p/q} (\|g * f\|_\infty)^{1-p/q},$$

from which it is easily seen that T is a continuous mapping from $L_p(G)$ to $L_q(G)$. Thus $T \in M(L_p(G), L_q(G))$.

When G is compact and $1 \leqq q < p < \infty$ then $L_p(G) \subset L_q(G)$, and it is evident that $Tf = g * f \in C(G) \subset L_p(G)$, $f \in L_p(G)$. Moreover $\|Tf\|_q = \|g * f\|_q \leqq \|g * f\|_p \leqq \|g\|_1 \|f\|_p$ and so $T \in M(L_p(G), L_q(G))$. □

Theorem 5.6.1. *Let G be either an arbitrary locally compact Abelian group and $1 \leqq p \leqq q < \infty$ or a compact Abelian group and $1 \leqq q < p < \infty$. If $T \in M(L_p(G), L_q(G))$ then there exists a net of functions $\{g_\alpha\} \subset C_c(G)$ such that if $T_\alpha f = g_\alpha * f$, $f \in L_p(G)$, then for each $f \in L_p(G)$,*

(i) $\|T_\alpha f\|_q \leqq \|T\|_{p,q} \|f\|_p$ *for each* α.

(ii) $\lim_\alpha \|T_\alpha f - Tf\|_q = 0$.

Proof. We consider three separate cases depending on the values of q and whether or not G is compact.

Suppose first that G is arbitrary, $1 \leqq p \leqq q$ and $q \neq 1$. We claim that it is sufficient, in order to establish the desired conclusion, to show that there exists a net $\{g_\alpha\} \subset C_c(G)$ such that $\lim_\alpha \langle T_\alpha f, g \rangle = \langle Tf, g \rangle$, $f \in L_p(G)$, $g \subset L_{q'}(G)$, that is, that $\{T_\alpha f\}$ converges weakly in $L_q(G)$ to Tf, and $\|T_\alpha f\|_q \leqq \|T\|_{p,q} \|f\|_p$, $f \in L_p(G)$. As suppose this were true. Then for any nonnegative numbers a_i, $i = 1, 2, \ldots, n$ for which $\sum_{i=1}^n a_i = 1$ it is apparent that

$$\left\| \sum_{i=1}^n a_i T_{\alpha_i} f \right\|_q \leqq \sum_{i=1}^n a_i \|T_{\alpha_i} f\|_q$$

$$\leqq \sum_{i=1}^n a_i \|T\|_{p,q} \|f\|_p = \|T\|_{p,q} \|f\|_p,$$

and

$$\lim_\alpha \left\langle \sum_{i=1}^n a_i T_{\alpha_i} f, g \right\rangle = \lim_\alpha \sum_{i=1}^n a_i \langle T_{\alpha_i} f, g \rangle$$

$$= \sum_{i=1}^n a_i \langle Tf, g \rangle = \langle Tf, g \rangle$$

for any choice of the a_i and each $f \in L_p(G)$, $g \in L_{q'}(G)$. Hence fixed convex combinations of $\{T_\alpha\}$ have the same properties as the $\{T_\alpha\}$ themselves. But since the closures of a convex set in the weak and strong operator topologies are identical (Theorem D.7.4), we conclude that there exists a net of functions $\{g_\alpha\} \subset C_c(G)$ such that $\{T_\alpha\}$ has the required properties with respect to the strong operator topology, which is what we wish to establish.

Now let $\{u_\beta\}$ be an approximate identity for $L_1(G)$ such that $\{u_\beta\} \subset C_c(G) * C_c(G)$, $u_\beta \geqq 0$, all the u_β are zero off of some common compact set, and $\|u_\beta\|_1 = 1$ (F.7d). And let $\{v_\delta\}$ be an approximate identity for $L_1(\hat{G})$ such that $\|v_\delta\|_1 = 1$ and $\{\hat{v}_\delta\} \subset C_c(G)$ (F.7d). Since T is continuous and commutes with translation it can be easily shown, by using essentially the same argument as the one given in the proof of Theorem 4.1.1, that $Tf * g = f * Tg = T(f * g)$, $f, g \in C_c(G)$. Thus, since $\{u_\beta\} \subset C_c(G) * C_c(G)$ we see that $\{Tu_\beta\} \subset C_0(G)$, as $L_q(G) * L_{q'}(G) \subset C_0(G)$, $1/q + 1/q' = 1$. Hence $\hat{v}_\delta\, Tu_\beta \in C_c(G)$ for each δ and β. Ordering the product space of the index spaces of the nets $\{u_\beta\}$ and $\{v_\delta\}$ lexicographically, we obtain a net $\{g_\alpha\} \subset C_c(G)$ upon setting $g_\alpha = \hat{v}_\delta\, Tu_\beta$ whenever $\alpha = (\beta, \delta)$. We claim that the net of multipliers $\{T_\alpha\} \subset M(L_p(G), L_q(G))$ defined by $T_\alpha f = g_\alpha * f$ satisfies the inequality (i) and converges in the weak operator topology to T.

Indeed, for $f, g \in C_c(G)$, and each α we have

$$
\begin{aligned}
|\langle T_\alpha f, g\rangle| & \\
&= |\langle g_\alpha, f * g\rangle| = \Big|\int_G \hat{v}_\delta\, Tu_\beta(s^{-1}) f * g(s)\, d\lambda(s)\Big| \\
&= \Big|\int_G \int_G \int_{\hat{G}} v_\delta(\gamma)(s^{-1}, \gamma^{-1})\, Tu_\beta(s^{-1}) f(s t^{-1})\, g(t)\, d\eta(\gamma)\, d\lambda(t)\, d\lambda(s)\Big| \\
&\leqq \int_{\hat{G}} |v_\delta(\gamma)| \Big\{\Big|\int_G \int_G Tu_\beta(s^{-1}) f(s t^{-1})\, g(t)(s, \gamma)\, d\lambda(t)\, d\lambda(s)\Big|\Big\}\, d\eta(\gamma) \\
&= \int_{\hat{G}} |v_\delta(\gamma)| \Big\{\Big|\int_G \int_G Tu_\beta(s^{-1})(s t^{-1}, \gamma) f(s t^{-1})(t, \gamma)\, g(t) \\
&\qquad \cdot d\lambda(t)\, d\lambda(s)\Big|\Big\}\, d\eta(\gamma) \\
&= \int_{\hat{G}} |v_\delta(\gamma)| \Big\{\Big|\int_G Tu_\beta(s^{-1})\, \gamma f * \gamma g(s)\, d\lambda(s)\Big|\Big\}\, d\eta(\gamma) \\
&\leqq \sup_{\gamma \in \hat{G}} |\langle Tu_\beta, \gamma f * \gamma g\rangle|\, \|v_\delta\|_1 = \sup_{\gamma \in \hat{G}} |\langle Tu_\beta * \gamma f, \gamma g\rangle| \\
&= \sup_{\gamma \in \hat{G}} |\langle T(u_\beta * \gamma f), \gamma g\rangle| \leqq \sup_{\gamma \in \hat{G}} \|T(u_\beta * \gamma f)\|_q\, \|\gamma g\|_{q'} \\
&\leqq \|T\|_{p,q} \sup_{\gamma \in \hat{G}} \|u_\beta * \gamma f\|_p\, \|\gamma g\|_{q'} \\
&\leqq \|T\|_{p,q}\, \|u_\beta\|_1 \sup_{\gamma \in \hat{G}} \|\gamma f\|_p\, \|\gamma g\|_{q'} = \|T\|_{p,q}\, \|f\|_p\, \|g\|_{q'},
\end{aligned}
$$

where, as usual, $1/q + 1/q' = 1$. Since $q' \neq \infty$, $C_c(G)$ is norm dense in $L_{q'}(G)$ and so we deduce from the previous inequalities for each α that

$$\|T_\alpha f\|_q \leqq \|T\|_{p,q} \|f\|_p \qquad (f \in C_c(G)).$$

Similarly, since $p \neq \infty$ and T_α is continuous, we conclude for each α that

$$\|T_\alpha f\|_q \leqq \|T\|_{p,q} \|f\|_p \qquad (f \in L_p(G)).$$

In particular, then we have for each α that $\|T_\alpha\|_{p,q} \leqq \|T\|_{p,q}$. Since $q \neq 1$, we note by the reflexivity of $L_q(G)$ and Alaoglu's Theorem (Theorem D.4.3) that closed balls in $L_q(G)$ are compact in the weak topology induced on $L_q(G)$ by $L_{q'}(G)$, and hence closed balls in $M(L_p(G), L_q(G))$ are compact in the weak operator topology. Thus the net $\{T_\alpha\}$ has a subnet $\{T_\gamma\}$ which converges in the weak operator topology to some $S \in M(L_p(G), L_q(G))$ such that $\|S\|_{p,q} \leqq \|T\|_{p,q}$. We claim that $S = T$. As suppose $\varepsilon > 0$, then for each $f, g \in C_c(G)$,

$$\begin{aligned}
&|\langle Tf, g\rangle - \langle T_\gamma f, g\rangle| \\
&\quad \leqq |\langle Tf, g\rangle - \langle u_\beta * Tf, g\rangle| + |\langle u_\beta * Tf, g\rangle - \langle \hat{v}_\delta\, Tu_\beta * f, g\rangle| \\
&\quad = |\langle Tf, g - u_\beta * g\rangle| + |\langle Tu_\beta - \hat{v}_\delta\, Tu_\beta, f * g\rangle| \\
&\quad \leqq \|Tf\|_q \|g - u_\beta * g\|_{q'} + \sup_{s \in K^{-1}} |1 - \hat{v}_\delta(s)| \, \|Tu_\beta\|_q \, \|f * g\|_{q'},
\end{aligned}$$

where K denotes the compact support of $f * g$. Because the supports of the u_β are contained in a common compact set it is easily seen that there exists some β_0 such that $\|g - u_{\beta_0} * g\|_{q'} < \varepsilon/2\, \|Tf\|_q$. For this β_0, since $\{\hat{v}_\delta\}$ converges uniformly to one on compact subsets of G, there exists a δ_0 such that

$$\sup_{s \in K^{-1}} |1 - \hat{v}_{\delta_0}(s)| < \varepsilon/2\, \|Tu_{\beta_0}\|_q \, \|f * g\|_{q'}.$$

Moreover, it is obvious that we may choose (β_0, δ_0) such that $T_{(\beta_0, \delta_0)}$ belongs to the subnet $\{T_\gamma\}$. Combining these inequalities we see at once that for $\gamma_0 = (\beta_0, \delta_0)$, $|\langle Tf, g\rangle - \langle T_{\gamma_0} f, g\rangle| < \varepsilon$. Consequently, since ε is arbitrary, we conclude that $\langle Tf, g\rangle = \langle Sf, g\rangle$ for $f, g \in C_c(G)$, from which we deduce, by means of the continuity of T and S and the norm denseness of $C_c(G)$ in $L_p(G)$ and $L_{q'}(G)$, that $T = S$.

Therefore, we have shown for each α that

$$\|T_\alpha f\|_q \leqq \|T\|_{p,q} \|f\|_p \qquad (f \in L_p(G)),$$

and

$$\lim_\alpha \langle T_\alpha f, g\rangle = \langle Tf, g\rangle \qquad (f \in L_p(G),\ g \in L_{q'}(G)).$$

In view of our introductory remarks, this is sufficient to prove the theorem in the case $q \neq 1$ and G arbitrary.

Now consider the case of a noncompact locally compact Abelian G and $p=q=1$. Then by Theorem 0.1.1 or Theorem 3.1.1 there exists a unique $\mu \in M(G)$ such that $Tf=\mu * f$, $f \in L_1(G)$. Let $\{u_\beta\}$ be an approximate identity for $L_1(G)$ as before, and for each compact set $K \subset G$ denote by μ_K the measure in $M(G)$ defined by

$$\mu_K(E)=\int_E \chi_K(s)\, d\mu(s).$$

Consider the compact subsets of G as directed by inclusion, that is, $K_1 \succ K_2$ if and only if $K_1 \supset K_2$. Then define $g_\alpha \in C_c(G)$ by $g_\alpha = g_{(\beta, K)} = u_\beta * \mu_K$, where we again order the pairs (β, K) lexicographically.

Then for each $f \in L_1(G)$,

$$\begin{aligned} \|T_\alpha f - Tf\|_1 &\leqq \|u_\beta * \mu_K * f - u_\beta * \mu * f\|_1 + \|u_\beta * \mu * f - \mu * f\|_1 \\ &\leqq \|\mu_K - \mu\| \, \|u_\beta * f\|_1 + \|u_\beta * \mu * f - \mu * f\|_1 \\ &\leqq \|\mu_K - \mu\| \, \|f\|_1 + \|u_\beta * \mu * f - \mu * f\|_1 . \end{aligned}$$

It is then apparent, since $\{u_\beta\}$ is an approximate identity for $L_1(G)$ and μ is regular, that $\lim_\alpha \|T_\alpha f - Tf\|_1 = 0$. Furthermore, for each α,

$$\|T_\alpha f\|_1 = \|u_\beta * \mu_K * f\|_1 \leqq \|\mu_K\| \, \|u_\beta * f\|_1 \leqq \|\mu\| \, \|f\|_1 = \|T\|_{1,1} \|f\|_1,$$

by Theorem 3.1.1. Thus the theorem is established for noncompact G and $p=q=1$.

Finally, suppose G is compact and $q=1$, $1 \leqq p < \infty$. This time choose $\{u_\beta\}$ to be an approximate identity for $L_1(G)$ consisting of trigonometric polynomials (F.7d), and set $g_\alpha = g_\beta = Tu_\beta$. Then $\{g_\alpha\}$ is again a net of trigonometric polynomials. This follows immediately from the linearity of T upon noting, since G is compact, that $T(\cdot,\gamma) = T[(\cdot,\gamma)*(\cdot,\gamma)] = [T(\cdot,\gamma)]*(\cdot,\gamma) = [T(\cdot,\gamma)]\hat{\ }(\gamma)(\cdot,\gamma)$. Thus $\{g_\alpha\} \subset C(G)$. Moreover in a similar fashion as before, we can show that now $Tf * g = f * Tg$, $f, g \in L_p(G)$.

Hence for each α,

$$\|T_\alpha f\|_1 = \|Tu_\alpha * f\|_1 = \|u_\alpha * Tf\|_1 \leqq \|u_\alpha\|_1 \, \|Tf\|_1 \leqq \|T\|_{p,1} \, \|f\|_p \quad (f \in L_p(G)),$$

and

$$\lim_\alpha \|T_\alpha f - Tf\|_1 = \lim_\alpha \|u_\alpha * Tf - Tf\|_1 = 0,$$

as $\{u_\alpha\}$ is an approximate identity for $L_1(G)$.

Therefore the theorem is valid when G is compact and $q=1$, $1 \leqq p < \infty$, which completes the proof. ▯

We shall shortly make use of this theorem. First, however, we need to introduce some new spaces which we shall denote by $A_p^q(G)$. Let G be an arbitrary locally compact Abelian group and $1 \leqq p \leqq q < \infty$. If $p < q$ we set $1/r = 1/p - 1/q$, and define $A_p^q(G)$ to be that subset of $L_r(G)$ consisting of the elements h which can be expressed as $h = \sum_{i=1}^{\infty} f_i * g_i$ almost everywhere with respect to Haar measure λ, where $f_i, g_i \in C_c(G), i = 1, 2, \ldots$ and $\sum_{i=1}^{\infty} \|f_i\|_p \|g_i\|_{q'} < \infty$. We recall when $1/r = 1/p - 1/q$ that $\|f * g\|_r \leqq \|f\|_p \|g\|_{q'}, f \in L_p(G), g \in L_{q'}(G)$ (F.4), so that the functions h defined above indeed lie in $L_r(G)$, and $\lim_n \left\| h - \sum_{i=1}^{n} f_i * g_i \right\|_r = 0$. If $p = q$ then $A_p^q(G)$ is the subset of $C_0(G)$ consisting of

$$h = \sum_{i=1}^{\infty} f_i * g_i, \qquad f_i, g_i \in C_c(G), \quad \text{and} \quad \sum_{i=1}^{\infty} \|f_i\|_p \|g_i\|_{p'} < \infty.$$

That $A_p^p(G) \subset C_0(G)$ is evident from the fact that $\|f_i * g_i\|_\infty \leqq \|f_i\|_p \|g_i\|_{p'}$, $i = 1, 2, 3, \ldots$, and so $\lim_n \left\| h - \sum_{i=1}^{n} f_i * g_i \right\|_\infty = 0$. In this case we also assume that $h = \sum_{i=1}^{\infty} f_i * g_i$ pointwise. Finally, let G be a compact Abelian group and $1 \leqq q < p < \infty$. Then $A_p^q(G)$ is the subset of $C(G)$ consisting of functions $h = \sum_{i=1}^{\infty} f_i * g_i$ where the equality holds pointwise, $f_i, g_i \in C(g), i = 1, 2, 3, \ldots$, and $\sum_{i=1}^{\infty} \|f_i\|_p \|g_i\|_{q'} < \infty$. Since $\|f_i * g_i\|_\infty \leqq \|f_i\|_p \|g_i\|_{p'} \leqq \|f_i\|_p \|g_i\|_{q'}$, $i = 1, 2, 3, \ldots$, as $p' < q'$ and G is compact, we see once again that

$$\lim_n \left\| h - \sum_{i=1}^{n} f_i * g_i \right\|_\infty = 0.$$

It is obvious that each of the $A_p^q(G)$, as defined, is a linear space. If $h \in A_p^q(G)$ we set

$$|||h||| = \inf \left\{ \sum_{i=1}^{\infty} \|f_i\|_p \|g_i\|_{q'} \right\}$$

where the infimum is taken over all representations $h = \sum_{i=1}^{\infty} f_i * g_i$, $f_i, g_i \in C_c(G)$, with $\sum_{i=1}^{\infty} \|f_i\|_p \|g_i\|_{q'} < \infty$. In view of the preceding discussion, it is apparent that if G is an arbitrary locally compact Abelian group and $1 \leqq p \leqq q < \infty$ then when $p < q$, $\|h\|_r \leqq |||h|||$, when $p = q$, $\|h\|_\infty \leqq |||h|||$, and if G is a compact Abelian group and $1 \leqq q < p < \infty$ then $\|h\|_\infty \leqq |||h|||$. It is now easily seen that $|||\cdot|||$ defines a norm for $A_p^q(G)$. Furthermore by essentially the same proof, *mutatis mutandis*, as that given for Theorem 5.1.1, one can show that $|||\cdot|||$ is a complete norm for $A_p^q(G)$. We shall omit the details.

We can summarize the preceding discussion in the following theorem.

Theorem 5.6.2. *Let G be either an arbitrary locally compact Abelian group and $1 \leqq p \leqq q < \infty$ or a compact Abelian group and $1 \leqq q < p < \infty$. Then $A_p^q(G)$ with the norm $|||\cdot|||$ is a Banach space.*

We are now in a position to establish the main result of this section, namely that $M(L_p(G), L_q(G))$ can be identified with the space of continuous linear functionals on $A_p^q(G)$.

Theorem 5.6.3. *Let G be either an arbitrary locally compact Abelian group and $1 \leqq p \leqq q < \infty$ or a compact Abelian group and $1 \leqq q < p < \infty$. Then there exists an isometric linear isomorphism of $M(L_p(G), L_q(G))$ onto $[A_p^q(G)]^*$, the Banach space of continuous linear functionals on $A_p^q(G)$.*

Proof. Let $T \in M(L_p(G), L_q(G))$. If $h \in A_p^q(G)$ has a representation $h = \sum_{i=1}^{\infty} f_i * g_i$ then set

$$\alpha(T)(h) = \sum_{i=1}^{\infty} \langle Tf_i, g_i \rangle .$$

First we note, since $Tf_i \in L_q(G)$ and $g_i \in C_c(G) \subset L_{q'}(G)$ for each i, that

$$\begin{aligned} |\alpha(T)(h)| &= \left| \sum_{i=1}^{\infty} \langle Tf_i, g_i \rangle \right| \leqq \sum_{i=1}^{\infty} \|Tf_i\|_q \, \|g_i\|_{q'} \\ &\leqq \|T\|_{p,q} \sum_{i=1}^{\infty} \|f_i\|_p \, \|g_i\|_{q'} < \infty . \end{aligned}$$

It is apparent that $\alpha(T)$ is linear. Moreover $\alpha(T)(h)$ is independent of the representation of h.

Indeed, suppose $h \in A_p^q(G)$ and $h = 0$. Let $\{g_\alpha\} \subset C_c(G)$ be a net of functions such that if $T_\alpha f = g_\alpha * f$, $f \in L_p(G)$, then $\|T_\alpha f\|_q \leqq \|T\|_{p,q} \|f\|_p$ for each α and $\lim_\alpha \|T_\alpha f = Tf\|_q = 0$, $f \in L_p(G)$. Such a net $\{g_\alpha\}$ exists by Theorem 5.6.1. Suppose $h = \sum_{i=1}^{\infty} f_i * g_i$, $f_i, g_i \in C_c(G)$. Then for each α we have by Lemma 5.6.1 that

$$\sum_{i=1}^{\infty} \|T_\alpha f_i * g_i\|_\infty \leqq \sum_{i=1}^{\infty} \|T_\alpha f_i\|_q \, \|g_i\|_{q'} \leqq \|T\|_{p,q} \sum_{i=1}^{\infty} \|f_i\|_p \, \|g_i\|_{q'} < \infty .$$

Hence $\sum_{i=1}^{\infty} T_\alpha f_i * g_i$ converges in the supremum norm uniformly with respect to α. Thus

$$\lim_\alpha \sum_{i=1}^{\infty} \langle T_\alpha f_i, g_i \rangle = \sum_{i=1}^{\infty} \lim_\alpha \langle T_\alpha f_i, g_i \rangle = \sum_{i=1}^{\infty} \langle T f_i, g_i \rangle,$$

the interchange of limits being justified by the uniformity of the convergence of the series with respect to α.

However, for each α we also have

$$\sum_{i=1}^{\infty} \langle T_\alpha f_i, g_i \rangle = \sum_{i=1}^{\infty} \langle g_\alpha * f_i, g_i \rangle = \sum_{i=1}^{\infty} \langle g_\alpha, f_i * g_i \rangle$$
$$= \langle g_\alpha, \sum_{i=1}^{\infty} f_i * g_i \rangle = 0.$$

The penultimate equality here is valid since in all cases $g_\alpha \in C_c(G)$ defines a continuous linear functional on either $L_r(G)$ or $C_0(G)$ and $\left\{ \sum_{i=1}^{n} f_i * g_i \right\}$ converges to $\sum_{i=1}^{\infty} f_i * g_i$ in $L_r(G)$ or $C_0(G)$, respectively. Consequently, if $h \in A_p^q(G)$ is such that $h = 0$ then $\sum_{i=1}^{\infty} \langle T f_i, g_i \rangle = 0$ for each representation $h = \sum_{i=1}^{\infty} f_i * g_i$.

Thus $\alpha(T)$ is a well defined linear functional on $A_p^q(G)$.

Moreover, $\alpha(T)$ is continuous since

$$|\alpha(T)(h)| = \left| \sum_{i=1}^{\infty} \langle T f_i, g_i \rangle \right| \leqq \| T \|_{p,q} \sum_{i=1}^{\infty} \| f_i \|_p \, \| g_i \|_{q'}$$

for each representation $h = \sum_{i=1}^{\infty} f_i * g_i$. Hence $|\alpha(T)(h)| \leqq \| T \|_{p,q} \, |||h|||$, $h \in A_p^q(G)$, and so $\alpha(T) \in [A_p^q(G)]^*$ with $\| \alpha(T) \| \leqq \| T \|_{p,q}$. Furthermore, it is evident that

$$\begin{aligned} \| T \|_{p,q} &= \sup \{ |\langle T f, g \rangle| \,|\, f, g \in C_c(G), \| f \|_p \leqq 1, \| g \|_{q'} \leqq 1 \} \\ &\leqq \sup \{ |\alpha(T)(h)| \,|\, h = f * g, f, g \in C_c(G), \| f \|_p \leqq 1, \| g \|_{q'} \leqq 1 \} \\ &\leqq \| \alpha(T) \|. \end{aligned}$$

Thus $\| T \|_{p,q} = \| \alpha(T) \|$.

Clearly then α: $M(L_p(G), L_q(G)) \to [A_p^q(G)]^*$ is an isometric linear isomorphism of $M(L_p(G), L_q(G))$ into $[A_p^q(G)]^*$. The proof will be complete once we have shown that the mapping is surjective.

Let $F\in[A_p^q(G)]^*$. For each $f\in C_c(G)$ define $F_f(g)=F(f*g)$, $g\in C_c(G)$. Clearly F_f defines a linear functional on $C_c(G)$, and

$$|F_f(g)|=|F(f*g)|\leqq\|F\|\,\|f\|_p\,\|g\|_{q'} \qquad (g\in C_c(G)).$$

Suppose that $q\neq 1$. The previous inequality shows that F_f defines a continuous linear functional on $C_c(G)$ considered as a subspace of $L_{q'}(G)$. Since $C_c(G)$ is norm dense in $L_{q'}(G)$ and $[L_{q'}(G)]^*=L_q(G)$, we conclude that there exists a unique $T'f\in L_q(G)$ such that $F_f(g)=F(f*g)=\langle T'f, g\rangle$, $g\in C_c(G)$, and $\|T'f\|_q\leqq\|F\|\,\|f\|_p$. Thus we obtain a continuous linear mapping T' from $C_c(G)$ considered as a subspace of $L_p(G)$ to $L_q(G)$. Since $C_c(G)$ is norm dense in $L_p(G)$, we can uniquely extend T' to a continuous linear mapping $T\colon L_p(G)\to L_q(G)$ for which $\|Tf\|_q\leqq\|F\|\,\|f\|_p$, $f\in L_p(G)$. Moreover, for each $s\in G$,

$$\langle T\tau_s f, g\rangle=F(\tau_s f*g)=F(f*\tau_s g)=\langle Tf, \tau_s g\rangle=\langle\tau_s Tf, g\rangle \qquad (f, g\in C_c(G)).$$

Hence, from the continuity of T and the density of $C_c(G)$ in $L_p(G)$ and $L_q(G)$, we conclude that $T\in M(L_p(G), L_q(G))$.

If $q=1$, then by similar reasoning as above we deduce the existence of a continuous linear mapping $T\colon L_p(G)\to M(G)$ which commutes with translation such that $\|Tf\|\leqq\|F\|\,\|f\|_p$, $f\in L_p(G)$, and $\langle Tf, g\rangle=F(f*g)$, $f, g\in C_c(G)$. However, for each $s\in G$, the estimates

$$\|\tau_s Tf-Tf\|=\|T(\tau_s f)-Tf\|\leqq\|F\|\,\|\tau_s f-f\|_p \qquad (f\in L_p(G)),$$

coupled with the continuity of translation in $L_p(G)$ (F.7a), reveal that the mapping $s\to\tau_s Tf$ is continuous from G to $M(G)$. Hence $Tf\in L_1(G)$, $f\in L_p(G)$ (F.7a). Therefore $T\in M(L_p(G), L_1(G))$.

In any case then, given $F\in[A_p^q(G)]^*$ there exists a unique $T\in M(L_p(G), L_q(G))$ such that $\langle Tf, g\rangle=F(f*g)$, $f, g\in C_c(G)$. But if $h\in A_p^q(G)$ has representation $h=\sum_{i=1}^{\infty} f_i*g_i$ then

$$\alpha(T)(h)=\sum_{i=1}^{\infty}\langle Tf_i, g_i\rangle=\sum_{i=1}^{\infty}F(f_i*g_i)=F(h),$$

since $\left\{\sum_{i=1}^{n} f_i*g_i\right\}$ converges to h in $A_p^q(G)$. That is, $\alpha(T)=F$.

Therefore α is surjective and the proof is complete. □

It should perhaps be noted explicitly that if G is arbitrary and $p=q$ or G is compact and $1\leqq q<p<\infty$ then the $A_p^q(G)$ are Banach spaces of continuous functions.

Combining the previous theorem with Theorem 3.1.1 we immediately obtain the following corollary.

Corollary 5.6.1. *Let G be a locally compact Abelian group. Then:*

(i) *There exists an isometric linear isomorphism of $M(G)$ onto $[A_1^1(G)]^*$.*

(ii) *There exists an isometric linear isomorphism of $L_q(G)$ onto $[A_1^q(G)]^*$.*

It can also be shown in this case that $A_1^1(G) = C_0(G)$ and $A_1^q(G) = L_{q'}(G)$, $1 < q < \infty$, $1/q + 1/q' = 1$, as is to be suspected from the corollary. We shall not provide the proofs here but refer the reader to the notes at the end of the chapter and the references given there.

Another interesting result is available by utilizing Theorem 5.3.3 and the previous theorem.

Corollary 5.6.2. *Let G be an infinite locally compact Abelian group and $1 < p < q < \infty$. If $1/p - 1/q = 1 - 1/r$ then $L_p(G) * L_r(G)$ is a proper subset of $L_q(G)$.*

Proof. Evidently $1/p + 1/r - 1/q = 1$, and so we have that $L_p(G) * L_r(G) \subset L_q(G)$ and $\|f * g\|_q \leq \|f\|_p \|g\|_r$, $f \in L_p(G)$, $g \in L_r(G)$ (F.4). We claim that also $L_p(G) * L_r(G) \subset A_p^{r'}(G)$, $1/r + 1/r' = 1$. Indeed, let $f \in L_p(G)$, $g \in L_r(G)$. Choose $f_1 \in C_c(G)$ such that $\|f - f_1\|_p < \frac{1}{2}$. Assuming $f_1, f_2, \ldots, f_n \in C_c(G)$ have been chosen so that $\left\|f - \sum_{i=1}^{n} f_i\right\|_p < 1/2^n$, we choose $f_{n+1} \in C_c(G)$ such that

$$\left\|f - \sum_{i=1}^{n} f_i - f_{n+1}\right\|_p = \left\|f - \sum_{i=1}^{n+1} f_i\right\|_p \leq 1/2^{n+1}.$$

Such a choice of f_{n+1} is possible as $C_c(G)$ is norm dense in $L_p(G)$. In this way we obtain a sequence $\{f_i\} \subset C_c(G)$ such that $\lim_n \left\|f - \sum_{i=1}^{n} f_i\right\|_p = 0$, and for which

$$\begin{aligned}\|f_n\|_p &= \left\|f - \left(\sum_{i=1}^{n-1} f_i\right) - \left(f - \left(\sum_{i=1}^{n} f_i\right)\right)\right\|_p \\ &\leq \left\|f - \sum_{i=1}^{n-1} f_i\right\|_p + \left\|f - \sum_{i=1}^{n} f_i\right\|_p \\ &< 1/2^{n-1} + 1/2^n = 3/2^n\end{aligned}$$

for each positive integer n. In a similar manner we construct a sequence $\{g_j\} \subset C_c(G)$ such that $\lim_m \left\|g - \sum_{j=1}^{m} g_j\right\|_r = 0$ and $\|g_m\|_r < 3/2^m$ for each positive integer m.

But $h = \sum_{i=1}^{\infty} \sum_{j=1}^{\infty} f_i * g_j$ belongs to $A_p^{r'}(G)$ since

$$\sum_{i=1}^{\infty} \sum_{j=1}^{\infty} \|f_i * g_j\|_q \leq \sum_{i=1}^{\infty} \sum_{j=1}^{\infty} \|f_i\|_p \|g_j\|_r \leq \sum_{i=1}^{\infty} \left[3/2^i \sum_{j=1}^{\infty} 3/2^j\right] = 9.$$

Moreover, $h=f*g$ as an element of $L_q(G)$, as for each m, n we have

$$\begin{aligned}\left\|f*g-\sum_{i=1}^{n}\sum_{j=1}^{m} f_i*g_j\right\|_q &\leqq \left\|f*g-f*\sum_{j=1}^{m} g_j\right\|_q+\left\|f*\sum_{j=1}^{m} g_j-\sum_{i=1}^{n}\sum_{j=1}^{m} f_i*g_j\right\|_q\\ &\leqq \|f\|_p\left\|g-\sum_{j=1}^{m} g_j\right\|_r+\sum_{j=1}^{m}\left\|f-\sum_{i=1}^{n} f_i\right\|_p\|g_j\|_r\\ &\leqq \|f\|_p/2^m+\sum_{j=1}^{m}(1/2^n)\,3/2^j\leqq \|f\|_p/2^m+3/2^n.\end{aligned}$$

Thus $L_p(G)*L_r(G)\subset A_p^{r'}(G)$.

Now if $L_p(G)*L_r(G)=L_q(G)$ then $A_p^{r'}(G)=L_q(G)$ as linear spaces. Moreover, from the discussion of $A_p^{r'}(G)$ preceding Theorem 5.6.2, we see that $\|h\|_q\leqq |||h|||$, $h\in A_p^{r'}(G)=L_q(G)$. Since $A_p^{r'}(G)$ and $L_q(G)$ are both Banach spaces with their respective norms, we conclude, by the Two Norm Theorem (Theorem D.6.3), that the norms in $A_p^{r'}(G)$ and $L_q(G)$ are equivalent. Hence it is clear that there exists a continuous linear isomorphism of $L_{q'}(G)=L_q(G)^*$ onto $[A_p^{r'}(G)]^*$. Thus by Theorem 5.6.3 there exists a continuous linear isomorphism of $L_{q'}(G)$ onto $M(L_p(G), L_{r'}(G))$, which contradicts Theorem 5.3.3 since $1/p-1/q=1-1/r$ implies $1/p-1/r'=1-1/q'$.

Therefore $L_p(G)*L_r(G)$ is a proper subset of $L_q(G)$. ☐

5.7. Multipliers with Small Support. We have seen in the previous section that if G is a locally compact Abelian group and $1\leqq p\leqq q<\infty$ or a compact Abelian group and $1\leqq q<p<\infty$ then $M(L_p(G), L_q(G))$ is isometrically isomorphic to the dual space $[A_p^q(G)]^*$. Moreover the correspondence between a multiplier $T\in M(L_p(G), L_q(G))$ and a continuous linear functional on $A_p^q(G)$ is determined by the formula

$$\alpha(T)(h)=\sum_{i=1}^{\infty}\langle Tf_i, g_i\rangle$$

where $h=\sum_{i=1}^{\infty} f_i*g_i\in A_p^q(G)$. In this case we define the *support of a multiplier* T in $M(L_p(G), L_q(G))$ as the support of the continuous linear functional $\alpha(T)$. That is, the support of T is the closed subset $\operatorname{supp} T$ of G such that $\alpha(T)(h)=0$ for all $h\in A_p^q(G)$ whose support is contained in $G\sim\operatorname{supp} T$. The theorems of this section describe those $T\in M(L_p(G), L_q(G))$ whose support lies in a given closed subset of G.

Let $E \subset G$ be closed. If G is locally compact Abelian and $p=q$ or G is compact Abelian and $1 \leqq q < p < \infty$ then for each $\mu \in M(G)$, we denote by T_μ the element of $M(L_p(G), L_q(G))$ defined by $T_\mu f = \mu * f$, $f \in L_p(G)$. As we have noted previously we have $\|T_\mu\|_{p,q} \leqq \|\mu\|$. Similarly if G is locally compact Abelian and $1 \leqq p < q < \infty$ then for each $g \in L_r(G)$, $1/p - 1/q = 1 - 1/r$, we denote by T_g the element of $M(L_p(G), L_q(G))$ defined by $T_g f = g * f$, $f \in L_p(G)$. Theorems 3.1.1 and 5.3.3 show that such a definition makes sense. Moreover $\|T_g\|_{p,q} \leqq \|g\|_r$. If E is a closed subset of G then we denote by $L_r(E)$, $1 < r < \infty$, the Banach space of functions on E whose r-th powers are absolutely integrable with respect to Haar measure.

Finally, we observe an elementary fact about the spaces $A_p^q(G)$, namely that they are norm dense in $C_0(G)$ and in $L_r(G)$, $1 \leqq r < \infty$. This is immediate from Theorem 5.1.2 upon noting that $D(G) \subset A_p^q(G)$. Indeed, if $h \in D(G)$ then suppose $h = \sum_{i=1}^{\infty} f_i * g_i$ where $f_i, g_i \in C_c^K(G)$ for some fixed compact set $K \subset G$ and $\sum_{i=1}^{\infty} \|f_i\|_\infty \|g_i\|_\infty < \infty$. But then we have

$$\sum_{i=1}^{\infty} \|f_i\|_p \|g_i\|_{q'} \leqq [\lambda(K)]^{1/p + 1/q'} \sum_{i=1}^{\infty} \|f_i\|_\infty \|g_i\|_\infty,$$

which shows that h defines an element of $A_p^q(G)$.

Theorem 5.7.1. *Let G be a locally compact Abelian group and $p=q$ or a compact Abelian group and $1 \leqq q < p < \infty$. If E is a closed subset of G then the following are equivalent:*

(i) *If $f \in C_0(E)$ then there exists $h \in A_p^q(G)$ such that $f(t) = h(t)$, $t \in E$.*

(ii) *There exists a constant $B > 0$ depending only on p, q and E such that*

$$B\|\mu\| \leqq \|T_\mu\|_{p,q} \leqq \|\mu\| \qquad (\mu \in M(E)).$$

(iii) *If $T \in M(L_p(G), L_q(G))$ and* supp $T \subset E$ *then there exists a unique $\mu \in M(E)$ such that $T = T_\mu$.*

Proof. Consider the injection mapping ι: $A_p^q(G) \to C_0(G)$ defined by $\iota(h) = h$, $h \in A_p^q(G)$. From the remarks preceding Theorem 5.6.2 it is clear that ι is a continuous linear injective mapping from $A_p^q(G)$ into $C_0(G)$. The observation noted above shows that $\iota(A_p^q(G))$ is norm dense in $C_0(G)$. Let $I(E) = \{h \mid h \in A_p^q(G), h(t) = 0, t \in E\}$. Since ι is continuous, it is apparent that $I(E)$ is a closed linear subspace of $A_p^q(G)$. Similarly we set $J(E) = \{h \mid h \in C_0(G), h(t) = 0, t \in E\}$. Then $J(E)$ is a closed linear subspace of $C_0(G)$ and both $A_p^q(G)/I(E)$ and $C_0(G)/J(E)$ are Banach spaces with their respective quotient norms. Moreover, it is apparent from the properties of ι that the mapping β: $A_p^q(G)/I(E) \to C_0(G)/J(E)$ defined by $\beta(h + I(E)) = h + J(E)$ is a continuous linear injective mapping and $\beta(A_p^q(G)/I(E))$ is norm dense in $C_0(G)/J(E)$. Furthermore, $C_0(G)/J(E)$ is isometrically isomorphic to $C_0(E)$ (E.4).

The adjoint β^* of β maps $[C_0(E)]^* = M(E)$ into $[A_p^q(G)/I(E)]^*$. Moreover it is easily seen that the elements of $[A_p^q(G)/I(E)]^*$ are precisely those $\alpha(T) \in [A_p^q(G)]^*$ for which supp $T \subset E$, and that $\beta^* \mu = T_\mu$, $\mu \in M(E)$.

However, if X and Y are Banach spaces and $\beta: X \to Y$ is a continuous linear injective mapping and βX is dense in Y then a well known theorem (Theorem D.3.1) asserts that $\beta X = Y$ if and only if $\beta^* Y^* = X^*$ if and only if there exists a $B > 0$ such that $B \|y^*\| \leqq \|\beta^* y^*\|$, $y^* \in Y^*$. In view of the preceding development we immediately deduce the validity of Theorem 5.7.1 from this result and Theorem 5.6.3 upon setting $X = A_p^q(G)/I(E)$ and $Y = C_0(E)$. □

By essentially the same argument, which we shall omit, one can also establish the following result.

Theorem 5.7.2. *Let G be a locally compact Abelian group, $1 \leqq p < q < \infty$ and $1/p - 1/q = 1 - 1/r$. If E is a closed subset of G then the following are equivalent:*

(i) *If $f \in L_r(E)$ then there exists $h \in A_p^q(G)$ such that $f = h$ almost everywhere on E.*

(ii) *There exists a constant $B > 0$ depending only on p, q and E such that*

$$B \|g\|_r \leqq \|T_g\|_{p,q} \leqq \|g\|_r \qquad (g \in L_r(E)).$$

(iii) *If $T \in M(L_p(G), L_q(G))$ and* supp $T \subset E$ *then there exists a unique $g \in L_r(E)$ such that $T = T_g$.*

5.8. Notes. The development in this chapter is based primarily on the work of Figà-Talamanca, Gaudry and Hörmander. The discussion of quasimeasures is based essentially on Gaudry [1]. See also Gaudry [3], [6], VI. Theorem 5.1.5 appears in Gaudry [1], while proof of Theorem 5.1.6 is available in Edwards [5, 12].

The basic properties of the spaces $M(L_p(G), L_q(G))$ for $G = R^n$ are studied in Hörmander [1], and subsequent extensions to arbitrary locally compact Abelian groups are due mainly to Gaudry [1–3, 6, 9]. In particular proofs of Theorems 5.2.1, 5.2.2, and 5.2.5 are available in Gaudry [1, 2], [6], V and VI, and Hörmander [1], while Theorems 5.3.1, 5.3.2, 5.3.3 and 5.3.4 can be found in Gaudry [6], V and VI, [9] and Hörmander [1]. A proof of Theorem 5.3.4 for infinite compact Abelian groups is available in Gaudry [9], while a proof of Theorem 5.2.4 can be found in Gaudry [6], V. Some further inclusion results are available in Doss [2].

Hörmander [1] has shown that if $T \in M(L_\infty(R^n), L_q(R^n))$, $1 \leqq q < \infty$, then $Tf = 0$ for each $f \in L_\infty(R^n)$ which vanishes at infinity. Gaudry [6] defines $M(L_\infty(G), L_q(G))$ to be all the linear mappings $T: L_\infty(G) \to L_q(G)$ which commute with translations and are continuous from the weak* topology on $L_\infty(G)$ to $L_q(G)$ with the weak topology induced by $L_{q'}(G) = L_q(G)^*$, $1/q + 1/q' = 1$. In this way he avoids the type of difficulty indicated, for example, at the end of 3.4, and the standard results for $M(L_p(G), L_q(G))$ are valid. The reader is referred to Gaudry [6], V and VI, for the details.

The functions constructed in Lemma 5.3.1 are generalizations of the Rudin-Shapiro polynomials. See for example, Katznelson [2], pp. 33–34.

Discussions of the Fourier transform on $L_p(G)$, $p > 2$, can be found in Gaudry [2], [6], VI, and L. Schwartz [2], the latter for the groups $R^n \times \Gamma^m$, while $M(L_p(G), L_q(G))\hat{}$ is studied in Gaudry [2], [6], VI, and Hörmander [1]. Theorem 5.5.1 was proved for $G = R^n$ in Hörmander [1] and for noncompact locally compact Abelian groups in Gaudry [9]. Edwards and Price [1] have also given a "constructive" proof of this theorem utilizing the Rudin-Shapiro functions. They also show when G is a nondiscrete locally compact Abelian group, $1 < p \leqq q < \infty$ and $1/p + 1/q \geqq 1$ that given any open set $U \subset G$ with compact closure there exists some $T \in M(L_p(G), L_q(G))$ such that the support of T is contained in the closure of U and $\hat{T} \notin V(\hat{G})$. The support of T is, of course, the support of the unique quasimeasure σ such that $Tf = \sigma * f$, $f \in C_c(G)$. We refer the reader to Edwards and Price [1] for further details, and to Gaudry [1], [6], VI, for a discussion of the supports of quasimeasures.

Corollary 5.5.1 was proved for $G = R^n$ in Hörmander [1], and its extension to arbitrary noncompact groups and Theorem 5.5.2 are due to Gaudry [2], [6], VI. The proof in Gaudry [6] makes use of the structure theorem for locally compact Abelian groups (Hewitt and Ross [1], 24.30 and Rudin [5], 2.4.1).

A portion of Theorem 5.5.3, namely part (iv), was shown to hold for $G = R^n$ in Hörmander [1]. The theorem in its full generality was established by Gaudry [2]. Corollary 5.5.1 shows that Theorem 5.5.3 (iv) is valid when G is noncompact and this obviously then implies Corollary 4.6.2. The new proof of Theorem 4.6.2, that is, the proof of Theorem 5.5.4, is taken from Gaudry [2]. It is due to R.E. Edwards. Theorem 5.5.5 is due to J.F. Price in a private communication to the author. See also Price [3], 6. When G is an infinite compact Abelian group and $1 < p \leqq \infty$, $1 \leqq q < \infty$, then there always exist $T \in M(L_p(G), L_q(G))$ which do not correspond to convolution with an element of $M(G)$. This is established in Brainerd and Edwards [1_{I}] by essentially the same argument involving Sidon sets as those given after Theorem 3.6.1 for the multipliers in $M(L_p(G))$, $1 < p < \infty$.

Other results about $M(L_p(G), L_q(G))\hat{}$ are to be found in Gaudry [6], VI, and Hörmander [1]. We mention only two of these. Denote by $L_p^{\mathrm{loc}}(G)$ the space of Borel measurable functions on G whose p-th powers are integrable over every compact subset of G.

Theorem 5.8.1. *Let G be a noncompact locally compact Abelian group and* $1 \leqq p \leqq q < \infty$. *If* $p \geqq 2$ *then* $M(L_p(G), L_q(G))^\wedge \subset L_p^{\mathrm{loc}}(\hat{G})$, *and if* $q \leqq 2$ *then* $M(L_p(G), L_q(G))^\wedge \subset L_{q'}^{\mathrm{loc}}(\hat{G})$, $1/q + 1/q' = 1$.

Thus when $p \geqq 2$ or $q \leqq 2$ we see that each $T \in M(L_p(G), L_q(G))$ corresponds to a function $\varphi \in L_p^{\mathrm{loc}}(\hat{G})$ or $L_{q'}^{\mathrm{loc}}(\hat{G})$ such that $(Tf)^\wedge = \varphi \hat{f}$ for each $f \in C_c(G)$. Clearly one can then consider $M(L_p(G), L_q(G))^\wedge$ as a set of functions on $\hat{G}$ and with $\|\varphi\|_{p,q} = \|T\|_{p,q}$ this space becomes a Banach space. Furthermore, one has the following result.

Theorem 5.8.2. *Let G be a noncompact locally compact Abelian group and suppose that either* $2 \leqq p \leqq q \leqq r$ *or* $p \leqq q \leqq r \leqq 2$. *If* $\varphi_1 \in M(L_p(G), L_q(G))^\wedge$ *and* $\varphi_2 \in M(L_q(G), L_r(G))^\wedge$ *then the pointwise product* $\varphi_1 \varphi_2 \in M(L_p(G), L_r(G))^\wedge$ *and*

$$\|\varphi_1 \varphi_2\|_{p,r} \leqq \|\varphi_1\|_{p,q} \|\varphi_2\|_{q,r}.$$

In particular, this theorem shows that $M(L_p(G), L_p(G))^\wedge \subset L_\infty(\hat{G})$ is a Banach algebra under pointwise operations and the operator norm.

The characterization of $M(L_p(G), L_q(G))$ as a dual space is taken essentially from Figà-Talamanca and Gaudry [1]. The case of $p = q$ is also contained in Figà-Talamanca [1, 3]. A proof of Theorem 5.6.1 for $p = q$ is also available in Brainerd and Edwards [1_{I}]. A description of the spaces $A_p^q(G)$ and $M(L_p(G), L_q(G))$ in terms of tensor products is given in Rieffel [4]. In this connection see also Harte [1]. A general exposition of the characterization theorem can be found in Gaudry [6], V.

The spaces $A_p^q(G)$ have been used in a number of instances in connection with the problem of factorization in topological algebras. Corollary 5.6.2 comes from Gaudry [9]. A collection of references dealing with the factorization problem, in general, is as follows: Cohen [1], Curtis and Figà-Talamanca [1], Edwards [8], Figà-Talamanca and Gaudry [1, 2], Gaudry [6], III and V, Gulick, Liu and van Rooij [$1_{\mathrm{II,III,IV}}$], Hewitt [3], Koosis [1], Máté [7], Rieffel [3], Rudin [2], Taylor [1] and Yap [1, 2].

Theorems 5.7.1 and 5.7.2 are taken from Máté [6, 7]. He uses his results to also obtain the characterization of Helson sets given in Rudin [4], 5.6.3. The reader is referred to these papers and to the notes to Chapter 4 for other references on such results.

A partial list of other sources dealing with $M(L_p(G), L_q(G))$ is the following: Edwards [14_{II}], 16.4, Gaudry [3], [6], V and VI, Hardy and Littlewood [4], Hewitt and Ross [2], 35, La Duke [1], Littman [1], Lizorkin [1], Price [3], Stein and Zygmund [1] and Taibleson [2_{III}].[1] In particular an excellent exposition of much of the material in this chapter, as well as the previous two chapters, is available in Gaudry [6], V and VI.

---

[1] See also Figà-Talamanca and Gaudry [4], Jodeit [1] and Lizorkin [2].

Chapter 6

The Multipliers for Functions with Fourier Transforms in $L_p(\hat{G})$

6.0. Introduction. In this chapter we shall discuss the Banach algebras $A_p(G)$ of elements in $L_1(G)$ whose Fourier transforms belong to $L_p(\hat{G})$ and the multipliers for these algebras. These algebras are similar to the group algebra $L_1(G)$ in a great many ways. In particular for noncompact groups we shall see that the algebras $A_p(G)$ and $L_1(G)$ have the same multipliers. However the algebras $A_p(G)$ are neither group nor QCG algebras. This leads to the observation that the nature of the multipliers of a commutative Banach algebra, by itself, contributes little information as to the character of the algebra. When G is a compact group it will be seen that the bounded measures $M(G)$ form a proper subset of the multipliers for $A_p(G)$. In this case we shall also characterize the multipliers of $A_p(G)$ as the dual space of a certain Banach space of continuous functions.

6.1. The Banach Algebras $A_p(G)$. Let G be a locally compact Abelian group and $1 \leqq p < \infty$. $A_p(G)$ will denote the translation invariant linear subspace of $L_1(G)$ consisting of all those $f \in L_1(G)$ such that $\hat{f} \in L_p(\hat{G})$. Since the Fourier transform of a function in $L_1(G)$ belongs to $C_0(\hat{G})$ it is evident that each $A_p(G)$ is an ideal in $L_1(G)$, and that $A_p(G) \subset A_r(G)$ if $p < r$. Moreover each $A_p(G)$ is norm dense in $L_1(G)$ because the elements of $L_1(G)$ whose Fourier transforms lie in $C_c(\hat{G})$ are norm dense in $L_1(G)$ (F.7c).

For each p, $1 \leqq p < \infty$, we define

$$\|f\|^p = \|f\|_1 + \|\hat{f}\|_p = \int_G |f(t)|\, d\lambda(t) + \Big(\int_{\hat{G}} |\hat{f}(\gamma)|^p\, d\eta(\gamma)\Big)^{1/p} \qquad (f \in A_p(G)).$$

The Haar measure η on $\hat{G}$ is chosen so that the Fourier Inversion Theorem (Theorem F.8.1) is valid. It is easy to see that $\|\cdot\|^p$ defines a norm on $A_p(G)$ and that $A_p(G)$ is a commutative algebra with convolution as multiplication. Our first result states that $A_p(G)$ is also a Banach algebra.

Theorem 6.1.1. *For each p,* $1 \leqq p < \infty$, $A_p(G)$ *is a commutative Banach algebra with the norm* $\|\cdot\|^p$ *and the usual convolution product.*

Proof. Let $\{f_n\} \subset A_p(G)$ be a Cauchy sequence in the norm $\|\cdot\|^p$. Clearly $\{f_n\}$ and $\{\hat{f}_n\}$ are Cauchy sequences in $L_1(G)$ and $L_p(\hat{G})$ respectively. Let $f \in L_1(G)$ and $g \in L_p(\hat{G})$ be such that $\lim_n \|f_n - f\|_1 = 0$ and $\lim_n \|\hat{f}_n - g\|_p = 0$. Since the Fourier transform is norm decreasing, the first assertion implies that $\lim_n \|\hat{f}_n - \hat{f}\|_\infty = 0$. Thus $\hat{f} = g$, since $\lim_n \|\hat{f}_n - g\|_p = 0$ implies that there exists a subsequence of $\{\hat{f}_n\}$ which converges almost everywhere to g.

Hence $f \in A_p(G)$ and $\lim_n \|f_n - f\|^p = 0$, that is, $A_p(G)$ is complete with the norm $\|\cdot\|^p$.

As indicated previously, $A_p(G)$ is a commutative algebra under convolution. To show $A_p(G)$ is a Banach algebra we observe that for each $f, g \in A_p(G)$,

$$\begin{aligned}\|f*g\|^p = \|f*g\|_1 + \|\hat{f}\hat{g}\|_p &\leqq \|f\|_1 \|g\|_1 + \|\hat{f}\|_\infty \|\hat{g}\|_p \\ &\leqq \|f\|_1 (\|g\|_1 + \|\hat{g}\|_p) \leqq \|f\|^p \|g\|^p.\end{aligned}$$

Therefore $A_p(G)$ is a commutative Banach algebra. □

The next theorem shows that we may identify the space of regular maximal ideals $\Delta(A_p(G))$ of $A_p(G)$ with $\hat{G}$.

Theorem 6.1.2. *For each p,* $1 \leqq p < \infty$, $\Delta(A_p(G))$ *is homeomorphic with* $\hat{G}$.

Proof. Let $f \in A_p(G)$. Then for each positive integer n we have

$$\begin{aligned}\|f^n\|^p = \|f^{n-1}*f\|^p &= \|f^{n-1}*f\|_1 + \|(\hat{f})^{n-1}\hat{f}\|_p \\ &\leqq \|f^{n-1}\|_1 \|f\|_1 + \|(\hat{f})^{n-1}\|_\infty \|\hat{f}\|_p \\ &\leqq \|f^{n-1}\|_1 \|f\|^p \leqq (\|f\|_1)^{n-1} \|f\|^p.\end{aligned}$$

Hence for each n,

$$(\|f^n\|^p)^{1/n} \leqq (\|f\|_1)^{1-1/n} (\|f\|^p)^{1/n}.$$

Letting n tend to infinity we conclude for each $f \in A_p(G)$, that

$$\|f\|^p_{\mathrm{Sp}} \leqq \|f\|_1$$

where $\|f\|^p_{\mathrm{Sp}} = \lim_n (\|f^n\|^p)^{1/n}$ is the spectral radius norm for the element $f \in A_p(G)$.

Now let F be any multiplicative linear functional on $A_p(G)$. Then

$$|F(f)| \leqq \|f\|^p_{\mathrm{Sp}} \leqq \|f\|_1 \qquad (f \in A_p(G)).$$

Hence F defines an L_1- bounded multiplicative linear functional on the norm dense subspace $A_p(G)$ of $L_1(G)$. Consequently F can be extended uniquely to a multiplicative linear functional on all of $L_1(G)$.

Thus there exists a unique element $\gamma\in\hat{G}$ such that

$$F(f)=\int_G (t^{-1},\gamma)f(t)\,d\lambda(t) \qquad (f\in A_p(G)).$$

It is obvious that each $\gamma\in\hat{G}$ defines a multiplicative linear functional on $A_p(G)$ by the previous formula. Hence the identity mapping on $\hat{G}$ defines a bijective mapping between $\hat{G}$ and $\Delta(A_p(G))$. A standard argument (Rudin [4], 1.2.6), *mutatis mutandis*, shows that the usual topology of $\hat{G}$ coincides with the Gelfand topology on $\hat{G}$ considered as the space of multiplicative linear functionals for $A_p(G)$.

Therefore $\Delta(A_p(G))$ is homeomorphic to $\hat{G}$. □

Corollary 6.1.1. *For each p, $1\leqq p<\infty$, $A_p(G)$ is a semi-simple commutative Banach algebra.*

When G is discrete then $\hat{G}$ is compact and $A_p(G)=L_1(G)$, $1\leqq p<\infty$. Obviously such a situation is of no interest and we shall assume throughout the remainder of this chapter that G is *a nondiscrete group*. It is fairly easy to see that the only instance when the $A_p(G)$ are not distinct for distinct p is the one just indicated.

For each p, $1\leqq p<\infty$, it is evident that the mapping ψ_p: $A_p(G)\to L_1(G)\times L_p(\hat{G})$ defined by $\psi_p(f)=(f,\hat{f})$ for each $f\in A_p(G)$ is a linear isometry of $A_p(G)$ into the Banach space $L_1(G)\times L_p(\hat{G})$ with the sum norm $\|(f,g)\|=\|f\|_1+\|g\|_p$. Thus we may consider $A_p(G)$ as a closed subspace of $L_1(G)\times L_p(\hat{G})$. Since the dual space of $L_1(G)\times L_p(\hat{G})$ is isomorphic with $L_\infty(G)\times L_{p'}(\hat{G})$, $1/p+1/p'=1$, it follows by an application of the Hahn-Banach Theorem (Theorem D.6.5) that every continuous linear functional F on $A_p(G)$ must be of the form

$$F(f)=\int_G f(t)\,a(t^{-1})\,d\lambda(t)+\int_{\hat{G}}\hat{f}(\gamma)\,b(\gamma^{-1})\,d\eta(\gamma) \qquad (f\in A_p(G))$$

where $(a,b)\in L_\infty(G)\times L_{p'}(\hat{G})$. Of course the pair (a,b) corresponding to a given functional may not be unique. This, however, is irrelevant for our purposes. We shall make use of this description of the continuous linear functionals of $A_p(G)$ in the following section.

The algebras $A_p(G)$ are similar in many respects to the group algebra $L_1(G)$. We have already seen that they have the same regular maximal ideal spaces. On the other hand, the algebras $A_p(G)$ are neither group algebras nor QCG algebras.

Indeed, we indicated in 1.6 that one necessary condition for a commutative Banach algebra to be a QCG algebra is that it possesses a minimal approximate identity. The next theorem shows that if G is nondiscrete then $A_p(G)$ has no minimal approximate identity, and hence cannot be a QCG algebra.

Theorem 6.1.3. *Let G be a locally compact Abelian group and $1 \leqq p < \infty$. Then the following are equivalent:*

(i) *$\hat{G}$ is compact.*

(ii) *$A_p(G)$ has a minimal approximate identity.*

Proof. If $\hat{G}$ is compact then $A_p(G) = L_1(G)$ and, as is easily verified,

$$\|f\|_1 \leqq \|f\|^p \leqq 2\|f\|_1 \qquad (f \in A_p(G)).$$

Thus any minimal approximate identity for $L_1(G)$ is also one for $A_p(G)$, and (i) implies (ii).

Conversely, suppose $\{u_\alpha\} \subset A_p(G)$ is a minimal approximate identity for $A_p(G)$. Without loss of generality we may assume that there exists a constant $B > 0$ such that $\|u_\alpha\|^p \leqq B$ for all α. Then for each $f \in A_p(G)$ and all α we would have

$$\begin{aligned} \|f * u_\alpha\|^p &= \|f * u_\alpha\|_1 + \|\hat{f}\hat{u}_\alpha\|_p \\ &\leqq \|f\|_1 \, \|u_\alpha\|_1 + \|\hat{f}\|_\infty \, \|\hat{u}_\alpha\|_p \\ &\leqq \|f\|_1 \, \|u_\alpha\|^p \\ &\leqq B\|f\|_1. \end{aligned}$$

Thus, since $\{u_\alpha\}$ is an approximate identity for $A_p(G)$, we conclude that

$$\|f\|^p \leqq B\|f\|_1 \qquad (f \in A_p(G)).$$

Consequently $A_p(G)$ is a closed subspace of $L_1(G)$, which implies that $A_p(G) = L_1(G)$ as $A_p(G)$ is norm dense in $L_1(G)$.

However, if $\hat{G}$ is not compact then for each positive integer n there exists a compact set $K_n \subset \hat{G}$ such that $\eta(K_n) > n^p$ because Haar measure η on $\hat{G}$ does not have finite total mass. Let $f_n \in L_1(G) = A_p(G)$ be such that $\hat{f}_n(\gamma) = 1$, $\gamma \in K_n$, and $\|f_n\|_1 \leqq 2$ (F.7e). Then we deduce at once from the inequality

$$\|\hat{f}\|_p \leqq (B-1)\|f\|_1 \qquad (f \in A_p(G))$$

that for each positive integer n,

$$n < \|\hat{f}_n\|_p \leqq (B-1)\|f_n\|_1 \leqq 2(B-1),$$

which is clearly absurd.

Therefore $\hat{G}$ is compact, and (ii) implies (i). □

Though $A_p(G)$ does not possess a minimal approximate identity when G is nondiscrete, it does always contain an approximate identity.

Theorem 6.1.4. *Let G be a nondiscrete locally compact Abelian group. Then $A_p(G)$, $1 \leqq p < \infty$, contains an approximate identity $\{u_\alpha\}$ such that $\{u_\alpha\} \subset A_1(G)$.*

Proof. Let $\{u_\alpha\}$ be any approximate identity for $L_1(G)$ for which $\|u_\alpha\|_1 \leqq B$ for all α and $\{\hat{u}_\alpha\} \subset C_c(\hat{G})$ (F.7d). Then $\{\hat{u}_\alpha\}$ converges uniformly to one on compact subsets of $\hat{G}$. Indeed, if $K \subset \hat{G}$ is compact then let $f \in L_1(G)$ be such that $\hat{f}(\gamma) = 1$, $\gamma \in K$. Since $\{u_\alpha\}$ is an approximate identity for $L_1(G)$, the desired conclusion is evident from the inequality

$$\begin{aligned} |\hat{u}_\alpha(\gamma) - 1| &= |\hat{u}_\alpha(\gamma)\hat{f}(\gamma) - \hat{f}(\gamma)| \\ &\leqq \|\hat{u}_\alpha \hat{f} - \hat{f}\|_\infty \\ &\leqq \|u_\alpha * f - f\|_1 \qquad (\gamma \in K). \end{aligned}$$

Now suppose $f \in A_p(G)$ and $\varepsilon > 0$. Then let $K \subset \hat{G}$ be a compact set such that

$$\int_{\hat{G} \sim K} |\hat{f}(\gamma)|^p \, d\eta(\gamma) < \frac{\varepsilon^p}{2(B+1)^p}.$$

Moreover, since $\{\hat{u}_\alpha\}$ converges uniformly to one on K it is easily seen that there exists an α_0 such that for $\alpha > \alpha_0$,

$$\int_K |\hat{f}\hat{u}_\alpha(\gamma) - \hat{f}(\gamma)|^p \, d\eta(\gamma) < \frac{\varepsilon^p}{2}.$$

Thus for $\alpha > \alpha_0$,

$$\begin{aligned} &\|\hat{f}\hat{u}_\alpha - \hat{f}\|_p \\ &\quad = \Big[\int_K |\hat{f}\hat{u}_\alpha(\gamma) - \hat{f}(\gamma)|^p \, d\eta(\gamma) + \int_{\hat{G} \sim K} |\hat{f}\hat{u}_\alpha(\gamma) - \hat{f}(\gamma)|^p \, d\eta(\gamma)\Big]^{1/p} \\ &\quad \leqq \left[\frac{\varepsilon^p}{2} + (B+1)^p \int_{\hat{G} \sim K} |\hat{f}(\gamma)|^p \, d\eta(\gamma)\right]^{1/p} \\ &\quad < \left[\frac{\varepsilon^p}{2} + \frac{\varepsilon^p}{2}\right]^{1/p} = \varepsilon. \end{aligned}$$

That is, $\lim_\alpha \|\hat{f}\hat{u}_\alpha - \hat{f}\|_p = 0$.

Combining this with the fact that $\lim_\alpha \|f * u_\alpha - f\|_1 = 0$ since $\{u_\alpha\}$ is an approximate identity for $L_1(G)$, we conclude that $\lim_\alpha \|f * u_\alpha - f\|^p = 0$. Hence $\{u_\alpha\}$ is an approximate identity for $A_p(G)$.

The assertion that $\{u_\alpha\} \subset A_1(G)$ is apparent as $\{\hat{u}_\alpha\} \subset C_c(\hat{G})$. $\square$

It should be noted that the existence of an approximate identity for $A_p(G)$ as in the preceding theorem reveals that each $A_p(G)$ is norm dense in $A_r(G)$, $r > p$, as the $A_p(G)$ are ideals in $L_1(G)$.

6.2. The Multipliers for $A_p(G)$ as Pseudomeasures. Having developed some of the basic facts concerning the commutative Banach algebras $A_p(G)$, we now wish to turn to an investigation of the multipliers for these algebras. Since $A_p(G)$ is a semi-simple commutative Banach algebra each multiplier for $A_p(G)$, that is, each mapping T from $A_p(G)$ to $A_p(G)$ such that $T(f*g)=Tf*g$, is automatically linear and continuous. Furthermore, to each $T\in M(A_p(G))$ there corresponds a unique bounded continuous function φ on $\hat{G}$ such that $(Tf)\hat{} = \varphi\hat{f}$ for each $f\in A_p(G)$. These assertions follow from Theorems 1.1.1 and 1.2.2.

Our first result shows that the elements of $M(A_p(G))$ can also be thought of as continuous linear operators which commute with translation.

Theorem 6.2.1. *Let G be a nondiscrete locally compact Abelian group. For each p, $1\leqq p<\infty$, if $T: A_p(G)\to A_p(G)$ then the following are equivalent:*

(i) *$T\in M(A_p(G))$.*

(ii) *T is a continuous linear operator such that $T\tau_s=\tau_s T$ for each $s\in G$.*

Proof. If $T\in M(A_p(G))$ then T is continuous and linear, and there exists a bounded continuous function φ on $\hat{G}$ such that $(Tf)\hat{} = \varphi\hat{f}$ for each $f\in A_p(G)$. From the properties of the Fourier transform and the semi-simplicity of $A_p(G)$, it follows at once that $T\tau_s=\tau_s T$ for each $s\in G$.

Suppose now that T is continuous, linear and commutes with translation. Let F be a continuous linear functional on $A_p(G)$. Then $F\circ T$ is also a continuous linear functional on $A_p(G)$. By the discussion of the preceding section there exists (a,b) and (α,β) in $L_\infty(G)\times L_{p'}(\hat{G})$, $1/p+1/p'=1$, such that for each $f\in A_p(G)$,

$$F(f)=\int_G f(t)\,a(t^{-1})\,d\lambda(t)+\int_{\hat{G}}\hat{f}(\gamma)\,b(\gamma^{-1})\,d\eta(\gamma)$$

and

$$F\circ T(f)=\int_G f(t)\,\alpha(t^{-1})\,d\lambda(t)+\int_{\hat{G}}\hat{f}(\gamma)\,\beta(\gamma^{-1})\,d\eta(\gamma).$$

Consequently, for $f, g\in A_p(G)$ we have

$$\begin{aligned}
&F(Tf*g)\\
&\quad=\int_G Tf*g(t)\,a(t^{-1})\,d\lambda(t)+\int_{\hat{G}}(Tf)\hat{}\,(\gamma)\,\hat{g}(\gamma)\,b(\gamma^{-1})\,d\eta(\gamma)\\
&\quad=\int_G\Big[\int_G Tf(t\,s^{-1})\,g(s)\,d\lambda(s)\Big]\,a(t^{-1})\,d\lambda(t)\\
&\qquad+\int_{\hat{G}}(Tf)\hat{}\,(\gamma)\Big[\int_G(s^{-1},\gamma)\,g(s)\,d\lambda(s)\Big]\,b(\gamma^{-1})\,d\eta(\gamma)
\end{aligned}$$

$$
\begin{aligned}
&= \int_G g(s)\Big[\int_G \tau_s Tf(t)\, a(t^{-1})\, d\lambda(t)\Big]\, d\lambda(s) \\
&\quad + \int_G g(s)\Big[\int_{\hat G} (\tau_s Tf)\hat{}\,(\gamma)\, b(\gamma^{-1})\, d\eta(\gamma)\Big]\, d\lambda(s) \\
&= \int_G g(s)\Big[\int_G T\tau_s f(t)\, a(t^{-1})\, d\lambda(t) + \int_{\hat G} (T\tau_s f)\hat{}\,(\gamma)\, b(\gamma^{-1})\, d\eta(\gamma)\Big]\, d\lambda(s) \\
&= \int_G g(s)\, F\circ T(\tau_s f)\, d\lambda(s) \\
&= \int_G g(s)\Big[\int_G \tau_s f(t)\, \alpha(t^{-1})\, d\lambda(t) + \int_{\hat G} (\tau_s f)\hat{}\,(\gamma)\, \beta(\gamma^{-1})\, d\eta(\gamma)\Big]\, d\lambda(s) \\
&= \int_G \Big[\int_G f(t s^{-1})\, g(s)\, d\lambda(s)\Big]\, \alpha(t^{-1})\, d\lambda(t) \\
&\quad + \int_{\hat G} \hat f(\gamma)\Big[\int_G (s^{-1},\gamma)\, g(s)\, d\lambda(s)\Big]\, \beta(\gamma^{-1})\, d\eta(\gamma) \\
&= \int_G f*g(t)\, \alpha(t^{-1})\, d\lambda(t) + \int_{\hat G} \hat f(\gamma)\, \hat g(\gamma)\, \beta(\gamma^{-1})\, d\eta(\gamma) \\
&= F\circ T(f*g) = F[T(f*g)].
\end{aligned}
$$

Since this holds for every continuous linear functional F it follows (D.6b) that $Tf*g = T(f*g)$ for each $f, g\in A_p(G)$.

Therefore T is a multiplier for $A_p(G)$. □

It is elementary to show that if $\mu\in M(G)$ then $Tf=\mu * f$ for $f\in A_p(G)$ defines an element of $M(A_p(G))$ such that $\|T\|^p \leqq \|\mu\|$, where $\|\cdot\|^p$ denotes the operator norm of $T\in M(A_p(G))$. Thus $M(G)$ can be continuously embedded into $M(A_p(G))$. It is not entirely clear under what conditions this embedding is surjective. In the following two sections we shall discuss instances where the mapping is surjective and others where it is not. However, at this point, we shall prove that every multiplier for $A_p(G)$ can be obtained by convolution with some pseudomeasure.

First we note that if $1\leqq p<q$ then $M(A_q(G))\subset M(A_p(G))$ as linear spaces. Because if $T\in M(A_q(G))$ then for each $f\in A_p(G)\subset A_q(G)$ we have $Tf\in A_q(G)$. Moreover, $(Tf)\hat{} = \varphi\, \hat f\in L_p(\hat G)$ as $f\in A_p(G)$ and φ is a bounded continuous function. Hence $Tf\in A_p(G)$. Clearly then T defines an element of $M(A_p(G))$, and $M(A_q(G))\subset M(A_p(G))$. We also observe the obvious fact that as linear spaces $A_2(G)$ and $L_1(G)\cap L_2(G)$ are identical.

Theorem 6.2.2. *Let G be a nondiscrete locally compact Abelian group and $1\leqq p<\infty$. If $T\in M(A_p(G))$ then there exists a unique pseudomeasure $\sigma\in P(G)$ such that $Tf=\sigma * f$ for each $f\in A_1(G)$. Moreover the correspondence betweeen T and σ defines a continuous algebra isomorphism from $M(A_p(G))$ into $P(G)$.*

Proof. Recall that $A_c(G)=A(G)\cap C_c(G)$ where $A(G)=\{\hat{f}\,|\,f\in L_1(\hat{G})\}$ with the norm $\|\hat{f}\|_A=\|f\|_1$. From the Fourier Inversion Theorem (Theorem F.8.1), it is apparent that $A_c(G)\subset A_1(G)\subset A(G)$. Suppose $p=1$ and $T\in M(A_1(G))$. Let $\varphi\in C(\hat{G})$ be such that $(Tf)\hat{}=\varphi\hat{f}$ for each $f\in A_1(G)$. For each $f\in A_c(G)$ define $F(f)=Tf(e)$. This is well defined as $Tf\in A_1(G)$ and hence is a continuous function. Since $Tf\in A_1(G)\subset A(G)$, the Fourier Inversion Theorem (Theorem F.8.1) reveals that for each $f\in A_1(G)$ we have $Tf=[(Tf)\hat{}]\check{}=([(Tf)\hat{}]\tilde{})\hat{}$. Consequently, for each $f\in A_c(G)$ we obtain

$$\begin{aligned}|F(f)|=|Tf(e)|&\leqq\|Tf\|_\infty\leqq\|[(Tf)\hat{}]\tilde{}\|_1=\|(Tf)\hat{}\|_1\\&=\|\varphi\hat{f}\|_1\leqq\|\varphi\|_\infty\|\hat{f}\|_1=\|\varphi\|_\infty\|f\|_A.\end{aligned}$$

Thus F defines a continuous linear functional on the subspace $A_c(G)$ of $A(G)$. Since by Theorem 5.1.2 the subspace $A_c(G)$ is dense in $A(G)$, we may uniquely extend F to a continuous linear functional on $A(G)$ without increasing the norm. From the definition of $P(G)$ as $A(G)^*$ we conclude that there exists a unique pseudomeasure $\sigma\in P(G)$ such that

$$F(f)=Tf(e)=\langle f,\sigma\rangle \qquad (f\in A_c(G)),$$

and $\|\sigma\|_P=\|F\|\leqq\|\varphi\|_\infty$.

If $f,g\in A_c(G)$ then $g*f\in A_c(G)$ and so

$$T(g*f)(e)=\langle g*f,\sigma\rangle=\langle g,\sigma*f\rangle.$$

On the other hand, since T is a multiplier for $A_1(G)$ we have

$$T(g*f)(e)=g*Tf(e)=\langle g,Tf\rangle.$$

Again using the denseness of $A_c(G)$ in $A(G)$, we conclude that $Tf=\sigma*f$ for each $f\in A_c(G)$.

To extend this relation to all $f\in A_1(G)$ we recall first that $A_1(G)\subset A_2(G)=L_1(G)\cap L_2(G)$.

Thus by Theorem 4.3.1 we see that if $f\in A_1(G)$ then $\sigma*f\in L_2(G)$. Let $\{v_\alpha\}\subset L_1(\hat{G})$ be an approximate identity for $L_1(\hat{G})$ such that $\|v_\alpha\|_1=1$ for all α and $\{\check{v}_\alpha\}\subset C_c(G)$ (F.7d). Clearly $\{\check{v}_\alpha\}\subset A_c(G)$. If $f\in A_1(G)$ then $f\check{v}_\alpha\in A_c(G)$ and so $T(f\check{v}_\alpha)=\sigma*(f\check{v}_\alpha)$. Since $\{\check{v}_\alpha\}$ converges uniformly to one on compact subsets of G and $\{v_\alpha\}$ is an approximate identity for $L_1(\hat{G})$, for each $f\in A_1(G)$ we can combine the estimates

$$\begin{aligned}\|\sigma*(f\check{v}_\alpha)-\sigma*(f\check{v}_\beta)\|^1&=\|T(f\check{v}_\alpha)-T(f\check{v}_\beta)\|^1\\&\leqq\|T\|^1(\|f\check{v}_\alpha-f\check{v}_\beta\|_1+\|\hat{f}*v_\alpha-\hat{f}*v_\beta\|_1),\end{aligned}$$

together with essentially the same argument as the one used in the proof of Theorem 6.1.4 to conclude that $\{\sigma*(f\check{v}_\alpha)\}$ is a Cauchy net in $A_1(G)$.

Thus there exists a $g\in A_1(G)$ for which $\lim_\alpha \|\sigma*(f\check{v}_\alpha)-g\|^1=0$. But then we deduce from

$$\begin{aligned}\|\hat{g}-\hat{\sigma}\hat{f}\|_\infty &\leqq \|\hat{g}-\hat{\sigma}(\hat{f}*v_\alpha)\|_\infty+\|\hat{\sigma}(\hat{f}*v_\alpha)-\hat{\sigma}\hat{f}\|_\infty\\ &\leqq \|g-\sigma*(f\check{v}_\alpha)\|_1+\|\hat{\sigma}\|_\infty\|f\check{v}_\alpha-f\|_1,\end{aligned}$$

and the fact that $\{\check{v}_\alpha\}$ converges uniformly to one on compact subsets of G, that $\hat{g}=\hat{\sigma}\hat{f}$. Consequently, since the Fourier-Plancherel transform is injective, we see that $\sigma*f=g\in A_1(G)$ by Corollary 4.2.2. Moreover, since $\{\check{v}_\alpha\}$ converges uniformly to one on compact subsets of G and $\{v_\alpha\}$ is an approximate identity for $L_1(\hat{G})$, we again see by the argument of Theorem 6.1.4 that $\lim_\alpha \|f\check{v}_\alpha-f\|^1=0$. This in turn implies that $\lim_\alpha \|T(f\check{v}_\alpha)-Tf\|^1=\lim_\alpha \|\sigma*(f\check{v}_\alpha)-Tf\|^1=0$. Therefore $Tf=\sigma*f$ for each $f\in A_1(G)$.

In general, if $1\leqq p<\infty$ and $T\in M(A_p(G))$ then $T\in M(A_1(G))$, as indicated before the statement of the theorem. The preceding argument then shows that there exists a unique pseudomeasure $\sigma\in P(G)$ such that $Tf=\sigma*f$ for each $f\in A_1(G)$.

If $1\leqq p<\infty$ and $T\in M(A_p(G))$ then the previous formula shows that $(Tf)\hat{}=\hat{\sigma}\hat{f}=\varphi\hat{f}$ for each $f\in A_1(G)$. Since $A_1(G)\hat{}$ separates points, we conclude that $\hat{\sigma}=\varphi$. Thus $\|\sigma\|_P=\|\hat{\sigma}\|_\infty=\|\varphi\|_\infty\leqq\|T\|^p$ by Theorem 1.2.2. Evidently the correspondence between T and σ defines a continuous linear mapping from $M(A_p(G))$ into $P(G)$. Clearly the mapping preserves products, and an elementary argument shows that it is injective.

Therefore the correspondence between T and σ defines a continuous algebra isomorphism from $M(A_p(G))$ into $P(G)$. □

Corollary 6.2.1. *Let G be a nondiscrete locally compact Abelian group and $1\leqq p<\infty$. Then there exists an isometric algebra isomorphism from $\mathscr{M}(A_p(G))$ into $P(G)$.*

We shall see in the last section that for certain values of p and compact groups the isomorphism given by Theorem 6.2.2 is not surjective. $\mathscr{M}(A_p(G))$ is defined as in 1.2, that is, $\mathscr{M}(A_p(G))$ is the normed algebra of all $\varphi\in C(\hat{G})$ such that $\varphi A_p(G)\hat{}\subset A_p(G)\hat{}$.

Before proceeding to the material of the next two sections we wish to give some results concerning the derived spaces of $A_p(G)$. Since the algebras $A_p(G)$ when G is nondiscrete do not possess minimal approximate identities, we cannot discuss the derived algebra for $A_p(G)$ as developed in Chapter 1. Instead, we shall restrict our attention to the set of elements in $A_p(G)$ whose transforms are mapped into $A_p(G)\hat{}$ upon multiplication by every function in $C_0(\hat{G})$. That is, for each p, $1\leqq p<\infty$, we consider $(A_p(G))_0=\{f\,|\,f\in A_p(G),\ \varphi\hat{f}\in A_p(G)\hat{}$ for all $\varphi\in C_0(\hat{G})\}$. This should be compared with the definition of the derived space for $L_p(G)$.

Theorem 6.2.3. *Let G be a nondiscrete locally compact Abelian group. If G is noncompact then $(A_p(G))_0=\{0\}$, $1\leqq p<\infty$, while if G is compact then $(A_p(G))_0=A_p(G)$, $1\leqq p\leqq 2$, and $(A_p(G))_0=L_2(G)$, $p>2$.*

The equality here is, of course, only as linear spaces.

Proof. For any group G it is obvious that $(A_p(G))_0\subset(L_1(G))_0$. The assertion for noncompact G is then immediate from Corollary 4.6.1.

If G is compact then $A_2(G)=L_1(G)\cap L_2(G)=L_2(G)$. Thus for $1\leqq p\leqq 2$ we have $A_p(G)\subset L_2(G)$. If $f\in A_p(G)$ and $\varphi\in C_0(\hat{G})$ then $\varphi\hat{f}\in L_2(\hat{G})\cap L_p(\hat{G})$. Consequently, by the Plancherel Theorem (Theorem F.8.2), there exists a $g\in L_2(G)$ such that $\hat{g}=\varphi\hat{f}$. Obviously $g\in A_p(G)$ and $f\in(A_p(G))_0$. Hence $(A_p(G))_0=A_p(G)$, $1\leqq p\leqq 2$.

If $p>2$ then $(A_p(G))_0\subset(L_1(G))_0=L_2(G)$ by Theorem 1.9.1 as G is compact. Conversely, $(A_2(G))_0=A_2(G)=L_2(G)\subset A_p(G)$ shows that $L_2(G)\subset(A_p(G))_0$.

Therefore $(A_p(G))_0=L_2(G)$ for $p>2$. □

6.3. The Multipliers for $A_p(G)$: G Noncompact. As mentioned previously there are a number of similarities between the algebras $A_p(G)$ and the group algebra $L_1(G)$. In this section we shall show that the multipliers for $A_p(G)$ can be identified with the bounded measures on G provided G is noncompact. Thus for noncompact locally compact Abelian groups G the group algebra $L_1(G)$ and the algebras $A_p(G)$ have the same multipliers. In particular, this result reveals that the multiplier algebra of a commutative Banach algebra does not uniquely determine the Banach algebra itself since $L_1(G)$ is a QCG algebra and the algebras $A_p(G)$ are not QCG algebras.

Theorem 6.3.1. *Let G be a noncompact nondiscrete locally compact Abelian group and $1\leqq p<\infty$. If T: $A_p(G)\to A_p(G)$ then the following are equivalent:*

(i) *$T\in M(A_p(G))$.*

(ii) *There exists a unique measure $\mu\in M(G)$ such that $Tf=\mu*f$ for each $f\in A_p(G)$.*

(iii) *There exists a unique function $\varphi\in C(\hat{G})$ such that $(Tf)\hat{}=\varphi\hat{f}$ for each $f\in A_p(G)$.*

Moreover the correspondence between T and μ defines an isometric algebra isomorphism from $M(A_p(G))$ onto $M(G)$.

Proof. It is evident that (ii) implies (iii) and (iii) implies (i). Suppose then that $T\in M(A_p(G))$. Then by Theorem 6.2.1 we see that T is a continuous linear operator from $A_p(G)$ to $A_p(G)$ which commutes with translation. Clearly

$$\|Tf\|_1 \leqq \|Tf\|^p \leqq \|T\|^p \|f\|^p \qquad (f \in A_p(G)).$$

Suppose $f \in A_c(G) \subset A_1(G) \subset A_p(G)$, and let $\varepsilon > 0$. Since $Tf \in L_1(G)$, there exists a compact subset $K \subset G$ which contains the support of f such that

$$\int_{G \sim K} |Tf(t)| \, d\lambda(t) < \frac{\varepsilon}{2}.$$

If $s_1, s_2, \ldots, s_n \in G$ have been chosen such that $K s_i \cap K s_j = \emptyset$, $i \neq j$, then let $s_{n+1} \in G \sim \bigcup_{i=1}^{n} K^{-1} K s_i$. Such an s_{n+1} exists as G is noncompact. Moreover $K s_{n+1} \cap K s_i = \emptyset$, $i = 1, 2, 3, \ldots, n$, as if $K s_{n+1} \cap K s_j \neq \emptyset$ then there exists some $s \in G$ such that $s = u s_{n+1} = v s_j$ where $u, v \in K$. Hence $s_{n+1} = u^{-1} v s_j \in K^{-1} K s_j$, contrary to the choice s_{n+1}. In this way we obtain a sequence $\{s_i\} \subset G$ such that $K s_i \cap K s_j = \emptyset$, $i \neq j$.

For each positive integer n, define $f_n = \left(\sum_{i=1}^{n} \tau_{s_i} f \right) \Big/ n$. Clearly $f_n \in A_c(G)$. Furthermore, by the choice of K, the fact that the sets $\{K s_i\}$ are pairwise disjoint and because T commutes with translation, we see that for each n,

$$\begin{aligned}
\|Tf_n\|_1 &= \left\| \left(\sum_{i=1}^{n} \tau_{s_i} Tf \right) \Big/ n \right\|_1 \\
&= \left\| \left(\sum_{i=1}^{n} \chi_{K s_i} \tau_{s_i} Tf \right) \Big/ n + \left(\sum_{i=1}^{n} \chi_{G \sim K s_i} \tau_{s_i} Tf \right) \Big/ n \right\|_1 \\
&\geqq \left\| \left(\sum_{i=1}^{n} \chi_{K s_i} \tau_{s_i} Tf \right) \Big/ n \right\|_1 - \left(\sum_{i=1}^{n} \|\chi_{G \sim K s_i} \tau_{s_i} Tf\|_1 \right) \Big/ n \\
&= \left(\sum_{i=1}^{n} \|\chi_{K s_i} \tau_{s_i} Tf\|_1 \right) \Big/ n - \left(\sum_{i=1}^{n} \|\chi_{G \sim K} Tf\|_1 \right) \Big/ n \\
&= \left(\sum_{i=1}^{n} \|\tau_{s_i} Tf - \chi_{G \sim K s_i} \tau_{s_i} Tf\|_1 \right) \Big/ n - \|\chi_{G \sim K} Tf\|_1 \\
&\geqq \left(\sum_{i=1}^{n} \|\tau_{s_i} Tf\|_1 \right) \Big/ n - \left(\sum_{i=1}^{n} \|\chi_{G \sim K s_i} \tau_{s_i} Tf\|_1 \right) \Big/ n - \|\chi_{G \sim K} Tf\|_1 \\
&= \|Tf\|_1 - 2 \|\chi_{G \sim K} Tf\|_1 \\
&> \|Tf\|_1 - \varepsilon.
\end{aligned}$$

It is apparent, since K contains the support of f, that $\lim_n \|f_n\|_\infty = 0$. Moreover, $\|f_n\|_1 \leqq \|f\|_1$, and so, by the convexity of $\log \|\cdot\|_{1/a}$, $0 \leqq a \leqq 1$

(Theorem D.11.3), we conclude that for each n,

$$\begin{aligned}\|f_n\|_2 &\leqq (\|f_n\|_1)^{\frac{1}{2}} (\|f_n\|_\infty)^{\frac{1}{2}} \\ &\leqq (\|f\|_1)^{\frac{1}{2}} (\|f_n\|_\infty)^{\frac{1}{2}}.\end{aligned}$$

Hence by the Plancherel Theorem (Theorem F.8.2) we see that $\lim_n \|f_n\|_2 = \lim_n \|\hat{f}_n\|_2 = 0$.

In particular, there exists a subsequence $\{\hat{f}_{n_k}\}$ of $\{\hat{f}_n\}$ such that $\lim_k \hat{f}_{n_k}(\gamma) = 0$ for almost all γ in $\hat{G}$. It is also easily seen that $\hat{f}_{n_k} = \varphi_k \hat{f}$ where φ_k is a sum of n_k characters of G divided by n_k. Thus $|\hat{f}_{n_k}| \leqq |\hat{f}|$ for each k. Consequently, we may apply the Lebesgue Dominated Convergence Theorem (Theorem C.5.2) to conclude that $\lim_k \|\hat{f}_{n_k}\|_p = 0$.

Hence for each k we have

$$\begin{aligned}\|Tf\|_1 - \varepsilon \leqq \|Tf_{n_k}\|_1 &\leqq \|T\|^p \|f_{n_k}\|^p \\ &= \|T\|^p (\|f_{n_k}\|_1 + \|\hat{f}_{n_k}\|_p) \\ &\leqq \|T\|^p (\|f\|_1 + \|\hat{f}_{n_k}\|_p),\end{aligned}$$

from which we immediately deduce that

$$\|Tf\|_1 - \varepsilon \leqq \|T\|^p \|f\|_1.$$

Since ε was arbitrary we see finally that

$$\|Tf\|_1 \leqq \|T\|^p \|f\|_1 \qquad (f \in A_c(G)).$$

Thus T defines a continuous linear mapping from $A_c(G)$ considered as a subspace of $L_1(G)$ to $L_1(G)$ which commutes with translation. Since $A_c(G)$ is norm dense in $L_1(G)$ by Theorem 5.1.2, we conclude that there exists a unique $T' \in M(L_1(G))$ such that $\|T'\| \leqq \|T\|^p$ and $T'f = Tf$ for each $f \in A_c(G)$. Hence, by Theorems 0.1.1 and 3.1.1, there exists a unique $\mu \in M(G)$ such that $\|\mu\| \leqq \|T\|^p$ and $Tf = \mu * f$ for each $f \in A_c(G)$.

If $f \in A_1(G)$ and $\{v_\alpha\}$ is a minimal approximate identity for $L_1(\hat{G})$ such that $\{\check{v}_\alpha\} \subset C_c(G)$, then from the argument at the end of the proof of Theorem 6.2.2 we know that $\lim_\alpha \|f\check{v}_\alpha - f\|^1 = 0$. Since $M(A_p(G)) \subset M(A_1(G))$ it follows that $Tf = \mu * f$ for each $f \in A_1(G)$. Consequently, $Tf = \mu * f$ for each $f \in A_p(G)$ as $A_1(G)$ is norm dense in $A_p(G)$.

Therefore (i) implies (ii), and the equivalence of (i), (ii) and (iii) is established.

Clearly the correspondence between $M(A_p(G))$ and $M(G)$ is an algebraic isomorphism, and by (ii) we have $\|T\|^p \leqq \|\mu\|$. However the argument used in showing that (i) implies (ii) reveals that $\|\mu\| \leqq \|T\|^p$. Thus the correspondence is isometric. □

It is perhaps worth while to point out that a very simple proof of this theorem is available for $p \geqq 2$ by means of Theorem 3.5.1. Indeed, suppose in this case that $T \in M(A_p(G))$. By the Plancherel Theorem and the Hausdorff-Young Theorem (Theorems F.8.2 and F.8.4) we have $L_1(G) \cap L_{p'}(G) \subset A_p(G)$, $1/p + 1/p' = 1$, and for each $f \in L_1(G) \cap L_{p'}(G)$,

$$\|Tf\|_1 \leqq \|T\|^p \|f\|^p = \|T\|^p (\|f\|_1 + \|\hat{f}\|_p)$$
$$\leqq \|T\|^p (\|f\|_1 + \|f\|_{p'}).$$

Thus T defines an element of $M(L_1(G) \cap L_{p'}(G), L_1(G))$, and so, by Theorem 3.5.1, there exists a unique $\mu \in M(G)$ such that $Tf = \mu * f$ for each $f \in L_1(G) \cap L_p(G)$. The density of $L_1(G) \cap L_{p'}(G)$ in $A_p(G)$ completes the proof.

6.4. The Multipliers for $A_p(G)$: G Compact. In the preceding section we discussed the multipliers for $A_p(G)$ when G is noncompact. We shall now examine the situation when G is an infinite compact Abelian group. The characterizations obtained in this case for the multipliers for $A_p(G)$ are in some instances more intricate than the ones we have previously discussed and not as satisfactory.

Since $A_p(G)$ is a semi-simple commutative Banach algebra, it is apparent that every function φ on $\hat{G}$ such that $\varphi \hat{f} \in A_p(G)\hat{}$ whenever $\hat{f} \in A_p(G)\hat{}$ defines a multiplier T for $A_p(G)$ by means of the formula $(Tf)\hat{} = \varphi \hat{f}$.

Thus if G is compact, $1 \leqq p \leqq 2$ and $\varphi \in C(\hat{G})$ then $\varphi \hat{f} \in L_2(\hat{G}) \cap L_p(\hat{G})$ for each $f \in A_p(G) \subset A_2(G)$. Hence, by the Plancherel Theorem (Theorem F.8.2) and the compactness of G, there exists $Tf \in A_p(G)$ for which $(Tf)\hat{} = \varphi \hat{f}$. Consequently each $\varphi \in C(\hat{G})$ defines a multiplier for $A_p(G)$. On the other hand, we know by Theorem 1.2.2 that if $T \in M(A_p(G))$, $1 \leqq p < \infty$, then there exists a $\varphi \in C(\hat{G})$ such that $(Tf)\hat{} = \varphi \hat{f}$ for each $f \in A_p(G)$, and $\|\varphi\|_\infty \leqq \|T\|^p$. From these observations we immediately deduce the following theorem.

Theorem 6.4.1. *Let G be an infinite compact Abelian group and $1 \leqq p \leqq 2$. Then there exists a continuous algebra isomorphism from $M(A_p(G))$ onto $C(\hat{G})$.*

Combining this with Theorems 6.2.2 and 4.2.2 we obtain the next corollary.

Corollary 6.4.1. *Let G be an infinite compact Abelian group and $1 \leqq p \leqq 2$. Then there exists a continuous algebra isomorphism of $M(A_p(G))$ onto $P(G)$.*

It should be noted that the isomorphism in the previous results does not appear to be isometric. However one can readily establish for infinite compact G that $\|\varphi\|_\infty \leqq \|T\|^2 \leqq 2\|\varphi\|_\infty$ for each $T \in M(A_2(G))$ where $(Tf)\hat{} = \varphi \hat{f}$.

When $p>2$ the situation is markedly different. First we observe that given $p>2$ there exist functions $\varphi \in C_0(\hat{G})$ which do not correspond to multipliers of $A_p(G)$. For if $\varphi A_p(G)\hat{} \subset A_p(G)\hat{}$ for each $\varphi \in C_0(\hat{G})$ then from Theorem 6.2.3 we would have $A_p(G) = (A_p(G))_0 = L_2(G) = A_2(G)$. But for infinite compact G this is impossible as for any $p>2$ there always exists $f \in L_1(G)$ such that $\hat{f} \in L_p(\hat{G})$ and $\hat{f} \notin L_2(\hat{G})$.

Indeed, suppose that if $f \in L_1(G)$ and $\hat{f} \in L_p(\hat{G})$ then $\hat{f} \in L_2(\hat{G})$. Then by the Hausdorff-Young Theorem (Theorem F.8.4) if $f \in L_{p'}(G) \subset L_1(G)$, $1/p + 1/p' = 1$, then $\hat{f} \in L_p(\hat{G})$ which implies $\hat{f} \in L_2(\hat{G})$, and hence $f \in L_2(G)$ by the Plancherel Theorem (Theorem F.8.2). Thus $L_{p'}(G) \subset L_2(G)$. Consequently, since G is compact, $L_{p'}(G) = L_2(G)$. However by Lemma 4.5.1 this is impossible.

Hence there exists at least one $f \in L_1(G)$ for which $\hat{f} \in L_p(\hat{G})$ and $\hat{f} \notin L_2(\hat{G})$, that is, $A_2(G)$ is a proper subset of $A_p(G)$, $p>2$.

Furthermore there exist $\varphi \in \mathcal{M}(A_p(G))$ which are not Fourier-Stieltje's transforms of any measure in $M(G)$. Indeed, let $p>2$, set $m = p/2$, $n = m/m-1$, and choose r such that $0<r<2$ and $rn>2$. Let $E \subset \hat{G}$ be any infinite Sidon set (F.11 a) and choose $\varphi \in C(\hat{G})$ such that: a) $\varphi(\gamma) = 0, \gamma \notin E$, b) $\sum_\gamma |\varphi(\gamma)|^2 = \infty$ and c) $\sum_\gamma |\varphi(\gamma)|^{rn} < \infty$. It is easily seen that such choices can always be made. If $f \in A_p(G)$ then using Hölder's inequality we have

$$\begin{aligned}
\sum_\gamma |\varphi \hat{f}(\gamma)|^2 &\leqq \Big(\sum_\gamma |\varphi(\gamma)|^{2p/p-2}\Big)^{1-2/p} \Big(\sum_\gamma |\hat{f}(\gamma)|^{2p/2}\Big)^{2/p} \\
&= \Big(\sum_\gamma |\varphi(\gamma)|^{2n}\Big)^{1/n} (\|\hat{f}\|_p)^2 \\
&\leqq (\|\varphi\|_\infty)^{2-r} \Big(\sum_\gamma |\varphi(\gamma)|^{rn}\Big)^{1/n} (\|\hat{f}\|_p)^2 .
\end{aligned}$$

Thus $\varphi \hat{f} \in L_2(\hat{G}) \subset L_p(\hat{G})$ since $\hat{G}$ is discrete. Hence φ defines a multiplier for $A_p(G)$. But $\varphi \neq \hat{\mu}$ for any $\mu \in M(G)$ because as seen from the discussion following Theorem 3.6.1 the function φ is a Fourier-Stieltjes transform if and only if $\sum_\gamma |\varphi(\gamma)|^2 < \infty$.

The preceding remarks show for infinite compact Abelian groups G and $p>2$ that we have $M(G) \subsetneqq M(A_p(G)) \subsetneqq P(G)$ since $P(G)\hat{} = L_\infty(\hat{G})$. Thus not every multiplier for $A_p(G)$ is obtained by convolution with a bounded measure, and not every pseudomeasure defines a multiplier for $A_p(G)$.

The characterization for $M(A_p(G))$ which we shall ultimately establish when G is compact and $p>2$ is analogous to the one given in Corollary 0.1.1 for $M(L_1(G))$, and the one discussed in 5.6 for $M(L_p(G), L_q(G))$. We shall show that $M(A_p(G))$ is continuously isomorphic to the dual space of a certain Banach space of continuous functions on G. We shall not be able, however, to show that the isomorphism involved is an isometry. The development will be fairly long and we shall establish a number of rather technical lemmas before stating and proving the central theorem.

First we wish to mention some elementary facts which it will be useful to keep in mind during the subsequent discussion. Namely, when G is compact, $p>2$, $1<p'<2$ and $1/p+1/p'=1$ then we have $A_1(G)\subset A_{p'}(G)\subset L_p(G)\subset L_2(G)=A_2(G)\subset L_{p'}(G)\subset A_p(G)\subset L_1(G)$. The proofs of certain of these inclusions depend on straightforward applications of the Hausdorff-Young Theorem (Theorem F.8.4). This theorem will also play an important role in the proof of the characterization theorem.

Moreover, when G is compact it is readily seen from the Fourier Inversion Theorem (Theorem F.8.1) that as linear spaces $A_1(G)=L_1(\hat{G})\hat{}$. In particular, then $A_1(G)\subset C(G)$ and is norm dense in $C(G)$ (F.7c).

Now consider a fixed $p>2$. For $T\in M(A_p(G))$ we shall denote by φ the function in $C(\hat{G})$ given by Theorem 1.2.2 such that $(Tf)\hat{}=\varphi\hat{f}$ for each $f\in A_p(G)$. If $f\in A_1(G)$ we set

$$\beta(T)(f)=\int_{\hat{G}}(Tf)\hat{}(\gamma)\,d\eta(\gamma)=\int_{\hat{G}}\varphi(\gamma)\hat{f}(\gamma)\,d\eta(\gamma),$$

and define

$$\|f\|_B=\sup_T\{|\beta(T)(f)|\,|\,T\in M(A_p(G)),\ \|T\|^p\leqq 1\}.$$

It is evident that these definitions make sense since if $T\in M(A_p(G))$ then $T\in M(A_1(G))$.

Routine arguments reveal that $\|\cdot\|_B$ is a norm on the linear space $A_1(G)$. We shall denote this normed linear space by $B_p(G)$. The preceding definitions also show for each $T\in M(A_p(G))$ that $\beta(T)$ defines a continuous linear functional on the normed linear space $B_p(G)$. Thus we obtain a mapping β: $M(A_p(G))\to B_p(G)^*$.

Lemma 6.4.1. *Let G be an infinite compact Abelian group. For each $p>2$ the mapping β is a continuous linear injective mapping from $M(A_p(G))$ to $B_p(G)^*$.*

Proof. β is clearly linear. If $\beta(T_1)=\beta(T_2)$ then for each $\delta\in\hat{G}$ we would have

$$\begin{aligned}\varphi_1(\delta)&=\int_{\hat{G}}\varphi_1(\gamma)(\cdot,\delta)\hat{}(\gamma)\,d\eta(\gamma)=\beta(T_1)[(\cdot,\delta)]\\&=\beta(T_2)[(\cdot,\delta)]=\int_{\hat{G}}\varphi_2(\gamma)(\cdot,\delta)\hat{}(\gamma)\,d\eta(\gamma)=\varphi_2(\delta).\end{aligned}$$

Hence $\varphi_1=\varphi_2$ and $T_1=T_2$. Thus β is injective.

If $f \in B_p(G)$ then

$$|\beta(T)(f)| = \Big|\int_{\hat{G}} (Tf)^\wedge(\gamma)\, d\eta(\gamma)\Big| = \Big|\|T\|^p \int_{\hat{G}} (Tf)^\wedge(\gamma)/\|T\|^p\, d\eta(\gamma)\Big|$$

$$= \big|\|T\|^p \beta(T/\|T\|^p)(f)\big| \leqq \|T\|^p \|f\|_B.$$

Therefore $\|\beta(T)\|_{B^*} \leqq \|T\|^p$ where $\|\cdot\|_{B^*}$ denotes the norm in $B_p(G)^*$. □

The remainder of the lemmas will be used in proving that β is surjective.

Lemma 6.4.2. *Let G be an infinite compact Abelian group, $p>2$ and $f, g \in B_p(G)$.*

(i) *If $1 \leqq r \leqq \infty$ and $1/r + 1/r' = 1$ then $\|f*g\|_B \leqq \|\hat{f}\|_r \|\hat{g}\|_{r'}$.*

(ii) $\|f*g\|_B \leqq \|f\|^p \|g\|_\infty$.

Proof. Clearly $f*g \in B_p(G)$ as $A_1(G)$ is an algebra under convolution. For each $T \in M(A_p(G))$ we have

$$|\beta(T)(f*g)| = \Big|\int_{\hat{G}} \varphi \hat{f} \hat{g}(\gamma)\, d\eta(\gamma)\Big| \leqq \|\varphi \hat{f}\|_r \|\hat{g}\|_{r'}$$

$$\leqq \|\varphi\|_\infty \|\hat{f}\|_r \|\hat{g}\|_{r'} \leqq \|T\|^p \|\hat{f}\|_r \|\hat{g}\|_{r'}.$$

The application of Hölder's inequality is valid since $L_1(\hat{G}) \subset L_r(\hat{G})$, $r \geqq 1$, as $\hat{G}$ is discrete.

Since this holds, in particular, when $\|T\|^p \leqq 1$, we conclude that $\|f*g\|_B \leqq \|\hat{f}\|_r \|\hat{g}\|_{r'}$, thereby proving (i).

To prove (ii) we observe that for $T \in M(A_p(G))$ we have

$$|\beta(T)(f*g)| = \Big|\int_{\hat{G}} [T(f*g)]^\wedge(\gamma)\, d\eta(\gamma)\Big| = \Big|\int_{\hat{G}} (Tf)^\wedge(\gamma)\, \hat{g}(\gamma)\, d\eta(\gamma)\Big|$$

$$= \Big|\int_{\hat{G}} (Tf)^\wedge(\gamma)\, \hat{\tilde{g}}(\gamma^{-1})\, d\eta(\gamma)\Big|.$$

However, $Tf, g \in A_1(G) \subset L_2(G)$ since $M(A_p(G)) \subset M(A_1(G))$. Thus we may apply Parseval's Formula (Theorem F.8.3) to obtain

$$|\beta(T)(f*g)| = \Big|\int_G Tf(t)\, \tilde{g}(t)\, d\lambda(t)\Big| \leqq \|Tf\|_1 \|\tilde{g}\|_\infty$$

$$\leqq \|Tf\|^p \|g\|_\infty \leqq \|T\|^p \|f\|^p \|g\|_\infty.$$

Consequently, if we restrict our attention to $T \in M(A_p(G))$ such that $\|T\|^p \leqq 1$, we deduce at once that $\|f*g\|_B \leqq \|f\|^p \|g\|_\infty$, proving (ii). □

Lemma 6.4.3. *Let C be an infinite compact Abelian group and $p>2$. Suppose $F \in B_p(G)^*$, $f \in B_p(G)$ and define $F_f(\hat{g}) = F(f*g)$ for each $g \in B_p(G)$. Then F_f defines a continuous linear functional on $L_p(\hat{G})$.*

Proof. It is evident that F_f defines a linear functional on $B_p(G)^\wedge$. Moreover, from the first portion of the preceding lemma, we see for each $g \in B_p(G)$ that

$$\begin{aligned}|F_f(\hat{g})| &= |F(f*g)| \leqq \|F\|_{B^*} \|f*g\|_B \\ &\leqq \|F\|_{B^*} \|\hat{f}\|_{p'} \|\hat{g}\|_p\end{aligned}$$

where $1/p + 1/p' = 1$. Thus F_f is a continuous linear functional on $B_p(G)^\wedge$ considered as a subspace of $L_p(\hat{G})$. Since $\hat{G}$ is discrete, $B_p(G)^\wedge = A_1(G)^\wedge \subset L_1(\hat{G})$ contains $C_c(\hat{G})$, and hence $B_p(G)^\wedge$ is norm dense in $L_p(\hat{G})$. Therefore F_f can be uniquely extended to a continuous linear functional on all of $L_p(\hat{G})$. □

If F_f is the functional defined in the previous lemma then denote by $\hat{h}$ the unique element of $L_{p'}(\hat{G})$, $1/p + 1/p' = 1$, such that

$$F_f(\hat{g}) = \langle \tilde{\hat{h}}, \hat{g} \rangle = \int_{\hat{G}} \hat{h}(\gamma)\, \hat{g}(\gamma)\, d\eta(\gamma) \qquad (\hat{g} \in B_p(G)^\wedge).$$

Since $1 < p' < 2$, the Hausdorff-Young Theorem (Theorem F.8.4) assures the existence of a unique $h \in L_p(G)$ whose Fourier transform is $\hat{h}$. Thus given $F \in B_p(G)^*$, for each $f \in B_p(G)$ we define $Tf = h$ where h is chosen as above. Clearly T is a linear transformation from the linear space $A_1(G) = B_p(G)$ to $A_{p'}(G) \subset A_p(G)$, $1/p + 1/p' = 1$.

Lemma 6.4.4. *Let G be an infinite compact Abelian group, $p > 2$ and $F \in B_p(G)^*$. If T is defined as above then T is a continuous linear operator from the subspace $A_1(G)$ of $A_p(G)$ to $A_p(G)$.*

Proof. Suppose $f \in A_1(G)$ and $1/p + 1/p' = 1$. Then since $B_p(G)^\wedge \subset L_{p'}(\hat{G}) \subset L_p(\hat{G})$ and $B_p(G)^\wedge$ is norm dense in $L_{p'}(\hat{G})$ we conclude that

$$\begin{aligned}\|(Tf)^\wedge\|_p &= \|\hat{h}\|_p = \|\tilde{\hat{h}}\|_p \\ &= \sup\{|\langle \tilde{\hat{h}}, \hat{g}\rangle| \mid \hat{g} \in L_{p'}(\hat{G}),\ \|\hat{g}\|_{p'} \leqq 1\} \\ &= \sup\{|\langle \tilde{\hat{h}}, \hat{g}\rangle| \mid \hat{g} \in B_p(G)^\wedge,\ \|\hat{g}\|_{p'} \leqq 1\} \\ &= \sup\{|F_f(\hat{g})| \mid \hat{g} \in B_p(G)^\wedge,\ \|\hat{g}\|_{p'} \leqq 1\} \\ &= \sup\{|F(f*g)| \mid \hat{g} \in B_p(G)^\wedge,\ \|\hat{g}\|_{p'} \leqq 1\} \\ &\leqq \sup\{\|F\|_{B^*} \|f*g\|_B \mid \hat{g} \in B_p(G)^\wedge,\ \|\hat{g}\|_{p'} \leqq 1\} \\ &\leqq \sup\{\|F\|_{B^*} \|\hat{f}\|_p \|\hat{g}\|_{p'} \mid \hat{g} \in B_p(G)^\wedge,\ \|\hat{g}\|_{p'} \leqq 1\} \\ &\leqq \|F\|_{B^*} \|\hat{f}\|_p.\end{aligned}$$

The penultimate inequality is valid because of Lemma 6.4.2 (i).

Furthermore, utilizing the fact that $A_1(G)$ is norm dense in $C(G)$ and Parseval's Formula (Theorem F.8.3), we have

$$\begin{aligned}
\|Tf\|_1 &= \sup\{|\langle Tf, g\rangle| \,|\, g\in C(G), \|g\|_\infty \leqq 1\} \\
&= \sup\{|\int_G h(t)\, g(t^{-1})\, d\lambda(t)| \,|\, g\in A_1(G), \|g\|_\infty \leqq 1\} \\
&= \sup\{|\int_{\hat{G}} \hat{h}(\gamma)\, \hat{g}(\gamma)\, d\eta(\gamma)| \,|\, g\in A_1(G), \|g\|_\infty \leqq 1\} \\
&= \sup\{|\langle \tilde{\hat{h}}, \hat{g}\rangle| \,|\, g\in A_1(G), \|g\|_\infty \leqq 1\} \\
&= \sup\{|F_f(\hat{g})| \,|\, g\in A_1(G), \|g\|_\infty \leqq 1\} \\
&= \sup\{|F(f*g)| \,|\, g\in A_1(G), \|g\|_\infty \leqq 1\} \\
&\leqq \sup\{\|F\|_{B^*} \|f*g\|_B \,|\, g\in A_1(G), \|g\|_\infty \leqq 1\} \\
&\leqq \sup\{\|F\|_{B^*} \|f\|^p \|g\|_\infty \,|\, g\in A_1(G), \|g\|_\infty \leqq 1\} \\
&\leqq \|F\|_{B^*} \|f\|^p.
\end{aligned}$$

The penultimate inequality is now due to Lemma 6.4.2(ii).

Combining these estimates we see that for each $f\in A_1(G)$,

$$\begin{aligned}
\|Tf\|^p &= \|Tf\|_1 + \|(Tf)^\wedge\|_p \\
&\leqq \|F\|_{B^*} \|f\|^p + \|F\|_{B^*} \|\hat{f}\|_p \\
&\leqq 2\, \|F\|_{B^*} \|f\|^p.
\end{aligned}$$

Therefore T is continuous from $A_1(G)\subset A_p(G)$ to $A_p(G)$. □

We are now in a position to state and prove the result mentioned before Lemma 6.4.2.

Theorem 6.4.2. *Let G be an infinite compact Abelian group. For each $p>2$ the mapping β: $M(A_p(G))\to B_p(G)^*$ defined by*

$$\beta(T)(f) = \int_{\hat{G}} (Tf)^\wedge(\gamma)\, d\eta(\gamma) \qquad (f\in B_p(G))$$

is a continuous linear isomorphism of $M(A_p(G))$ onto $B_p(G)^*$.

Proof. By Lemma 6.4.1 we need only show that β is surjective. Given $F\in B_p(G)^*$ let T be the operator constructed preceding Lemma 6.4.4. In view of this lemma T can be uniquely extended to a bounded linear operator on all of $A_p(G)$ since $A_1(G)$ is norm dense in $A_p(G)$. We shall also denote this extension by T.

Moreover, if $f, g \in A_1(G)$ and $s \in G$ then

$$\begin{aligned}
&\int_{\hat{G}} [T(\tau_s f)]^\wedge(\gamma)\,\hat{g}(\gamma)\,d\eta(\gamma)\\
&\quad = F(\tau_s f * g) = F(f * \tau_s g) = \int_{\hat{G}} (Tf)^\wedge(\gamma)(\tau_s g)^\wedge(\gamma)\,d\eta(\gamma)\\
&\quad = \int_{\hat{G}} [Tf * \tau_s g]^\wedge(\gamma)\,d\eta(\gamma) = \int_{\hat{G}} [\tau_s(Tf) * g]^\wedge(\gamma)\,d\eta(\gamma)\\
&\quad = \int_{\hat{G}} [\tau_s(Tf)]^\wedge(\gamma)\,\hat{g}(\gamma)\,d\eta(\gamma).
\end{aligned}$$

Since, as we have remarked previously, $A_1(G)^\wedge$ is norm dense in $L_{p'}(\hat{G})$, $1/p + 1/p' = 1$, and $(Tf)^\wedge \in L_p(\hat{G})$ for each $f \in A_1(G)$, we conclude that $[T(\tau_s f)]^\wedge = [\tau_s(Tf)]^\wedge$ for each $f \in A_1(G)$ and $s \in G$. The semi-simplicity of $A_1(G)$, the continuity of T and the norm denseness of $A_1(G)$ in $A_p(G)$ combine to imply that $T\tau_s = \tau_s T$ for each $s \in G$.

Therefore $T \in M(A_p(G))$ by Theorem 6.2.1.

Furthermore, if $f, g \in A_1(G)$ then

$$\begin{aligned}
\beta(T)(f*g) &= \int_{\hat{G}} [T(f*g)]^\wedge(\gamma)\,d\eta(\gamma) = \int_{\hat{G}} (Tf)^\wedge(\gamma)\,\hat{g}(\gamma)\,d\eta(\gamma)\\
&= F_f(\hat{g}) = F(f*g)
\end{aligned}$$

by the definition of T. But $\{f*g \mid f, g \in A_1(G)\}$ is norm dense in $B_p(G)$. Indeed, let $\{u_\alpha\} \subset A_1(G)$ be an approximate identity for the algebra $A_1(G)$. Then in particular we have for each $f \in A_1(G)$ that $\lim_\alpha \|\hat{f} - \hat{f}\hat{u}_\alpha\|_1 = 0$. However,

$$\begin{aligned}
\|f - f*u_\alpha\|_B &= \sup\{|\beta(T)(f - f*u_\alpha)| \mid T \in M(A_p(G)),\ \|T\|^p \leqq 1\}\\
&= \sup\Big\{\Big|\int_{\hat{G}} \varphi(\gamma)[\hat{f}(\gamma) - \hat{f}\hat{u}_\alpha(\gamma)]\,d\eta(\gamma)\Big| \,\Big|\, T \in M(A_p(G)),\ \|T\|^p \leqq 1\Big\}\\
&\leqq \sup\{\|\varphi\|_\infty \|\hat{f} - \hat{f}\hat{u}_\alpha\|_1 \mid T \in M(A_p(G)),\ \|T\|^p \leqq 1\}\\
&\leqq \|\hat{f} - \hat{f}\hat{u}_\alpha\|_1
\end{aligned}$$

as $\|\varphi\|_\infty \leqq \|T\|^p \leqq 1$. Thus we see at once that $\lim_\alpha \|f - f*u_\alpha\|_B = 0$, and so $\{f*g \mid f, g \in A_1(G)\}$ is norm dense in $B_p(G)$.

Hence $\beta(T) = F$ and β is surjective. ▯

From the previous lemmas and theorem it is apparent that $\|\beta(T)\|_{B^*} \leqq \|T\|^p \leqq 2\|\beta(T)\|_{B^*}$.

If we denote the completion of the normed linear space $B_p(G)$ by $\bar{B}_p(G)$ then the dual of $\bar{B}_p(G)$ is the same as that of $B_p(G)$. In particular, the preceding theorem establishes the existence of a continuous linear isomorphism between $M(A_p(G))$ and $\bar{B}_p(G)^*$. As indicated previously, considered as linear spaces, $B_p(G) = L_1(\hat{G})^\wedge \subset C(G)$. The next theorem shows that we may also consider $\bar{B}_p(G)$ as a subspace of $C(G)$.

Theorem 6.4.3. *Let G be an infinite compact Abelian group. For each $p>2$ there exists a continuous linear injective mapping $\imath$ of $\bar{B}_p(G)$ onto a subspace of $C(G)$.*

Proof. From the Fourier Inversion Theorem (Theorem F.8.1), if $f \in B_p(G)$ then for each $t \in G$,

$$\begin{aligned} |f(t)| &= \Big|\int_{\hat{G}} (t,\gamma) \hat{f}(\gamma)\, d\eta(\gamma)\Big| = \Big|\int_{\hat{G}} (\tau_{t^{-1}} f)^\wedge(\gamma)\, d\eta(\gamma)\Big| \\ &= |\beta(\tau_{t^{-1}})(f)| \leqq \sup\{|\beta(T)(f)| \,|\, T \in M(A_p),\ \|T\|^p \leqq 1\} \\ &= \|f\|_B. \end{aligned}$$

Hence $\|f\|_\infty \leqq \|f\|_B$ for each $f \in B_p(G)$.

Considering the elements of $\bar{B}_p(G)$ as Cauchy sequences of elements of $B_p(G)$, it is apparent from the preceding inequality that if $\{f_n\} \subset B_p(G)$ is a Cauchy sequence in $B_p(G)$ then there exists a unique function $f \in C(G)$ such that $\lim_n \|f_n - f\|_\infty = 0$. Setting $\imath(\{f_n\}) = f$ we obtain a well defined linear mapping from $\bar{B}_p(G)$ onto a subspace of $C(G)$. It follows at once from the previous inequality that $\imath$ is a continuous mapping.

To prove that $\imath$ is injective it is sufficient to show that if $\{f_n\} \subset B_p(G)$ is a Cauchy sequence such that $\lim_n \|f_n\|_\infty = 0$ then $\lim_n \|f_n\|_B = 0$. Now if $T \in M(A_p(G))$ then by Theorem 6.4.2 we have $\beta(T) \in B_p(G)^*$. Moreover, the inequality

$$|\beta(T)(f_n) - \beta(T)(f_m)| \leqq \|\beta(T)\|_{B^*} \|f_n - f_m\|_B$$

shows that $\{\beta(T)(f_n)\}$ is a Cauchy sequence of numbers. Define $G(T) = \lim_n \beta(T)(f_n)$. We claim that $G(T) = 0$ for each $T \in M(A_p(G))$, that is, $\{f_n\}$ converges weakly to zero.

Indeed, given $g \in L_1(G)$ we denote by T_g the multiplier for $A_p(G)$ defined by $T_g f = g * f$ for each $f \in A_p(G)$. Since $\{f_n\} \subset A_1(G)$ we may apply the Fourier Inversion Theorem (Theorem F.8.1) to deduce that for each n

$$\begin{aligned} |\beta(T_g)(f_n)| &= \Big|\int_{\hat{G}} (T_g f_n)^\wedge(\gamma)\, d\eta(\gamma)\Big| = \Big|\int_{\hat{G}} (g * f_n)^\wedge(\gamma)\, d\eta(\gamma)\Big| \\ &= |g * f_n(e)| \leqq \|g\|_1 \|f_n\|_\infty. \end{aligned}$$

Hence $G(T_g) = 0$ for each $g \in L_1(G)$.

Furthermore, suppose $\{u_\alpha\} \subset A_1(G)$ is an approximate identity for $A_1(G)$ such that $\|u_\alpha\|_1 \leqq 1$ for all α. If $T \in M(A_p(G))$ then, as indicated earlier, $T \in M(A_1(G))$, and by Theorem 1.1.6 we see that $\{T_{u_\alpha}\}$ converges to T in the strong operator topology on $M(A_1(G))$. Setting $T_\alpha = T_{u_\alpha}$, we have for each $f \in B_p(G)$ and each α that

$$|\beta(T)(f) - \beta(T_\alpha)(f)| = \left|\int_{\hat{G}} (Tf)^\wedge(\gamma)\, d\eta(\gamma) - \int_{\hat{G}} (T_\alpha f)^\wedge(\gamma)\, d\eta(\gamma)\right|$$
$$\leqq \|[(T-T_\alpha)f]^\wedge\|_1 \leqq \|(T-T_\alpha)f\|^1.$$

Consequently $\lim_\alpha \beta(T_\alpha)(f) = \beta(T)(f)$ for each $f \in B_p(G)$.

Thus suppose $T \in M(A_p(G))$ and $\varepsilon > 0$. Then since $G(T_\alpha) = 0$ for all α we have

$$\begin{aligned} |G(T)| &= |G(T) - G(T_\alpha)| \\ &\leqq |G(T) - \beta(T)(f_n)| + |\beta(T)(f_n) - \beta(T_\alpha)(f_n)| \\ &\qquad + |\beta(T_\alpha)(f_n) - G(T_\alpha)| \\ &\leqq |G(T) - \beta(T)(f_n)| + \|(T-T_\alpha)f_n\|^1 + \|f_n\|_\infty. \end{aligned}$$

Since $\lim_n \beta(T)(f_n) = G(T)$ and $\lim_n \|f_n\|_\infty = 0$ there exists an N such that $|G(T) - \beta(T)(f_N)| < \varepsilon/3$ and $\|f_N\|_\infty < \varepsilon/3$. For this N choose α_0 such that $\|(T - T_{\alpha_0})f_N\|^1 < \varepsilon/3$. Combining these estimates with the preceding one we conclude that $|G(T)| < \varepsilon$. But ε is arbitrary. Hence $G(T) = 0$ for each $T \in M(A_p(G))$.

Thus $\{f_n\}$ converges weakly to zero.

Finally, let $\varepsilon > 0$ and for each positive integer n choose $T_n \in M(A_p(G))$ such that $\|T_n\|^p \leqq 1$ and $\|f_n\|_B < |\beta(T_n)(f_n)| + \varepsilon/3$. This is possible by the definition of $\|\cdot\|_B$. Because $\{f_n\}$ is Cauchy in $B_p(G)$ there exists a positive integer N such that if $n, m \geqq N$ then $\|f_n - f_m\|_B < \varepsilon/3$. In particular, since $\|\beta(T_N)\|_{B^*} \leqq \|T_N\|^p$ for all $m > N$ we have

$$\begin{aligned} \|f_N\|_B &< |\beta(T_N)(f_N)| + \varepsilon/3 \\ &\leqq |\beta(T_N)(f_N - f_m)| + |\beta(T_N)(f_m)| + \varepsilon/3 \\ &\leqq \|T_N\|^p \|f_N - f_m\|_B + |\beta(T_N)(f_m)| + \varepsilon/3 \\ &\leqq 2\varepsilon/3 + |\beta(T_N)(f_m)|. \end{aligned}$$

However, since $\{f_m\}$ converges weakly to zero we see that $\lim_m |\beta(T_N)(f_m)| = 0$. Hence $\|f_N\|_B \leqq 2\varepsilon/3$. Moreover, if $n \geqq N$ then

$$\|f_n\|_B \leqq \|f_n - f_N\|_B + \|f_N\|_B < \varepsilon/3 + 2\varepsilon/3 = \varepsilon,$$

that is, $\lim_n \|f_n\|_B = 0$.

Therefore the mapping ι is injective. ☐

We summarize the previous two theorems in the next result.

Theorem 6.4.4. *Let G be an infinite compact Abelian group. For each $p>2$ there exists a continuous linear isomorphism of $M(A_p(G))$ onto the dual space of a Banach space of continuous functions.*

It should be recalled, however, that the norm in this Banach space is not the supremum norm.

The analog for $M(A_p(G))$, $p>2$, of Theorem 0.1.2 for compact groups takes the following form.

Theorem 6.4.5. *Let G be an infinite compact Abelian group. For each $p>2$ the space of finite linear combinations of the functionals $\{\beta(\tau_s)\,|\,s\in G\}$ is weak* dense in $B_p(G)^*$.*

Proof. Suppose $f\in B_p(G)$ and $\beta(\tau_s)(f)=0$ for each $s\in G$. Then for each $s\in G$,

$$\begin{aligned}0=\beta(\tau_s)(f)&=\int_{\hat{G}}(\tau_s f)^{\wedge}(\gamma)\,d\eta(\gamma)=\int_{\hat{G}}(s^{-1},\gamma)\hat{f}(\gamma)\,d\eta(\gamma)\\&=f(s^{-1}),\end{aligned}$$

by the Fourier Inversion Theorem (Theorem F.8.1). Hence $f=0$ and $\beta(T)(f)=0$ for each $T\in M(A_p(G))$. Consequently every weak* continuous linear functional which vanishes on $\{\beta(\tau_s)\,|\,s\in G\}$ vanishes on all of $B_p(G)^*$. Thus (Theorem D.4.1 and D.6f) we conclude that $\{\beta(\tau_s)\,|\,s\in G\}$ is weak* dense in $B_p(G)^*$. ▯

6.5. Notes. The spaces $A_p(G)$ were introduced in Larsen, Liu and Wang [1]. Subsequently several other authors have considered these spaces and their generalizations, namely, Figà-Talamanca and Gaudry [2], Gaudry [8], Lai [1–3], Larsen [1, 3], Liu and van Rooij [1], Martin and Yap [1], Reiter [1][1] and Warner [1].

Theorems 6.1.1 and 6.1.2 are taken from Larsen, Liu and Wang [1], while Theorem 6.1.3 comes from Martin and Yap [1]. The proof is based on a suggestion of J. Burnham. An alternative proof is available in Larsen [2], 6.1.3. The existence of approximate identities for the $A_p(G)$ as in Theorem 6.1.4 was used in Larsen, Liu and Wang [1], although the proof given here comes from Lai [1].

The ideal structure of $A_p(G)$ has been investigated in several papers. Larsen, Liu and Wang [1] and Reiter [1][1] showed that there exists a one-to-one correspondence between the closed ideals of $A_p(G)$ and $L_1(G)$, while Lai [1] showed that such a correspondence exists between the closed primary ideals of $A_p(G)$ and $L_1(G)$. In Martin and Yap [1], however, it was proved that there exist maximal ideals in $A_p(G)$ which are neither closed, primary nor regular, so that the ideal structures of $A_p(G)$

[1] See also Cigler [1] and Reiter [2].

and $L_1(G)$ are not completely identical. The latter authors also showed that there exist discontinuous positive linear functionals on $A_p(G)$.

Theorem 6.2.1 is taken from Larsen, Liu and Wang [1] and Theorem 6.2.2 from Larsen [3]. The characterization of $M(A_p(G))$ as $M(G)$ for noncompact G was asserted in Larsen, Liu and Wang [1] but the proof given there is faulty as pointed out by G.I. Gaudry. The proof of Theorem 6.3.1 is taken from Figà-Talamanca and Gaudry [2]. The proof of the theorem for $p \geqq 2$ noted at the end of 6.3 is to be found in Gaudry [8], as is a proof of the theorem for all $1 \leqq p < \infty$ when G is a noncompact group which contains a subgroup isomorphic to the integers.

The description of $M(A_p(G))$ for compact G is based on Larsen [3]. The proof and statement of Theorem 6.4.2 in Larsen [2] contains an error. Clearly the characterization of $M(A_p(G))$, $p > 2$, for compact G is closely related to that for $M(L_p(G), L_q(G))$ as discussed in 5.6, Figà-Talamanca [1] and Figà-Talamanca and Gaudry [1].

Chapter 7

The Multipliers for the Pair $(H_p(G), H_q(G))$, $1 \leqq p, q \leqq \infty$

7.0. Introduction. In this final chapter we shall discuss a few of the more elementary results concerning the multipliers for the Hardy spaces $H_p(G)$. Throughout we shall assume that G is a compact connected Abelian group and that its discrete dual group $\hat{G}$ has been given some fixed order (B.3). We shall denote the set of nonnegative elements in $\hat{G}$ with respect to the given order by $\hat{G}_+$. We define the space $H_p(G)$, $1 \leqq p \leqq \infty$, to be the closed ideal in the semi-simple Banach algebra $L_p(G)$, $1 \leqq p \leqq \infty$, consisting of all those $f \in L_p(G)$ such that $\hat{f}(\gamma) = 0$, $\gamma \in \hat{G} \sim \hat{G}_+ = \hat{G}_-$. Clearly each $H_p(G)$, $1 \leqq p \leqq \infty$, is a semi-simple Banach algebra with convolution multiplication.

It is not immediately obvious which of the various notions of multiplier discussed in the previous chapters is most appropriate when considering the pair $(H_p(G), H_q(G))$. Rather than take the approach used when discussing the pair $(L_p(G), L_q(G))$, we prefer to make use of the Banach algebra structure of the spaces $H_p(G)$. Thus we say that a mapping $T: H_p(G) \to H_q(G)$, $1 \leqq p, q \leqq \infty$, is a *multiplier for the pair* $(H_p(G), H_q(G))$ if $Tf * g = f * Tg$ for each $f, g \in H_p(G)$. This definition makes sense since $L_1(G) * L_r(G) \subset L_r(G)$, $1 \leqq r \leqq \infty$, (F.2). As usual we denote the linear space of all multipliers for the pair $(H_p(G), H_q(G))$ by $M(H_p(G), H_q(G))$. Note, that as in the general definition of multipliers given in Chapter 1, we have made no assumptions of linearity or continuity on T. These properties shall once again be seen to be consequences of the definition.

Besides some general results about $M(H_p(G), H_q(G))$, we shall content ourselves in the following sections with describing the spaces $M(H_p(G), H_q(G))$, $1 \leqq q \leqq 2 \leqq p \leqq \infty$, and $M(H_p(G), H_\infty(G))$, $1 \leqq p \leqq \infty$.

7.1. General Properties of $M(H_p(G), H_q(G))$, $1 \leqq p, q \leqq \infty$. We begin by showing that every element of $M(H_p(G), H_q(G))$ is a continuous linear operator from $H_p(G)$ to $H_q(G)$.

Theorem 7.1.1. *Let G be a compact connected Abelian group and $1 \leqq p, q \leqq \infty$. If $T \in M(H_p(G), H_q(G))$ then T is a continuous linear operator from $H_p(G)$ to $H_q(G)$.*

Proof. Let $f, g, h \in H_p(G)$. Then for any scalar $a, b \in C$ we have

$$\begin{aligned} h * T(af + bg) &= Th * (af + bg) \\ &= Th * af + Th * bg \\ &= a(Th * f) + b(Th * g) \\ &= h * (a\, Tf + b\, Tg). \end{aligned}$$

Hence

$$\hat{h}[T(af + bg)]\hat{} = \hat{h}(a\, Tf + b\, Tg)\hat{} \qquad (f, g, h \in H_p(G)),$$

from which we conclude via the semi-simplicity of $L_1(G)$ that

$$T(af + bg) = a\, Tf + b\, Tg \qquad (f, g \in H_p(G)).$$

Thus T is linear.

Further, suppose $\lim_n \|f_n - f\|_p = 0$ and $\lim_n \|Tf_n - g\|_q = 0$. By Hölder's inequality and the fact that $\|f * g\|_q \leqq \|f\|_1 \|g\|_q$, $f \in L_1(G)$, $g \in L_q(G)$, $1 \leqq q \leqq \infty$ (F.2), we see for each $h \in H_p(G)$ that

$$\begin{aligned} \|h * g - h * Tf\|_q &\leqq \|h * g - h * Tf_n\|_q + \|h * Tf_n - h * Tf\|_q \\ &= \|h * (g - Tf_n)\|_q + \|Th * (f_n - f)\|_q \\ &\leqq \|h\|_1 \|g - Tf_n\|_q + \|Th\|_q \|f_n - f\|_1 \\ &\leqq \|h\|_1 \|g - Tf_n\|_q + \|Th\|_q \|f_n - f\|_p. \end{aligned}$$

Hence $h * g = h * Tf$ for each $h \in H_p(G)$. Utilizing the semi-simplicity of $L_1(G)$ we again deduce that $g = Tf$.

Therefore T is a closed linear transformation, from whence it follows, by the Closed Graph Theorem (Theorem D.6.1), that T is continuous. □

Corollary 7.1.1. *Let G be a compact connected Abelian group and $1 \leqq p, q \leqq \infty$. Then $M(H_p(G), H_q(G))$ is a Banach space of continuous linear operators from $H_p(G)$ to $H_q(G)$.*

If $T \in M(H_p(G), H_q(G))$ we shall denote its norm by $\|T\|^{p,q}$.

Let $\varphi \in L_\infty(\hat{G}_+)$. Then φ can obviously be considered as an element of $L_\infty(\hat{G})$ by defining $\varphi(\gamma) = 0$, $\gamma \in \hat{G}_-$. Assuming such an identification, suppose that $\varphi \hat{f} \in H_q(G)\hat{}$ for each $f \in H_p(G)$. It is apparent that the equation $(Tf)\hat{} = \varphi \hat{f}$ then defines a unique linear mapping T: $H_p(G) \to H_q(G)$ such that $Tf * g = f * Tg$, $f, g \in H_p(G)$. That is, φ determines an element of $M(H_p(G), H_q(G))$. The converse of this assertion is also valid.

Theorem 7.1.2. *Let G be compact connected Abelian group and $1 \leqq p, q \leqq \infty$. If T: $H_p(G) \to H_q(G)$ then the following are equivalent:*

(i) $T \in M(H_p(G), H_q(G))$.

(ii) *There exists a unique* $\varphi \in L_\infty(\hat{G}_+)$ *such that* $(Tf)\hat{}\,(\gamma) = \varphi(\gamma)\hat{f}(\gamma)$, $\gamma \in \hat{G}_+$, *for each* $f \in H_p(G)$.

Moreover the correspondence between T *and* φ *defines a continuous linear isomorphism from* $M(H_p(G), H_q(G))$ *into* $L_\infty(\hat{G}_+)$.

Proof. The previous remarks show that (ii) implies (i). Suppose $T \in M(H_p(G), H_q(G))$. For each $\gamma \in \hat{G}_+$ let $g \in H_p(G)$ be such that $\hat{g}(\gamma) \neq 0$. Define

$$\varphi(\gamma) = \frac{(Tg)\hat{}\,(\gamma)}{\hat{g}(\gamma)}.$$

The definition of $\varphi(\gamma)$ is independent of the choice of g since $(Tf)\hat{}\,\hat{g} = \hat{f}(Tg)\hat{}$, $f, g \in H_p(G)$, as $T \in M(H_p(G), H_q(G))$. In particular, for each $\gamma \in \hat{G}_+$ it is apparent that $(\cdot, \gamma) \in H_p(G)$ and $[T(\cdot, \gamma)]\hat{}\,(\gamma) = [T(\cdot, \gamma)]\hat{}\,/\chi_{\{\gamma\}}(\gamma)$ $= [T(\cdot, \gamma)]\hat{}\,/[(\cdot, \gamma)]\hat{}\,(\gamma) = \varphi(\gamma)$. Consequently, for each $f \in H_p(G)$ we have

$$\begin{aligned}(Tf)\hat{}\,(\gamma) &= (Tf)\hat{}\,(\gamma)[(\cdot, \gamma)]\hat{}\,(\gamma)\\ &= [Tf * (\cdot, \gamma)]\hat{}\,(\gamma)\\ &= [f * T(\cdot, \gamma)]\hat{}\,(\gamma)\\ &= \hat{f}(\gamma)[T(\cdot, \gamma)]\hat{}\,(\gamma)\\ &= \varphi(\gamma)\hat{f}(\gamma) \qquad (\gamma \in \hat{G}_+).\end{aligned}$$

The uniqueness of φ is evident.

Therefore (i) implies (ii).

Moreover, we note for each $\gamma \in \hat{G}_+$ that by Hölder's inequality we have

$$\begin{aligned}|\varphi(\gamma)| &= |[T(\cdot, \gamma)]\hat{}\,(\gamma)|\\ &\leqq \|[T(\cdot, \gamma)]\hat{}\,\|_\infty\\ &\leqq \|T(\cdot, \gamma)\|_1\\ &\leqq \|T(\cdot, \gamma)\|_q\\ &\leqq \|T\|^{p,q} \|(\cdot, \gamma)\|_p\\ &= \|T\|^{p,q}.\end{aligned}$$

Hence $\|\varphi\|_\infty \leqq \|T\|^{p,q}$, and the mapping determined by the correspondence between T and φ is linear and continuous. ▯

It is of course easily seen that if $T \in M(H_p(G), H_q(G))$ then T is a continuous linear operator from $H_p(G)$ to $H_q(G)$ such that $T\tau_s = \tau_s T$, $s \in G$.

7.2. The Multipliers for the Pair $(H_p(G), H_q(G))$, $1 \leqq q \leqq 2 \leqq p \leqq \infty$. If the indices p and q satisfy the inequalities $1 \leqq q \leqq 2 \leqq p \leqq \infty$ then every function in $L_\infty(\hat{G}_+)$ determines a multiplier for the pair $(H_p(G), H_q(G))$.

Theorem 7.2.1. *Let G be a compact connected Abelian group and $1 \leqq q \leqq 2 \leqq p \leqq \infty$. Then there exists a continuous linear isomorphism of $M(H_p(G), H_q(G))$ onto $L_\infty(\hat{G}_+)$.*

Proof. By Theorem 7.1.2 it is sufficient to show that the mapping determined by the equation $(Tf)\hat{} = \varphi \hat{f}$ is surjective. Let $\varphi \in L_\infty(\hat{G}_+)$. Consider φ as an element of $L_\infty(\hat{G})$ by setting $\varphi(\gamma)=0$, $\gamma \in \hat{G}_-$. Then, since $H_p(G) \subset H_2(G)$ as $2 \leqq p \leqq \infty$, we see from Theorem 4.1.1 that $\varphi \hat{f} \in H_2(G)\hat{}$ for each $f \in H_p(G)$. Thus there exists a unique $Tf \in H_2(G)$ such that $(Tf)\hat{} = \varphi \hat{f}$. But $H_2(G) \subset H_q(G)$ as $1 \leqq q \leqq 2$. Hence $T\colon H_p(G) \to H_q(G)$ and $(Tf)\hat{}(\gamma) = \varphi(\gamma)\hat{f}(\gamma)$, $\gamma \in \hat{G}_+$, for each $f \in H_p(G)$.

Therefore by Theorem 7.1.2 we conclude that $T \in M(H_p(G), H_q(G))$, and the mapping from $M(H_p(G), H_q(G))$ to $L_\infty(\hat{G}_+)$ is surjective. ▯

Corollary 7.2.1. *Let G be a compact connected Abelian group. Then there exists an isometric algebra isomorphism of $M(H_2(G), H_2(G))$ onto $L_\infty(\hat{G}_+)$.*

Proof. From Corollary 7.1.1 and Theorems 7.1.2 and 7.2.1 it is apparent that $M(H_2(G), H_2(G))$ is a commutative Banach algebra and that the correspondence between T and φ defines a continuous algebra isomorphism from $M(H_2(G), H_2(G))$ onto $L_\infty(\hat{G}_+)$. Suppose $T \in M(H_2(G), H_2(G))$ and $\varphi \in L_\infty(\hat{G}_+)$ is such that $(Tf)\hat{}(\gamma) = \varphi(\gamma)\hat{f}(\gamma)$, $\gamma \in \hat{G}_+$, for each $f \in H_2(G)$. Then considering φ, as before, to be both an element of $L_\infty(\hat{G}_+)$ and of $L_\infty(\hat{G})$ we deduce via Theorems 7.1.2 and 4.1.1 and the Plancherel Theorem (Theorem F.8.2) that

$$\begin{aligned}
\sup_{\gamma \in \hat{G}_+} |\varphi(\gamma)| = \|\varphi\|_\infty \leqq \|T\|^{2,2}
&= \sup_{\substack{f \in H_2(G) \\ \|f\|_2 = 1}} \|Tf\|_2 \\
&= \sup_{\substack{f \in H_2(G) \\ \|f\|_2 = 1}} \|(Tf)\hat{}\|_2 \\
&= \sup_{\substack{f \in H_2(G) \\ \|f\|_2 = 1}} \|\varphi \hat{f}\|_2 \\
&\leqq \sup_{\substack{f \in L_2(G) \\ \|f\|_2 = 1}} \|\varphi \hat{f}\|_2 \\
&= \|T\|_{2,2} \\
&= \sup_{\gamma \in \hat{G}} |\varphi(\gamma)| \\
&= \sup_{\gamma \in \hat{G}_+} |\varphi(\gamma)|.
\end{aligned}$$

Thus $\|\varphi\|_\infty = \|T\|^{2,2}$, and the mapping is an isometry. ▯

7.3. The Multipliers for the Pair $(H_p(G), H_\infty(G))$, $1 \leqq p \leqq \infty$. Finally we wish to characterize the elements of $M(H_p(G), H_\infty(G))$, $1 \leqq p \leqq \infty$. These results should be compared with those for $M(L_p(G), L_\infty(G))$ given in Theorems 3.3.1 and 3.4.2. $M(G)\hat{}\,|_{\hat{G}_+}$ and $L_{p'}(G)\hat{}\,|_{\hat{G}_+}$, $1 < p' \leqq \infty$, will denote the restrictions to $\hat{G}_+$ of the elements of $M(G)\hat{}$ and $L_{p'}(G)\hat{}$, $1 < p' \leqq \infty$, respectively. Clearly $M(G)\hat{}\,|_{\hat{G}_+}$ and $L_{p'}(G)\hat{}\,|_{\hat{G}_+}$ are subspaces of $L_\infty(\hat{G}_+)$.

Theorem 7.3.1. *Let G be a compact connected Abelian group and $1 \leqq p \leqq \infty$. If $T: H_p(G) \to H_\infty(G)$ then the following are equivalent:*

(i) *$T \in M(H_p(G), H_\infty(G))$.*

(ii) *If $p = \infty$, then there exists a unique $\varphi \in M(G)\hat{}\,|_{\hat{G}_+}$ such that $(Tf)\hat{}\,(\gamma) = \varphi(\gamma)\hat{f}(\gamma)$, $\gamma \in \hat{G}_+$, for each $f \in H_p(G)$; and if $1 \leqq p < \infty$ then there exists a unique $\varphi \in L_{p'}(G)\hat{}\,|_{\hat{G}_+}$, $1/p + 1/p' = 1$, such that $(Tf)\hat{}\,(\gamma) = \varphi(\gamma)\hat{f}(\gamma)$, $\gamma \in \hat{G}_+$, for each $f \in H_p(G)$.*

Proof. Suppose $T \in M(H_p(G), H_\infty(G))$ and let $\varphi \in L_\infty(\hat{G}_+)$ be such that $(Tf)\hat{}\,(\gamma) = \varphi(\gamma)\hat{f}(\gamma)$, $\gamma \in \hat{G}_+$, for each $f \in H_p(G)$. From the relation $[T(\cdot, \gamma)]\hat{}$ $= \varphi(\gamma)$, $\gamma \in \hat{G}_+$, established in the proof of Theorem 7.1.2, it follows immediately that $[T(\cdot, \gamma)](\cdot) = \varphi(\gamma)(\cdot, \gamma)$ for each $\gamma \in \hat{G}_+$. Let A denote the linear subspace of $H_\infty(G)$ consisting of the trigonometric polynomials in $H_\infty(G)$. The preceding observation shows that $T: A \to A$. In particular, if $f \in A$ then $Tf \in A$ is continuous and we can define $F_T(f) = Tf(e)$. Clearly F_T is a linear functional on A.

If $p = \infty$ then we see that

$$|F_T(f)| = |Tf(e)| \leqq \|Tf\|_\infty \leqq \|T\|^{\infty, \infty} \|f\|_\infty \qquad (f \in A).$$

Thus F_T defines a continuous linear functional on A considered as a subspace of $C(G)$. Extending F_T to all of $C(G)$, we deduce from the fact that $C(G)^* = M(G)$ (Theorem D.9.4) the existence of a $\mu \in M(G)$ such that

$$F_T(f) = \langle f, \mu \rangle = \int_G f(t^{-1})\, d\mu(t) \qquad (f \in A).$$

Hence

$$\begin{aligned} \varphi(\gamma) &= \varphi(\gamma)(e, \gamma) = [T(\cdot, \gamma)](e) \\ &= \int_G (t^{-1}, \gamma)\, d\mu(t) = \hat{\mu}(\gamma) \end{aligned} \qquad (\gamma \in \hat{G}_+).$$

That is, $\varphi \in M(G)\hat{}\,|_{\hat{G}_+}$.

Similarly if $1 \leqq p < \infty$ then, utilizing the estimate

$$|F_T(f)| \leqq \|Tf\|_\infty \leqq \|T\|^{p, \infty} \|f\|_p,$$

we deduce the existence of a $g \in L_{p'}(G)$ such that

$$Tf(e) = \int_G f(t^{-1})\, g(t)\, d\lambda(t) \qquad (f \in A).$$

Then, as before,

$$\varphi(\gamma) = \int_G (t^{-1}, \gamma)\, g(t)\, d\lambda(t) = \hat{g}(\gamma) \qquad (\gamma \in \hat{G}_+),$$

and $\varphi \in L_{p'}(G)\hat{}\,|_{\hat{G}_+}$.

The uniqueness of the φ is a consequence of Theorem 7.1.2.

Thus (i) implies (ii).

The converse assertion follows immediately from the facts that if $\mu \in M(G)$ then $\mu * f \in L_\infty(G)$ whenever $f \in L_\infty(G)$ (F.2), and if $g \in L_{p'}(G)$, $1/p + 1/p' = 1$, then $g * f \in C_0(G) \subset L_\infty(G)$ whenever $f \in L_p(G)$ (F.3). □

We should note that if $q < \infty$ then $M(H_p(G), H_q(G))$ can not, in general, be described as in Theorem 7.3.1. This is of course obvious from Theorem 7.2.1. Even without the restriction $1 \leqq q \leqq 2 \leqq p \leqq \infty$ on the indices p and q it is not generally the case that Theorem 7.3.1 is valid when $q < \infty$. For example, let $G = \Gamma = \{z \mid |z| = 1\}$ and $1 \leqq p \leqq \infty$, $1 \leqq q < \infty$. Then $\hat{G} = Z$, the group of integers, and we shall take $\hat{G}_+ = Z_+$, the set of nonnegative integers. Suppose $\{n_k\} \subset Z_+$ is a sequence such that $n_{k+1}/n_k \geqq r > 1$, that is, $\{n_k\}$ is a Hadamard sequence in Z_+. Then $\{n_k\}$ is an infinite Sidon set in Z_+ (F.11e). Define $\varphi \in L_\infty(Z_+)$ by setting $\varphi(n_k) = c_k$, $k = 1, 2, 3, \ldots$, and $\varphi(n) = 0$, $n \neq n_k$, $n \in Z_+$, where $\{c_k\}$ is so chosen that $\|\varphi\|_\infty < \infty$ and $\lim_k \varphi(n_k) \neq 0$. We claim that φ determines an element of $M(H_p(\Gamma), H_q(\Gamma))$ but that $\varphi \notin M(\Gamma)\hat{}\,|_{Z_+}$.

Indeed, if $f \in H_p(\Gamma) \subset H_1(\Gamma)$ then since $\{n_k\}$ is a Hadamard sequence we have that $\sum_{k=1}^{\infty} |\hat{f}(n_k)|^2 < \infty$ (F.11f). Hence $\varphi \hat{f} \in L_2(Z)$ for each $f \in H_p(\Gamma)$, where we here are considering φ as an element of $L_\infty(Z)$ by defining $\varphi(n) = 0$, $n \in Z \sim Z_+$. Thus, by the Plancherel Theorem (Theorem F.8.2), there exists a unique $Tf \in L_2(\Gamma)$ such that $(Tf)\hat{} = \varphi \hat{f}$ for each $f \in H_p(\Gamma)$. However $(Tf)\hat{}(n) = 0$, $n \neq n_k$, implies that $Tf \in L_q(\Gamma)$, $1 \leqq q < \infty$, as $\{n_k\}$ is a Hadamard sequence (F.11e). Hence $T: H_p(G) \to H_q(G)$ and $(Tf)\hat{}(\gamma) = \varphi(\gamma) \hat{f}(\gamma)$, $\gamma \in Z_+$, for each $f \in H_p(G)$. That is, $T \in M(H_p(G), H_q(G))$.

But suppose that $\varphi \in M(\Gamma)\hat{}\,|_{Z_+}$. Then there exists a $\mu \in M(\Gamma)$ such that $\hat{\mu}(n) = \varphi(n)$, $n \in Z_+$. In particular, $\hat{\mu}(n) = 0$, $n \in Z_+ \sim \{n_k\}$, from which we conclude, via an extension of the F. and M. Riesz Theorem (Theorem F.12.2), that μ is absolutely continuous with respect to λ. Consequently $\lim_k \varphi(n_k) = \lim_k \hat{\mu}(n_k) = 0$, contrary to the choice of φ. Therefore $\varphi \notin M(\Gamma)\hat{}\,|_{Z_+}$.

In particular χ_E where $E = \{n_k\}$, a Hadamard sequence in Z_+, defines a multiplier for the pair $(H_p(\Gamma), H_q(\Gamma))$, $1 \leqq p \leqq \infty$, $1 \leqq q < \infty$, and $\chi_E \notin M(\Gamma)\hat{}\,|_{Z_+}$.

7.4. Notes. The majority of the development in this chapter is based on Gaudry [5]. In particular Theorem 7.3.1 and the example following that theorem come from this paper. A special case of the example was given in Meyer [2], who showed also that there exist functions φ on the real line R which determine multipliers for the pair $(H_1(R), H_1(R))$ but for which the restriction of φ to $[0, +\infty)$ does not belong to $M(R)\hat{}\,|_{[0,+\infty)}$. Rudin [4] contains examples of characteristic functions χ_E which determine multipliers for the pairs $(H_p(G), H_2(G))$, $1<p\leqq 2$, but not for $(H_1(G), H_2(G))$.

Gaudry [5] also contains a characterization of the multipliers for the pair $(H_1(G), A(G)\cap H_1(G))$ where, as usual, $A(G)=L_1(\hat{G})\hat{}$ and for $f\in A(G)\cap H_1(G)$ we set $\|f\|=\|f\|_A=\|\hat{f}\|_1$. The theorem is as follows.

Theorem 7.4.1. *Let G be a compact connected Abelian group. If T: $H_1(G)\to A(G)\cap H_1(G)$ then the following are equivalent:*

(i) *$T\in M(H_1(G), A(G)\cap H_1(G))$.*

(ii) *There exists a unique $\varphi\in L_\infty(\hat{G}_+)$ such that $(Tf)\hat{}\,(\gamma)=\varphi(\gamma)\hat{f}(\gamma)$, $\gamma\in\hat{G}_+$, for each $f\in H_1(G)$ and both φ and $|\varphi|$ belong to $L_\infty(G)\hat{}\,|_{\hat{G}_+}$.*

This result has also been established by Boas [1] and Caveny [1].

Many other authors have considered the question of obtaining either necessary or sufficient conditions for a function φ to determine a multiplier for the pair $(H_p(G), H_q(G))$. We mention Duren [1], Duren and Shields [1, 2], Hardy and Littlewood [1, 3, 4], Hedlund [1, 2], Stein [4–6], Stein and Zygmund [1], and Wells [3]. The results of Hardy and Littlewood [3] and Stein and Zygmund [1] provide a characterization of $M(H_1(\Gamma), H_2(\Gamma))$.

Theorem 7.4.2. *Let $G=\Gamma$ and T: $H_1(\Gamma)\to H_2(\Gamma)$. Then the following are equivalent:*

(i) *$T\in M(H_1(\Gamma), H_2(\Gamma))$.*

(ii) *There exists a unique $\varphi\in L_\infty(Z_+)$ such that $(Tf)\hat{}\,(n)=\varphi(n)\hat{f}(n)$, $n\in Z_+$, for each $f\in H_1(\Gamma)$ and*

$$\sum_{n=1}^{N} n|\varphi(n)|^2=O(N).$$

In this connection, see also Benedek and Panzone [1] and Hedlund [2].

Multipliers for the pair $(H_p(\Gamma), L_q(Z_+))$ have been studied in several papers, among which we mention Duren and Shields [1, 2], Hardy and Littlewood [3] and Hedlund [1].

The connection between $M(H_p(G), H_q(G))$ and interpolation in these spaces is considered in Gaudry [5] and Hedlund [2], while some results on the spectra of multipliers of $H_p(G)$ can be found in Widom [1, 2].

Some other references dealing with various aspects of multipliers of H_p-spaces and other related spaces are: Bohr [1], Duren, Romberg and Shields [1], Fournier [1], Helson [2], Meyer [3], Paley [1], Rider [1], Rudin [1], [5], 8.7.8, and Zygmund [5].

Appendices

In order to facilitate the reading of the book, we have collected in these appendices many of the concepts and results cited in the body of the exposition. The presentation is not complete, but rather consists of those items which we felt to be most valuable as a supplement to the material in the book proper. No proofs are given, although sources of the different results are generally indicated. More complete and thorough discussions of the material in the various appendices are available in the following references: Appendix A: Dugundji [1] and Kelley [1]; Appendix B: Hewitt and Ross [1], Husain [1], Pontryagin [1], Rudin [5] and Weil [1]; Appendix C: Bourbaki [1], Dunford and Schwartz [1], Gaudry [6], Halmos [1], Hewitt and Ross [1], Nachbin [1], Pontryagin [1], Royden [1], Rudin [5, 7] and Weil [1]; Appendix D: Bourbaki [1], Dunford and Schwartz [1], Edwards [11], Hewitt and Ross [1], Kelley and Namioka [1], Loomis [1], Naimark [1], Royden [1], Rudin [7] and Zygmund [6]; Appendix E: Edwards [14_{II}], Gelfand, Naimark and Šilov [1], Hille and Phillips [1], Hoffman [1], Katznelson [2], Loomis [1], Naimark [1], Rickart [1], Wang [2] and Wermer [2]; Appendix F: Edwards [54], Gaudry [6], Hewitt [2], Hewitt and Ross [1, 2], Hoffman [1], Katznelson [2], Loomis [1], Rickart [1], Rudin [5], Weil [1] and Zygmund [6].

Appendix A: Topology

A.1. We mention only a few topological results. For the fundamental ideas and theorems of topology we refer the readers to Dugundji [1] and Kelley [1]. Let S be a topological space. A family of sets $\{U_\alpha\}$ is called a *locally finite cover* of S if $S = \bigcup_\alpha U_\alpha$ and each point $t \in S$ has a neighborhood U_t which intersects only finitely many of the U_α. A Hausdorff topological space is *paracompact* if each open covering has a refinement which is an open locally finite cover of S. Every paracompact space is normal (Dugundji [1], VII.2.2). Moreover, we have the following theorem.

Theorem A.1.1. (Dugundji [1], VII.6.1). *Let S be a topological space. Then the following are equivalent:*

(i) *S is normal.*

(ii) *If $\{U_\alpha\}$ is an open locally finite cover of S then there exists an open cover $\{V_\alpha\}$ of S such that $\overline{V}_\alpha \subset U_\alpha$, where $\overline{V}_\alpha$ denotes the closure of V_α, and $V_\alpha \neq \emptyset$ whenever $U_\alpha \neq \emptyset$.*

Appendix B: Topological Groups

B.1. A *topological group* G is a group equipped with a Hausdorff topology such that the mapping $(s,t) \to st$ is continuous from the product $G \times G$ onto G, and the mapping $t \to t^{-1}$ is continuous from G onto G. Then the mappings $t \to st$ and $t \to ts$ for a given $s \in G$, and the mapping $t \to t^{-1}$ are homeomorphisms of G onto itself. In particular, if E and F are subsets of G and $EF = \{st \mid s \in E, t \in F\}$ then EF is open whenever E is open, EF is compact if both E and F are compact, EF is closed if one set is compact and the other is closed, and $E^{-1} = \{t^{-1} \mid t \in E\}$ is open if E is open. If $\mathscr{U} = \{U\}$ is a neighborhood basis at the identity of G then the families $\{sU \mid s \in G, U \in \mathscr{U}\}$ and $\{Us \mid s \in G, U \in \mathscr{U}\}$ form bases for the topology of G. A set E is *symmetric* if $E = E^{-1}$. Every topological group has a basis at the identity consisting of open symmetric neighborhoods. And if U is any neighborhood of the identity, there exists symmetric neighborhoods W and V of the identity such that $W^2 \subset U$ and $\overline{V} \subset U$, where $\overline{V}$ denotes the closure of V. If the topology of G is locally compact then G is a *locally compact group*. U is a *compact neighborhood* in G if it is a compact set with a nonempty interior. Every locally compact group has a neighborhood basis at the identity consisting of symmetric compact neighborhoods.

Expositions of these basic concepts and results for topological groups can be found in Hewitt and Ross [1], 4, and Rudin [5], Appendix B.

B.2. Let Γ denote the set of complex numbers of unit modulus. Then with the usual multiplication of complex numbers and the topology inherited from the usual topology of the complex plane, Γ is a compact Abelian group. If G is any locally compact Abelian group then a continuous homomorphism of G into Γ is called a *continuous character of* G. We denote the collection of all such homomorphisms by $\hat{G}$. If $\gamma \in \hat{G}$ we denote its value at $t \in G$ by (t, γ). The continuous functions $\{(\cdot, \gamma) \mid \gamma \in \hat{G}\}$ separate the points of G (Hewitt and Ross [1], 22.17). For $\gamma_1, \gamma_2 \in \hat{G}$ we define

$\gamma_1\gamma_2$ by $(t, \gamma_1\gamma_2) = (t, \gamma_1)(t, \gamma_2)$, $t \in G$. Then with this binary operation $\hat{G}$ becomes an Abelian group. For each positive real number r let $U_r = \{z \mid z \in \Gamma, |1-z| < r\}$. If K is a compact subset of G then we set $N(K, r) = (\gamma \mid \gamma \in \hat{G}, (t, \gamma) \in U_r$ for all $t \in K\}$. Then the family of sets $\{N(K, r)\}$ and their translates form a basis for a topology on $\hat{G}$. With this topology $\hat{G}$ becomes a locally compact Abelian group (Rudin [5], 1.2.6). The topology on $\hat{G}$ is also equivalent to the weak* topology on $\hat{G}$ considered as the space of multiplicative linear functionals of the Banach algebra $L_1(G)$. (See Appendix F.) $\hat{G}$ as a locally compact Abelian group will be called the *dual group* of G. The most important relation between G and $\hat{G}$ is given in the following theorem. For $t \in G$ define $\alpha(t)$: $\hat{G} \to \Gamma$ by $\alpha(t)(\gamma) = (t, \gamma)$, $\gamma \in \hat{G}$. Then $\alpha(t)$ is a continuous homomorphism of $\hat{G}$ into Γ, that is, $\alpha(t) \in \hat{\hat{G}}$.

Theorem B.2.1. [Pontryagin-van Kampen Duality Theorem (Hewitt and Ross [1], 24.8, Rudin [5], 1.7)]. *Let G be a locally compact Abelian group and $\hat{G}$ its dual group. Then α: $G \to \hat{\hat{G}}$ is a topological isomorphism from G onto $\hat{\hat{G}}$, the dual group of $\hat{G}$.*

Clearly this theorem allows us to identify the dual group of the dual group of G with G itself. One consequence of this duality theorem is that a locally compact Abelian group is compact (discrete) if and only if its dual group is discrete (compact) (Hewitt and Ross [1], 23.17).

If H is a closed subgroup of a locally compact Abelian group G then $\hat{H}$ consists of precisely the restriction of the elements of $\hat{G}$ to H (Rudin [5], 2.1.4). Furthermore, given such a subgroup H of G we denote by $A(\hat{G}, H)$ the collection of all $\gamma \in \hat{G}$ for which $(t, \gamma) = 1$, $t \in H$. $A(\hat{G}, H)$ is called the *annihilator of H in $\hat{G}$*. $A(\hat{G}, H)$ is a closed subgroup of $\hat{G}$ (Hewitt and Ross [1], 23.24) and $\hat{H}$ is topologically isomorphic with $\hat{G}/A(\hat{G}, H)$ (Hewitt and Ross [1], 24.11).

Given a locally compact Abelian group G with dual group $\hat{G}$ it is evident that $\hat{G}_d$, the group $\hat{G}$ in the discrete topology, is again a locally compact Abelian group. Its dual group is then a compact Abelian group which we denote by $\beta(G)$. $\beta(G)$ is called the *Bohr compactification* of G, and there exists a continuous isomorphism of G onto a dense subgroup of $\beta(G)$ (Gaudry [6], VII.2, Hewitt and Ross [1], 26.13, and Rudin [5], 1.8).

B.3. Finally, we note that if G is a compact connected Abelian group then $\hat{G}$ is discrete, Abelian and contains no elements of finite order (Rudin [5], 2.5.6). Hence $\hat{G}$ can be ordered (Rudin [5], 8.1.2).

B.4. Every locally compact group G is paracompact (Hewitt and Ross [1], 8.13) and hence normal (A.1). In particular then, every locally compact topological group has an open locally finite cover and Theorem A.1.1 applies to such groups. Furthermore, we have the following result.

Theorem B.4.1. (Gaudry [1], p. 463). *Let G be a locally compact group. Then G has an open locally finite cover $\{U_\alpha\}$ such that the closure $\overline{U}_\alpha$ of U_α is compact for each α.*

Appendix C: Measure and Integration

C.1. Let S be a locally compact Hausdorff space. A *nonnegative Borel measure* ν on S is a countably additive set function defined on the σ-algebra of Borel subsets of S whose values lie in the interval $[0, \infty]$ and for which $\nu(E) < \infty$ whenever E is a Borel set with compact closure. A *signed Borel measure* ν is a countably additive set function on the Borel subsets of S whose values lie in $[-\infty, \infty]$ and such that (i) $\nu(\emptyset)=0$, (ii) $|\nu(E)| < \infty$ whenever E is a Borel set with compact closure, (iii) ν takes at most one of the values $+\infty$ or $-\infty$ and (iv) if $\{E_i\}$ is a sequence of pairwise disjoint Borel sets then either $\sum_{i=1}^{\infty} \nu(E_i)$ converges absolutely if $\nu\left(\bigcup_{i=1}^{\infty} E_i\right)$ is finite, or diverges to $+\infty$ or $-\infty$. If ν is a nonnegative or a signed Borel measure on S and S' is any Borel subset of S then $\nu_{S'}$ denotes the nonnegative or signed Borel measure defined by $\nu_{S'}(E) = \nu(E \cap S')$. A Borel subset S' of S is said to be *positive (negative)* with respect to ν if $\nu_{S'}(E) \geqq 0$ $\left(\nu_{S'}(E) \leqq 0\right)$ for each Borel set E. Evidently $\nu_{S'}$ $(-\nu_{S'})$ is a nonnegative Borel measure on S whenever S' is a positive (negative) set. Such positive and negative sets always exist.

Theorem C.1.1. [Hahn Decomposition Theorem (Halmos [1], 29A, Royden [1], 11.5.21)]. *Let ν be a signed Borel measure on S. Then there exists disjoint Borel sets S^+ and S^- in S such that $S = S^+ \cup S^-$, S^+ is a positive set and S^- is a negative set.*

This decomposition is not unique. However, if S^+ and S^- are positive and negative sets then the nonnegative Borel measures on S defined by $\nu^+ = \nu_{S^+}$ and $\nu^- = -\nu_{S^-}$ provide a unique decomposition of ν into the difference of nonnegative measures.

Theorem C.1.2. [Jordan Decomposition Theorem (Halmos [1], 29B, Royden [1], 11.5.22)]. *Let ν be a signed Borel measure on S. Then there exist unique nonnegative Borel measures ν^+ and ν^- on S such that*

$$\nu(E) = \nu^+(E) - \nu^-(E)$$

for each Borel set E.

Since ν takes at most one of the values $+\infty$ or $-\infty$, it is clear that either ν^+ or ν^- takes values only in some finite interval. ν^+ and ν^- are called the *upper and lower variations* of ν. The *total variation* of a signed Borel measure ν is the nonnegative Borel measure $|\nu|$ defined on each Borel set E by

$$|\nu|(E)=\nu^+(E)+\nu^-(E).$$

A *complex valued Borel measure* ν on S is a complex valued set function on the Borel subsets of S for which there exist unique signed Borel measures ν_1 and ν_2 on S such that for each Borel set E one has

$$\nu(E)=\nu_1(E)+i\,\nu_2(E).$$

In view of the Jordan Decomposition Theorem we see that there then exists unique nonnegative Borel measures $\nu_1^+, \nu_1^-, \nu_2^+$ and ν_2^- such that

$$\nu(E)=\nu_1^+(E)-\nu_1^-(E)+i\,\nu_2^+(E)-i\,\nu_2^-(E)$$

for each Borel set E. The total variation of a complex Borel measure ν is the nonnegative Borel measure $|\nu|$ defined on each Borel set E by

$$|\nu|(E)=|\nu_1|(E)+|\nu_2|(E)=\nu_1^+(E)+\nu_1^-(E)+\nu_2^+(E)+\nu_2^-(E).$$

ν is said to have *finite total mass* if $|\nu|(S)<\infty$.

A complex valued Borel measure ν is said to be *regular* if for each Borel set E we have

$$|\nu|(E)=\sup_K\{|\nu|(K)\,|\,K\subset E,\, K \text{ compact}\}=\inf_U\{|\nu|(U)\,|\,U\supset E,\, U \text{ open}\}.$$

A Borel set E is *σ-finite* with respect to ν if there exists a sequence of Borel sets $\{E_i\}$ such that $E=\bigcup_{i=1}^{\infty} E_i$ and $|\nu|(E_i)<\infty$, $i=1,2,3,\ldots$. If μ is a nonnegative Borel measure on S then ν is *absolutely continuous with respect to* μ if whenever E is a Borel set such that $\mu(E)=0$ then $|\nu|(E)=0$. ν is *singular with respect to* μ if there exist disjoint Borel sets S' and S'' in S such that $\nu=\nu_{S'}$ and $\mu=\mu_{S''}$, where as before $\nu_{S'}(E)=\nu(E\cap S')$ for each Borel set E in S. Given a regular nonnegative Borel measure μ and a regular complex valued Borel measure ν with finite total mass then ν can always be decomposed into absolutely continuous and singular parts with respect to μ.

Theorem C.1.3. [Lebesgue Decomposition Theorem (Hewitt and Ross [1], 14.22)]. *Let μ be a regular nonnegative Borel measure on S and ν a regular complex valued Borel measure on S with finite total mass. Then there exists unique regular complex valued Borel measures ν_a and ν_s on S such that:*

(i) $\nu=\nu_a+\nu_s$.
(ii) ν_a *is absolutely continuous with respect to* μ.
(iii) ν_s *is singular with respect to* μ.

A real valued function f on S is *Borel measurable* if the inverse image of every open set of real numbers is a Borel set. A complex valued function is Borel measurable if its real and imaginary parts are Borel measurable. A property for a function f on S is said to hold for *ν-almost all t in S* if the set of points where f fails to have the property has ν-measure zero. Thus a sequence of functions $\{f_n\}$ converges *ν-almost everywhere* to a function f if it does so for ν-almost all t in S. The ν-almost everywhere limit of a sequence of Borel measurable functions is a Borel measurable function (Royden [1], 11.2.6).

C.2. If S is a locally compact Hausdorff space then the *Baire sets* is the σ-algebra generated by the compact G_δ-subsets of S. A *nonnegative Baire measure* ν on S is a countably additive set function defined on the Baire sets whose values lie in $[0,\infty]$ and such that $\nu(E)<\infty$ whenever E is a Baire set with compact closure. A nonnegative Baire measure is *regular* if for each Baire set E

$$\begin{aligned}\nu(E)&=\sup_K\{\nu(E)\,|\,K\subset E,\ K \text{ is a compact } G_\delta\text{-set}\}\\&=\inf_U\{\nu(U)\,|\,U\supset E,\ U \text{ is an open Baire set}\}.\end{aligned}$$

Since S is locally compact every Baire set is a Borel set. The converse is not generally valid but does hold if S is a locally compact metric space (Royden [1], 14.1).

C.3. It will be necessary to be able to define a measure on a locally compact Hausdorff space from a family of measures defined on open subsets. The following theorem assures that this can be done.

Theorem C.3.1. (Bourbaki [1], III.3.1). *Let S be a locally compact Hausdorff space and $\{U_\alpha\}$ an open cover of S. Suppose for each α that ν^α is a regular complex valued Borel measure on U_α and that $\nu^\alpha_{U_\alpha\cap U_\beta}=\nu^\beta_{U_\alpha\cap U_\beta}$ for each α and β such that $U_\alpha\cap U_\beta\neq\emptyset$. Then there exists a unique regular complex valued Borel measure ν on S such that $\nu_{U_\alpha}=\nu^\alpha$ for each α.*

C.4. Let G be a locally compact Abelian group. If ν is a regular complex valued Borel measure on G and $s\in G$ then we denote by $\tau_s\nu$ that complex valued Borel measure on G defined on each Borel set E by $\tau_s\nu(E)=\nu(Es^{-1})$. A measure is said to be *translation invariant* if $\tau_s\nu=\nu$ for each $s\in G$. Nonzero translation invariant measures on locally compact Abelian groups always exist.

Theorem C.4.1. (Hewitt and Ross [1], 15.8, Nachbin [1], II.4, II.8 and II.9, Pontryagin [1], 25, and Weil [1], 7). *Let G be a locally compact Abelian group. Then there exists a nonzero regular translation invariant nonnegative Borel measure λ on G. λ is unique up to multiplication by a positive constant.*

Such a measure on a locally compact Abelian group is called a *Haar measure.* A Haar measure λ has finite total mass if and only if G is compact (Hewitt and Ross [1], 15.9). Generally when G is compact one normalizes Haar measure so that $\lambda(G)=1$. If G is infinite and discrete then one normalizes λ so that $\lambda(\{t\})=1$ for each $t\in G$. If G is nondiscrete then $\lambda(\{t\})=0$ for each $t\in G$ (Nachbin [1], p. 76). For any group if E is a Borel set in G then $\lambda(E)=\lambda(E^{-1})$ and if E is open then $\lambda(E)>0$.

Theorem C.4.2. *Let G be a nondiscrete locally compact Abelian group and λ a Haar measure on G. If E is a Borel set in G such that $0<\lambda(E)<\infty$ then there exists a Borel set $E'\subset E$ such that $\lambda(E')=\lambda(E)/2$.*

This result follows from (Halmos [1], p. 174). Another important property of Haar measure is the next result.

Theorem C.4.3. (Rudin [5], 2.6.7). *Let G be a locally compact Abelian group and λ a Haar measure on G. If K is a compact subset of G and $\varepsilon>0$ then there exists a Borel set V in G with compact closure such that $\lambda(VK)<(1+\varepsilon)\,\lambda(V)$.*

General discussions of the existence and properties of Haar measures are available in Hewitt and Ross [1], 15, Nachbin [1], II.4, II.5, II.8 and II.9, Pontryagin [1], 25, and Weil [1], 7 and 8.

C.5. Let ν be a regular complex valued Borel measure on a locally compact Hausdorff space S. A Borel measurable function f on S is said to be *absolutely integrable* (or just *integrable*) if

$$\int_S |f(t)|\, d|\nu|(t)<\infty .$$

In view of this definition we can restrict our attention to nonnegative measures.

We state two standard convergence theorems.

Theorem C.5.1. [Monotone Convergence Theorem (Royden [1], 11.3.12)]. *Let ν be a regular nonnegative Borel measure on a locally compact Hausdorff space S. Suppose $\{f_n\}$ is a sequence of nonnegative Borel measurable functions which converges ν-almost everywhere to the function f. If $f_n(t)\leqq f(t)$ for ν-almost all t in S and each n then*

$$\lim_n \int_S f_n(t)\, d\nu(t)=\int_S f(t)\, d\nu(t).$$

The conclusion here is to be understood in the sense that both sides of the equality are ∞ if f is not integrable.

Theorem C.5.2. [Lebesgue Dominated Convergence Theorem (Dunford and Schwartz [1], III.6.16, and Royden [1], 11.3.16)]. *Let ν be a regular nonnegative Borel measure on a locally compact Hausdorff space S. Suppose $\{f_n\}$ is a sequence of integrable functions which converges ν-almost everywhere to the function f. If there exists an integrable function g such that $|f_n(t)| \leqq |g(t)|$ for ν-almost all t in S and each n then f is an integrable function and*

$$\lim_n \int_S f_n(t)\, d\nu(t) = \int_S f(t)\, d\nu(t).$$

C.6. Suppose S and S' are locally compact Hausdorff spaces and ν and ν' are regular complex valued Borel measures on S and S', respectively. The set function $\nu \times \nu'$ defined on the sets $E \times E'$ in the product space $S \times S'$, where both E and E' are Borel sets, by

$$\nu \times \nu'(E \times E') = \nu(E)\, \nu'(E'),$$

can be uniquely extended to a regular complex valued Borel measure on $S \times S'$. We denote this measure by $\nu \times \nu'$.

Theorem C.6.1. [Fubini's Theorem (Dunford and Schwartz [1], III.11.9 and III.11.13, Hewitt and Ross [1], 13.8 and 14.25, and Royden [1], 12.4.19 and 12.4.20)]. *Let S and S' be locally compact Hausdorff spaces and suppose that either ν and ν' are regular nonnegative Borel measures on S and S', respectively, and f is a nonnegative Borel measurable function on $S \times S'$, or ν and ν' are regular complex valued Borel measures on S and S', respectively, which have finite total mass and f is a Borel measurable function on $S \times S'$ such that*

$$\int_S \int_{S'} |f(t,s)|\, d|\nu'|(s)\, d|\nu|(t) < \infty$$

or

$$\int_{S'} \int_S |f(t,s)|\, d|\nu|(t)\, d|\nu'|(s) < \infty.$$

Then

$$\int_{S \times S'} f(t,s)\, d(\nu \times \nu')(t,s) = \int_S \int_{S'} f(t,s)\, d\nu'(s)\, d\nu(t) = \int_{S'} \int_S f(t,s)\, d\nu(t)\, d\nu'(s).$$

C.7. The form of the Radon-Nikodym theorem which we require is the following one.

Theorem C.7.1. [Radon-Nikodym Theorem (Hewitt and Ross [1], 14.19)]. *Let S be a locally compact Hausdorff space and μ a regular nonnegative Borel measure on S. If ν is a regular complex valued Borel measure on S with finite total mass which is absolutely continuous with respect to μ*

then there exists a Borel measurable function f which is integrable with respect to μ and such that

$$\nu(E) = \int_E f(t)\, d\mu(t)$$

for each Borel set E. Moreover

$$|\nu|(S) = \int_S |f(t)|\, d\mu(t).$$

C.8. A real valued function f on a locally compact Hausdorff space S is *upper semi-continuous* if for every real number α the set $\{t \mid f(t) < \alpha\}$ is open. f is *lower semi-continuous* if $-f$ is upper semi-continuous. If f and g are upper (lower) semi-continuous then $f+g$ is upper (lower) semi-continuous. A continuous function is clearly both upper and lower semi-continuous.

Theorem C.8.1. [Dini's Theorem (Royden [1], 9.2.11)]. *Let S be a compact Hausdorff space and suppose $\{f_n\}$ is a sequence of upper semi-continuous real valued functions on S such that for each $t \in S$ the sequence of real numbers $\{f_n(t)\}$ decreases monotonically to zero. Then the sequence $\{f_n\}$ converges uniformly to zero.*

Appendix D: Functional Analysis

A.1. A *topological vector space* X is a vector space X over the complex numbers equipped with a topology such that X is an Abelian topological group, if the group operation is taken to be vector space addition, and such that the mapping $(a, x) \to ax$ is continuous from the product $C \times X \to X$ where C is the field of the complex numbers. X is a *locally convex topological vector space* if it is a topological vector space which possesses a neighborhood basis at the identity consisting of convex sets. A topological vector space in which the topology is given by a complete invariant metric is called an *F-space*. A locally convex F-space is a *Fréchet space*. A vector space X is a *normed linear space* if there exists a mapping $x \to \|x\|$ such that for $x, y \in X$ and scalar a we have (i) $\|x\| \geqq 0$, (ii) $\|x\| = 0$ if and only if $x = 0$, (iii) $\|ax\| = |a|\,\|x\|$ and (iv) $\|x+y\| \leqq \|x\| + \|y\|$. A normed linear space X is a *Banach space* if the invariant metric defined by $\rho(x, y) = \|x - y\|$ is complete. Clearly every Banach space is a Fréchet space, but the converse need not be valid (Dunford and Schwartz [1], IV.2.27 and IV.2.28, and Edwards [11], 6.1.3 and 6.1.4). For each α in a directed set A let X_α be a Banach space with norm $\|\cdot\|_\alpha$. Furthermore, assume that if $\alpha, \beta \in A$, $\alpha < \beta$ then $X_\alpha \subset X_\beta$ and $\|x\|_\alpha = \|x\|_\beta$, $x \in X_\alpha$. Since

A is a directed set, it is evident that $X=\bigcup_{\alpha} X_\alpha$ is a vector space. For each $\varepsilon>0$ define $U_\varepsilon=\bigcup_{\alpha}\{x \mid x\in X_\alpha,\ \|x\|_\alpha<\varepsilon\}$. Then the family of convex sets $\{U_\varepsilon \mid \varepsilon>0\}$ in X forms a neighborhood basis at the identity in X which defines a locally convex topology on X. The locally convex topological vector space so obtained will be called the *internal inductive limit* of the Banach spaces X_α. This is a special instance of the more general notion of internal inductive limit of locally convex topological vector spaces as discussed for example in Edwards [11], 6.3. The foregoing is however sufficient for our purposes.

D.2. We state the Baire Category Theorem in the form most useful for our development.

Theorem D.2.1. [Baire Category Theorem (Dunford and Schwartz [1], I.6.9)]. *Let X be a complete metric space and suppose $X=\bigcup_{i=1}^{\infty} X_i$ where each X_i is a closed subset of X. Then there exists an i_0 such that X_{i_0} contains a nonempty open subset of X.*

D.3. If X is a topological vector space then X^* is the vector space of all continuous linear functionals on X. If $x^*\in X^*$ then we denote its value at $x\in X$ by $\langle x, x^*\rangle$. X^* is called the *dual space* of X. Given topological vector spaces X and Y and a continuous linear mapping $T\colon X\to Y$ we can define a linear mapping $T^*\colon Y^*\to X^*$ by the formula $\langle x, T^* y^*\rangle = \langle Tx, y^*\rangle$, $x\in X$, $y^*\in Y^*$. T^* is called the *adjoint* of T. We denote by $E(X, Y)$ the linear space of all continuous linear mappings from X to Y. If X is a normed linear space and Y is a Banach space then $E(X, Y)$ is a Banach space where if $T\in E(X, Y)$ then $\|T\|=\sup_{\|x\|=1}\|Tx\|$ (Dunford and Schwartz [1], II.3.8). In particular, X^* is a Banach space whenever X is a normed linear space. Moreover, $E(Y^*, X^*)$ is a Banach space and if $T\in E(X, Y)$ then $T^*\in E(Y^*, X^*)$ and $\|T\|=\|T^*\|$ (Dunford and Schwartz [1], VI.2.2). If X is a Banach space then X^{**} is the Banach space of continuous linear functionals on X^*. The formula $\langle x, x^*\rangle=\langle \tau(x), x^*\rangle$ clearly defines a linear injective mapping from X into X^{**}. If this mapping is surjective then X is said to be *reflexive*. A useful theorem is the following result.

Theorem D.3.1. (Rudin [5], Appendix C.11). *Let X, Y be Banach spaces and suppose $T\in E(X, Y)$ is injective and the range of T is dense in Y. Then the following are equivalent:*

(i) *T is surjective.*
(ii) *T^* is surjective.*
(iii) *There exists $\delta>0$ such that $\|T^* y^*\|\geqq\delta\|y^*\|$ for each $y^*\in Y^*$.*

D.4. If X is a topological vector space then a net $\{x_\alpha\} \subset X$ is *weakly convergent* to $x \in X$ if for each $x^* \in X^*$ we have $\lim_\alpha \langle x_\alpha, x^* \rangle = \langle x, x^* \rangle$. This notion of convergence induces a topology on X, called the *weak topology*, which is weaker than the original topology on X. If X is a locally convex topological vector space then so is X with the weak topology (Kelley and Namioka [1], 5.17). Given a topological vector space and its dual X^* we say that a net $\{x_\alpha^*\} \subset X^*$ is *weak* convergent* to $x^* \in X^*$ if for each $x \in X$ we have $\lim_\alpha \langle x, x_\alpha^* \rangle = \langle x, x^* \rangle$. X^* equipped with the topology induced by this notion of convergence is a locally convex topological vector space. The topology is called the *weak* topology* (Kelley and Namioka [1], 5.17). In this case it is evident that for each $x \in X$ the mapping $x^* \to \langle x, x^* \rangle$ defines a continuous linear functional on X^*. The converse is also valid.

Theorem D.4.1. (Kelley and Namioka [1], 5.17.6). *Let X be a topological vector space and X^* its dual space. Then F is a weak* continuous linear functional on X^* if and only if there exists a unique $x \in X$ such that $F(x^*) = \langle x, x^* \rangle$ for each $x^* \in X^*$.*

We note also that a linear functional F on X^* is weak* continuous if and only if $F^{-1}(0)$ is a weak* closed subset of X^* (Hille and Phillips [1], p. 28).

Theorem D.4.1 combined with V.5.6 in Dunford and Schwartz [1] easily yields the next result.

Theorem D.4.2. *Let X be a Banach space and suppose T: $X^* \to X^*$ is linear. Then T is continuous in the weak* topology on X^* if and only if T is continuous in the weak* topology on X^* when restricted to norm bounded subsets of X^*.*

A final result concerning the weak* topology is Alaoglu's Theorem. If X is a Banach space then a closed norm bounded ball in X^* is a set of the form $\{x^* \mid \|x^*\| \leqq B\}$ for some constant $B > 0$.

Theorem D.4.3. [Alaoglu's Theorem (Dunford and Schwartz [1], V.4.2, and Royden [1], 10.6.17)]. *Let X be a Banach space. Then any closed norm bounded ball in X^* is compact in the weak* topology.*

The theorem, of course, implies that if $\{x_\alpha^*\}$ is a net in some closed norm bounded ball in X^* then it has a weak* convergent subnet.

We shall also have need of a compactness result for the weak topology. A subset Y in a Banach space X is said to be *weakly sequentially compact* if every sequence $\{x_n\} \subset Y$ has a subsequence which converges weakly to a point in X.

Theorem D.4.4. (Dunford and Schwartz [1], II.3.28). *Let X be a reflexive Banach space and suppose $Y \subset X$. Then the following are equivalent:*

(i) *Y is norm bounded.*
(ii) *Y is weakly sequentially compact.*

D.5. A *Hilbert space* is a vector space X together with a function $\langle \cdot, \cdot \rangle$ from $X \times X$ to the complex numbers such that for $x, y, z \in X$ and $a \in C$ we have (i) $\langle x, x \rangle = 0$ if and only if $x = 0$, (ii) $\langle x, x \rangle \geqq 0$, (iii) $\langle x+y, z \rangle = \langle x, z \rangle + \langle y, z \rangle$, (iv) $\langle a x, y \rangle = a \langle x, y \rangle$, (v) $\langle x, y \rangle = \overline{\langle y, x \rangle}$, where the bar denotes complex conjugation, and (vi) $\|x\| = \sqrt{\langle x, x \rangle}$ defines a norm for X under which X is a Banach space. A family of elements $\{x_\alpha\} \subset X$ is said to be *orthogonal* if $\langle x_\alpha, x_\beta \rangle = 0$, $\alpha \neq \beta$. An orthogonal family in a Hilbert space is *orthonormal* if $\langle x_\alpha, x_\alpha \rangle = 1$. Every Hilbert space contains orthonormal families, and indeed contains orthonormal families $\{x_\alpha\}$ such that $x = \sum_\alpha \langle x, x_\alpha \rangle x_\alpha$ for each $x \in X$. An orthonormal family with this property is called *complete*. For such a complete orthonormal family one has that $\|x\|^2 = \sum_\alpha |\langle x, x_\alpha \rangle|^2$ for each $x \in X$. Moreover, if $\{a_\alpha\}$ are scalars then $\sum_\alpha a_\alpha x_\alpha$ converges in X if and only if $\sum_\alpha |a_\alpha|^2 < \infty$ (Dunford and Schwartz [1], IV.4.9, IV.4.12 and IV.4.13).

Two important inequalities are valid in Hilbert space.

Theorem D.5.1. [Cauchy-Schwarz Inequality (Dunford and Schwartz [1], IV.4.1)]. *Let X be a Hilbert space. Then $|\langle x, y \rangle| \leqq \|x\| \, \|y\|$ for each $x, y \in X$.*

Theorem D.5.2. [Bessel's Inequality (Dunford and Schwartz [1], p. 252)]. *Let X be a Hilbert space and suppose $\{x_\alpha\}$ is an orthonormal family in X. Then $\sum_\alpha |\langle x, x_\alpha \rangle|^2 \leqq \|x\|^2$ for each $x \in X$.*

If Y is a closed linear subspace of a Hilbert space X then we set $Y^\perp = \{z \mid z \in X, \langle y, z \rangle = 0, \ y \in Y\}$. $Y^\perp$ is called the *orthogonal complement* of Y.

Theorem D.5.3. (Dunford and Schwartz [1], IV.4.4). *Let X be a Hilbert space and Y a closed linear subspace of X. Then $Y^\perp$ is a closed linear subspace of X and $X = Y \oplus Y^\perp$ where $\oplus$ denotes the vector space direct sum of Y and $Y^\perp$.*

Thus, given a closed linear subspace Y of a Hilbert space X, we can uniquely write each $x \in X$ as $x = y + z$ where $y \in Y$ and $z \in Y^\perp$. The *projection* of X *onto* Y is the linear mapping defined by $Tx = y$. $T \in E(X, Y)$ and $\|T\| = 1$.

D.6. Next we wish to state some of the fundamental theorems of functional analysis. We give them in the forms which seem most appropriate for our purposes.

Let X and Y be F-spaces and suppose T is a linear mapping from X to Y. T is said to be *closed* if whenever $\{x_\alpha\}$ converges to x and $\{Tx_\alpha\}$ converges to y then $Tx=y$. Equivalently, the *graph of* T consisting of all the points (x, Tx) in the product space $X\times Y$ is closed in $X\times Y$.

Theorem D.6.1. [Closed Graph Theorem (Dunford and Schwartz [1], II.2.4)]. *Let X and Y be F-spaces and suppose T is a closed linear mapping from X to Y. Then T is continuous from X to Y.*

Closely related to the Closed Graph Theorem is the Open Mapping Theorem. A mapping T from one F-space to another is called *open* if the image of every open set is open.

Theorem D.6.2. [Open Mapping Theorem (Dunford and Schwartz [1], II.2.1)]. *Let X and Y be F-spaces and suppose T is a continuous linear mapping of X onto Y. Then T is an open mapping of X onto Y.*

An immediate consequence of this theorem is that a continuous linear injective mapping of an F-space X onto an F-space Y has a continuous linear inverse.

An application of this observation or the Closed Graph Theorem to the identity mapping provides us with the next result.

Theorem D.6.3. [Two Norm Theorem]. *Let X be a vector space and suppose that $\|\cdot\|_1$ and $\|\cdot\|_2$ are norms under which X is a Banach space. If there exists a constant $B>0$ such that $\|x\|_1\leqq B\|x\|_2$ for each $x\in X$, then there exists a constant $B'>0$ such that $\|x\|_2\leqq B'\|x\|_1$ for each $x\in X$.*

We also state the next results for Banach spaces although they are more generally valid.

Theorem D.6.4. [Uniform Boundedness Theorem (Dunford and Schwartz [1], II.1.18)]. *Let X and Y* be *Banach spaces and $\{T_\alpha\}$ a net in $E(X, Y)$. Suppose for each $x\in X$ that there exists some $Tx\in Y$ and some constant $B_x>0$ such that:*

(i) $\lim_\alpha T_\alpha x=Tx$.

(ii) $\|T_\alpha x\|\leqq B_x$ *for all* α.

Then (i) *defines a linear mapping $T\in E(X, Y)$.*

Theorem D.6.5. [Hahn-Banach Theorem (Dunford and Schwartz [1], II.3.11)]. *Let Y be a subspace of the normed linear space X. If $y^*\in Y^*$ then there exists an $x^*\in X^*$ such that $\|y^*\|=\|x^*\|$ and $\langle y, y^*\rangle=\langle y, x^*\rangle$ for each $y\in Y$.*

Thus every continuous linear functional on a subspace of a normed linear space X can be extended to a continuous linear functional on all of X without increasing the norm. This theorem has a number of important consequences which we list below. Throughout, X is a normed linear space.

a) X^* separates the points of X.

b) If $\langle x, x^*\rangle = \langle y, x^*\rangle$ for all $x^* \in X^*$ then $x = y$.

c) If Y is a norm dense subspace of X and $\langle y, x^*\rangle = 0$ for all $y \in Y$ then $x^* = 0$.

d) If $x \in X$ then $\|x\| = \sup_{\|x^*\|=1} |\langle x, x^*\rangle| = \sup_{\|x^*\| \leqq 1} |\langle x, x^*\rangle|$.

e) The mapping $\tau\colon X \to X^{**}$ defined in D.3 is an isometric linear isomorphism, and $\tau(X)$ is weak* dense in X^{**}.

f) Let Y^* be a linear subspace of X^*. Then Y^* is weak* dense in X^* if and only if given $x \in X$ for which $\langle x, y^*\rangle = 0$ for all $y^* \in Y^*$, implies $x = 0$.

These results can be found in Dunford and Schwartz [1], II.3 and V.4.6, and Hille and Phillips [1], 2.10.

We shall also have need of an extension of the Hahn-Banach theorem which allows one to extend continuous linear functionals on subspaces which are invariant under certain mappings to functionals on the whole space which possess the same invariance property.

Theorem D.6.6. (Edwards [11], 3.3.1). *Let X be a normed linear space and suppose G is an Abelian group of linear mappings of X into itself such that $\|\tau(x)\| = \|x\|$ for all $x \in X$ and each $\tau \in G$. If Y is a subspace of X such that $\tau(Y) \subset Y$ for each $\tau \in G$ and $y^* \in Y^*$ is such that $\langle \tau(y), y^*\rangle = \langle y, y^*\rangle$ for all $y \in Y$ and all $\tau \in G$ then there exists an $x^* \in X^*$ such that $\|y^*\| = \|x^*\|$, $\langle y, y^*\rangle = \langle y, x^*\rangle$ for all $y \in Y$ and $\langle \tau(x), x^*\rangle = \langle x, x^*\rangle$ for all $x \in X$ and all $\tau \in G$.*

D.7. Next we wish to recall some results dealing with convex sets.

Theorem D.7.1. [Krein-Šmulian Theorem (Dunford and Schwartz [1], V.5.7)]. *Let X be a Banach space and X^* its dual. A convex set $Y^* \subset X^*$ is closed in the weak* topology on X^* if and only if $X^* \cap S_r^*$ is closed in the weak* topology on X^* for every positive real number r, where $S_r^* = \{x^* \mid x^* \in X^*, \|x^*\| \leqq r\}$.*

If X is a vector space and $Y \subset X$ is convex then $y \in Y$ is an *extreme point* of Y whenever $y = \alpha y_1 + (1-\alpha) y_2$, $0 < \alpha < 1$, $y_1, y_2 \in Y$, implies $y = y_1 = y_2$. For any $Y \subset X$ the set of all elements of X of the form $y = \sum_{i=1}^{n} \alpha_i y_i$, $y_i \in Y$, $0 \leqq \alpha_i \leqq 1$ and $\sum_{i=1}^{n} \alpha_i = 1$ is called the *convex hull* of Y.

It is denoted by $\mathrm{co}(Y)$. If X is a topological vector space then the closure of the convex hull of a set $Y \subset X$ is denoted by $\overline{\mathrm{co}}(Y)$.

Theorem D.7.2. [Krein-Milman Theorem (Dunford and Schwartz [1], V.8.4)]. *Let X be a locally convex topological vector space and Y a convex compact subset of X. Let $Y' = \{y \mid y \in Y,\ y$ is an extreme point of $Y\}$. Then $\overline{\mathrm{co}}(Y') = Y$.*

We also have need of relationships between the closures of convex sets in various topologies.

Theorem D.7.3. (Dunford and Schwartz [1], V.3.14). *Let X be a locally convex topological vector space and Y a convex subset of X. Then Y is closed in X if and only if Y is closed in the weak topology on X.*

Consider Banach spaces X and Y and $E(X, Y)$, the space of continuous linear transformations from X to Y. Besides the norm topology on $E(X, Y)$ we also wish to introduce the strong and weak operator topologies. The *strong operator topology* on $E(X, Y)$ is that topology in which a net $\{T_\alpha\} \subset E(X, Y)$ converges to $T \in E(X, Y)$ if and only if for each $x \in X$ we have $\lim_\alpha \|T_\alpha x - Tx\|_Y = 0$, where $\|\cdot\|_Y$ denotes the norm in Y. The *weak operator topology* is that where $\{T_\alpha\}$ converges to T provided for each $x \in X$ and $y^* \in Y^*$ we have $\lim_\alpha \langle T_\alpha x, y^* \rangle = \langle Tx, y^* \rangle$. With either of these topologies $E(X, Y)$ is a locally convex topological space.

Theorem D.7.4. (Dunford and Schwartz [1], VI.1.5 and VI.9.3, and Greenleaf [3], p. 275). *Let X and Y be Banach spaces.*

(i) *If $W \subset E(X, Y)$ is convex then W is closed in the strong operator topology if and only if W is closed in the weak operator topology.*

(ii) *If $W \subset E(X, Y)$ is compact in the strong operator topology then $\overline{\mathrm{co}}(W)$ is compact where the closure is taken with respect to the strong operator topology.*

D.8. Another result dealing with the strong operator topology is the characterization of strong operator continuous linear functionals on $E(X, Y)$.

Theorem D.8.1. (Dunford and Schwartz [1], VI.1.4). *Let X and Y be Banach spaces. If F is a strong operator continuous linear functional on $E(X, Y)$ then there exists $x_i \in X$ and $y_i^* \in Y^*$, $i = 1, 2, \ldots, n$ such that*

$$F(T) = \sum_{i=1}^{n} \langle T x_i, y_i^* \rangle \qquad (T \in E(X, Y)).$$

A general discussion of these results on convexity and the weak and strong operator topologies can be found in Dunford and Schwartz [1], V.3, V.5, V.8 and VI.1.

D.9. We turn now to some particular spaces. If v is a regular nonnegative Borel measure on a locally compact Hausdorff space S then $L_p(v)$, $1 \leqq p < \infty$, denotes the linear space of equivalence classes of complex valued functions on S whose p-th powers are absolutely integrable with respect to v. $L_p(v)$ is a Banach space (Royden [1], 11.7.25) under the norm

$$\|f\|_p = \Big(\int_S |f(t)|^p \, dv(t)\Big)^{1/p} \qquad (f \in L_p(v)).$$

$L_\infty(v)$ denotes the linear space of equivalence classes of v-essentially bounded measurable complex valued functions on S. It is a Banach space (Royden [1], 11.7.25) with the norm

$$\|f\|_\infty = \operatorname*{ess\,sup}_{t \in S} |f(t)| \qquad (f \in L_\infty(v)).$$

The dual spaces of these Banach spaces are described in the following theorems.

Theorem D.9.1. (Dunford and Schwartz [1], IV.8.1). *Let v be a regular nonnegative Borel measure on S. If $1 < p < \infty$ and $1/p + 1/p' = 1$ then to each $x^* \in L_p(v)^*$ there exists a unique $g \in L_{p'}(v)$ such that*

$$\langle f, x^* \rangle = \int_S f(t)\, g(t)\, dv(t) \qquad (f \in L_p(v)).$$

Moreover the correspondence between x^* and g defines an isometric linear isomorphism of $L_p(v)^*$ onto $L_{p'}(v)$.

Theorem D.9.2. (Dunford and Schwartz [1], IV.8.5). *Let v be a regular nonnegative Borel measure on S and suppose $S = \bigcup_\alpha S_\alpha$ where* (i) $S_\alpha \cap S_\beta = \emptyset$, $\alpha \neq \beta$, (ii) *S_α is σ-finite and* (iii) *for each measurable $E \subset S$, $v(E) < \infty$, there exists at most a countable number of the S_α for which $E \cap S_\alpha \neq 0$. Then to each $x^* \in L_1(v)^*$ there exists a unique $g \in L_\infty(v)$ such that*

$$\langle f, x^* \rangle = \int_S f(t)\, g(t)\, dv(t) \qquad (f \in L_1(v)).$$

Moreover the correspondence between x^* and g defines an isometric linear isomorphism from $L_1(v)^*$ onto $L_\infty(v)$.

We note that every locally compact group G satisfies the hypotheses of the theorem (Dunford and Schwartz [1], p. 290, and Rudin [5], Appendix E.10). For $1 < p < \infty$ it follows from Theorem D.9.1 that $L_p(v)$ is reflexive, that is, $L_p(v)^{**} = L_p(v)$. When $S = G$ is a locally compact Abelian group and $v = \lambda$ is Haar measure then to insure a simplification in notation we shall make the convention that

$$\langle f, x^* \rangle = \int_G f(t^{-1})\, g(t)\, d\lambda(t).$$

That is, the function given by the previous theorems is $\tilde{g}(t) = g(t^{-1})$. We shall, however, write $\langle f, x^* \rangle = \langle f, g \rangle$.

For the spaces $L_p(v)$ we have the following inequality.

Theorem D.9.3. [Hölder's Inequality (Dunford and Schwartz [1], III.3.2)]. *Let v be a regular nonnegative Borel measure on S and suppose $1 \leqq p < \infty$, $1/p + 1/p' = 1$. If $f \in L_p(v)$ and $g \in L_{p'}(v)$ then $fg \in L_1(v)$ and $\|fg\|_1 \leqq \|f\|_p \|g\|_{p'}$.*

When $p = 2$ this is referred to also as the *Cauchy-Schwarz inequality*.

A finite linear combination of characteristic functions of v-measurable sets is called a *simple function*. The integrable simple functions $\mathscr{S}(v)$ are norm dense in $L_p(v)$, $1 \leqq p < \infty$ (Dunford and Schwartz [1], III.3.8).

If S is a locally compact Hausdorff topological space then $C_0(S)$ and $C_c(S)$ denote, respectively, the linear spaces of complex valued continuous functions which vanish at infinity and which have compact support. If for $f \in C_0(S)$ we set

$$\|f\|_\infty = \sup_{t \in S} |f(t)|$$

then $C_0(S)$ is a Banach space with this supremum norm. With this norm $C_c(S)$ is a normed linear space but not a Banach space unless S is compact. In general, $C_c(S)$ is norm dense in $C_0(S)$ (Rudin [7], 3.17). If $K \subset S$ is compact then $C_c^K(S)$ denotes the elements of $C_c(S)$ whose supports lie in K. Clearly $C_c^K(S)$ is a Banach space with the supremum norm

$$\|f\|_K = \sup_{t \in K} |f(t)| \qquad (f \in C_c^K(S)).$$

Clearly $C_c(S) = \bigcup_K C_c^K(S)$, where the union is taken over all compact subsets K of S. We thus can consider $C_c(S)$ as the internal inductive limit of the Banach spaces $C_c^K(S)$ (D.1). $C_c(S)$ so considered is a locally convex topological vector space.

We wish to describe the dual spaces of $C_0(S)$ and $C_c(S)$. For this purpose we denote by $M(S)$ the linear space of all regular complex valued Borel measures μ on S which have finite total mass, that is, for which $|\mu|(S) < \infty$. $M(S)$ is a Banach space with the norm $\|\mu\| = |\mu|(S)$. By $V(S)$ we shall mean the linear space of all regular complex valued Borel measures on S.

Theorem D.9.4. [Riesz Representation Theorem (Hewitt and Ross [1], 14.10, and Rudin [7], 6.19)]. *Let S be a locally compact Hausdorff* *topological space. If* $x^* \in C_0(S)^*$ *then there exists a unique* $\mu \in M(S)$ *such that*

$$\langle f, x^* \rangle = \int_S f(t)\, d\mu(t) \qquad (f \in C_0(S)).$$

Moreover the correspondence between x^* *and* μ *defines an isometric linear isomorphism from* $C_0(S)^*$ *onto* $M(S)$.

Theorem D.9.5. (Edwards [11], p. 430). *Let S be a locally compact noncompact Hausdorff topological space and consider* $C_c(S)$ *as an interval inductive limit of the Banach spaces* $C_c^K(S)$, *K compact. If* $x^* \in C_c(S)^*$ *then there exists a unique* $\mu \in V(S)$ *such that*

$$\langle f, x^* \rangle = \int_S f(t)\, d\mu(t) \qquad (f \in C_0(S)).$$

Again, as with the L_p-spaces, when $S = G$ is a locally compact Abelian group we shall make the convention that the measure μ in the previous theorems is chosen such that

$$\langle f, x^* \rangle = \int_G f(t^{-1})\, d\mu(t).$$

We shall also write $\langle f, x^* \rangle = \langle f, \mu \rangle$ in this case.

Finally we note that $C_c(S)$ is always norm dense in $L_p(v)$, $1 \leqq p < \infty$ (Hewitt and Ross [1], 12.10).

D.10. A topological vector space X is said to be *weakly complete* if every sequence $\{x_n\} \subset X$ which is Cauchy in the weak topology on X converges in the weak topology to some element of X. The spaces $L_p(v)$, $1 \leqq p < \infty$ are weakly complete, while the spaces $L_\infty(v)$ and $C_0(S)$ are not (Dunford and Schwartz [1], IV.15).

D.11. We state the Riesz-Thorin Convexity Theorem in several different forms for easy reference. A real valued function f of n-real variables is said to be *convex* if given any two n-tuples of real numbers $(u_1, u_2, \ldots, u_n)$ and $(v_1, v_2, \ldots, v_n)$ and $0 \leqq \alpha \leqq 1$ we have

$$f(w_1, w_2, \ldots, w_n) \leqq \alpha f(u_1, u_2, \ldots, u_n) + (1-\alpha) f(v_1, v_2, \ldots, v_n)$$

where $w_i = \alpha u_i + (1-\alpha) v_i$, $i = 1, 2, \ldots, n$.

A mapping T from the *n-fold* product of vector spaces $X = X_1 \times X_2 \times \cdots \times X_n$ to a vector space Y is said to be *multilinear* if T restricted to each of the component spaces X_i is a linear mapping from X_i to Y, $i = 1, 2, \ldots, n$.

Theorem D.11.1. [Multilinear Riesz-Thorin Theorem (Zygmund [6], XII.3.3)]. *Let* v *and* v_i, $i = 1, 2, \ldots, n$, *be regular nonnegative Borel measures on the locally compact Hausdorff spaces* S *and* S_i, $i = 1, 2, \ldots, n$, *respectively. Suppose T is a multilinear mapping from the product space* $\mathscr{S}(v_1)\times\mathscr{S}(v_2)\times\cdots\times\mathscr{S}(v_n)$ *to the vector space of v-measurable functions on S where* $\mathscr{S}(v_i)$ *denotes the integrable simple functions with respect to* v_i, $i=1,2,\ldots,n$. *If for a given n+1-tuple of real numbers* $(p_1,p_2,\ldots,p_n,q)$, $1\leqq p_i\leqq\infty$, $i=1,2,\ldots,n$, $1\leqq q\leqq\infty$, *the mapping T has an extension to a continuous multilinear mapping from the Banach space* $L_{p_1}(v_1)\times L_{p_2}(v_2)\times\cdots\times L_{p_n}(v_n)$ *to* $L_q(v)$ *then denote by* $\|T\|_{(p_1,p_2,\ldots,p_n,q)}$ *the norm of this extension. If no such extension exists then set* $\|T\|_{(p_1,p_2,\ldots,p_n,q)}=\infty$. *Then* $\log\|T\|_{(1/a_1,1/a_2,\ldots,1/a_n,1/b)}$ *is a convex function on the set* $0\leqq a_i$, $b\leqq 1$, $i=1,2,\ldots,n$ *in* R^{n+1}.

If only two spaces are involved the theorem takes the following form.

Theorem D.11.2. [Riesz-Thorin Convexity Theorem (Dunford and Schwartz [1], VI.10.11, and Zygmund [6], XII.1.11)]. *Let* v_1 *and* v_2 *be regular nonnegative Borel measures on the locally compact Hausdorff spaces* S_1 *and* S_2. *Suppose T is a linear mapping from* $\mathscr{S}(v_1)$ *to the vector space of* v_2*-measurable functions on* S_2. *If for a given pair of real numbers* (p,q), $1\leqq p$, $q\leqq\infty$, *the mapping T has an extension to a continuous linear mapping from* $L_p(v_1)$ *to* $L_q(v_2)$ *then denote by* $\|T\|_{p,q}$ *the norm of this extension. If no such extension exists then set* $\|T\|_{p,q}=\infty$. *Then* $\log\|T\|_{1/a,1/b}$ *is a convex function on the rectangle* $0\leqq a$, $b\leqq 1$.

And if only one space is involved we have the following result.

Theorem D.11.3. (Dunford and Schwartz [1], VI.10.8). *Let v be a regular nonnegative Borel measure on the locally compact Hausdorff space S. Suppose T is a linear mapping from* $\mathscr{S}(v)$ *to the v-measurable functions on S. If* $1\leqq p\leqq\infty$ *and the mapping T has an extension to a continuous linear mapping from* $L_p(v)$ *to* $L_p(v)$ *then denote by* $\|T\|_p$ *the norm of this extension. If no such extension exists then set* $\|T\|_p=\infty$. *Then* $\log\|T\|_{1/a}$ *is a convex function on the interval* $0\leqq a\leqq 1$.

In particular this last result says something about the convexity of $\log\|\cdot\|_p$ in $L_p(v)$. Indeed, let f be a v-measurable function on S and let $\|f\|_p$ denote the norm of f if $f\in L_p(v)$, and set $\|f\|_p=\infty$ if f is not an element of $L_p(v)$. Then $\log\|f\|_{1/a}$ is a convex function for $0\leqq a\leqq 1$.

These convexity theorems are of course valid in a more general context than given here. The reader is refered to Dunford and Schwartz [1], VI.10, and Zygmund [6], XII.1, XII.2 and XII.3, for further details.

D.12. The last result of this appendix is a theorem concerning vector valued integrals.

Theorem D.12.1. (Bourbaki [1], p. 81). *Let S be a compact Hausdorff space and X a locally convex topological vector space which satifies the* *property that the closed convex hull of any compact set is compact in the weak topology. If v is a regular complex valued Borel measure on S and f is a continuous vector valued function from S to X then the vector valued integral*

$$\int_S f(t)\, dv(t)$$

exists and belongs to X.

Since in a Banach space the closed convex hull of a compact set is always weakly compact (Dunford and Schwartz [1], V.6.4), we see that the theorem is valid whenever X is a Banach space. Indeed, in a Banach space the closed convex hull of a compact set is compact (Dunford and Schwartz [1], V.2.6).

Theorem D.12.2. (Bourbaki [1], p. 85). *Let S be a compact Hausdorff space and X a locally convex topological vector space which satifies the property that the closed convex hull of any compact set is compact in the weak topology. If v is a regular complex valued Borel measure on S and f is a continuous vector valued function from S to X then*

$$\langle \int_S f(t)\, dv(t), x^* \rangle = \int_S \langle f(t), x^* \rangle\, dv(t) \qquad (x^* \in X^*).$$

Appendix E: Banach Algebras

E.1. A *Banach algebra A* is an algebra over the complex numbers which is a Banach space and such that $\|x\,y\| \leqq \|x\|\,\|y\|$ for each $x, y \in A$. We shall restrict our attention here, as in the bulk of the book itself, to *commutative* Banach algebras. We wish to discuss Banach algebras with and without units, that is, with and without an element e such that $x\,e = x$ for all $x \in A$. If A has a unit we shall assume without loss of generality that $\|e\| = 1$ (Katznelson [2], VIII.1.4).

A subalgebra I of A is called an *ideal* if $x\,y \in I$ whenever $x \in I$ and $y \in A$. If I is an ideal and $I \neq A$ or $I \neq \{0\}$ then I is a *proper ideal*. An ideal I is *maximal* if it is proper and contained in no other proper ideal. If I is an ideal in A then in the quotient space A/I of the vector spaces A and I we may introduce a multiplication under which A/I becomes an algebra. Namely, one defines

$$(x+I)(y+I) = x\,y + I \qquad (x, y \in A).$$

If I is a closed ideal then the quotient algebra A/I becomes a Banach algebra (Katznelson [2], VIII.1.5) under the norm defined by

$$\|x+I\| = \inf_{y \in I} \|x+y\| \qquad (x \in A).$$

A closed ideal I is said to be *regular* if the quotient algebra A/I has a unit. Clearly if A has a unit then every closed ideal is regular.

Theorem E.1.1. (Katznelson [2], VIII.2.6 and VIII.2.7). *Let A be a commutative Banach algebra. Then very regular closed proper ideal is contained in a regular maximal ideal, and every regular maximal ideal is closed.*

E.2. A homomorphism μ of a commutative Banach algebra onto the complex numbers is called a *multiplicative linear functional* on A. Every such functional is continuous and of norm one (Katznelson [2], VIII.2.8). If μ is a multiplicative linear functional on A then $\mu^{-1}(0)=\{x \mid x\in A,\ \mu(x)=0\}$ is a regular maximal ideal in A. To avoid triviality we shall always assume that it is not the case that $xy=0$ for all $x, y\in A$.

Theorem E.2.1. (Katznelson [2], VIII.2.8). *Let A be a commutative Banach algebra. Then the correspondence $\mu \to \mu^{-1}(0)$ defines an injective mapping from the set of all multiplicative linear functionals on A onto the set of all regular maximal ideals in A.*

Denote by $\Delta(A)$ the set of all regular maximal ideals in A. The previous theorem shows that $\Delta(A)$ can be identified with the subset of the unit ball $\{x^* \mid x^*\in A^*,\ \|x^*\|\leqq 1\}$ in A^* consisting of the multiplicative linear functionals on A. The weak* topology on the unit ball in A^* relativized to the subset of multiplicative linear functionals induces a locally compact Hausdorff topology on this subset. This topology is called the *Gelfand topology*. Thus by this process, $\Delta(A)$ can be made into a locally compact Hausdorff space. $\Delta(A)$ so considered is called the *regular maximal ideal space* of A. If A has a unit then $\Delta(A)$ is compact (Katznelson [2], VIII.3.1).

On the locally compact Hausdorff space $\Delta(A)$ we can define certain functions corresponding to the elements of A by setting $\hat{x}(\mu^{-1}(0))=\langle x, \mu\rangle$ for each multiplicative linear functional μ.

Theorem E.2.2. (Hille and Phillips [1], 4.15, and Katznelson [2], VIII.3.2). *Let A be a commutative Banach algebra. Then the correspondence $x\to\hat{x}$ defines a continuous homomorphism of A onto a subalgebra of $C_0(\Delta(A))$ which separates the points of $\Delta(A)$ and such that if $m\in\Delta(A)$ then there exists some $x\in A$ for which $\hat{x}(m)\neq 0$. Moreover $\|\hat{x}\|_\infty \leqq \|x\|$ for each $x\in A$.*

If A has a unit then the mapping is obviously into $C(\Delta(A))$ as $\Delta(A)$ is compact. Moreover, we note that for $m\in\Delta(A)$ we have $x\in m$ if and only if $\hat{x}(m)=0$.

Thus every commutative Banach algebra A can be represented homomorphically as an algebra of functions in $C_0(\Delta(A))$. Let $\hat{A}=\{\hat{x} \mid x\in A\}$.

Then $\hat{A}$ is called the *Gelfand representation* of A, and $\hat{x}$ is called the *Gelfand transform* of $x \in A$.

If the Gelfand transform is an injective mapping, that is, $\hat{x}=0$ implies $x=0$, then A is said to be *semi-simple*. In this case A is represented isomorphically as an algebra of functions in $C_0(\Delta(A))$. If the Gelfand transform is an isometric isomorphism then we say that A is a *supremum norm algebra*. Obviously every supremum norm algebra is semi-simple. Also it is easily seen when A is semi-simple that $\hat{A}=\{\hat{x} \mid x \in A\}$ is a commutative Banach algebra provided the operations of addition and multiplication are defined pointwise and the norm is defined by $\|\hat{x}\|=\|x\|$.

In general $\|\hat{x}\|_\infty \neq \|x\|$. However, one does have (Katznelson [2], 3.6) for each $x \in A$ that

$$\|\hat{x}\|_\infty = \lim_n \|x^n\|^{1/n}.$$

One also defines $\|x\|_{s_p} = \lim_n \|x^n\|^{1/n}$, $x \in A$, and calls it the *spectral radius norm* of x. Clearly $\|x\|_{s_p} \leqq \|x\|$.

E.3. A commutative Banach algebra A is said to have an *involution* if there exists a mapping $x \to x^+$ from A to A such that (i) $(a\,x+b\,y)^+ = \bar{a}\,x^+ + \bar{b}\,y^+$, (ii) $(x\,y)^+ = x^+\,y^+$ and (iii) $x^{++}=x$ for any x, $y \in A$ and scalars a, $b \in C$. A is said to be *self-adjoint* if there exists an *involution*$^+$ on A such that $\hat{x}^+(\mu^{-1}(0)) = \overline{\hat{x}(\mu^{-1}(0))}$ for each multiplicative linear functional μ. The bar, of course, denotes complex conjugation. Not all algebras with involution are self-adjoint (Naimark [1], III.14.1).

E.4. Let I be an ideal in a commutative Banach algebra A. Then the *hull of* I, denoted by $h(I)$, is the collection of all the regular maximal ideals which contain I. Equivalently, $h(I)$ is the intersection of the sets of regular maximal ideals in $\Delta(A)$ where $\hat{x}$ vanishes, $x \in I$. Thus it is evident that $h(I)$ is always a closed subset of $\Delta(A)$. If E is a subset of $\Delta(A)$ the the *kernel of* E, denoted by $k(E)$, is the intersection of all the elements of E. Clearly $k(E)$ is a closed ideal in A as it is the intersection of the regular maximal ideals in E. Equivalently, $k(E)$ is the set of all $x \in A$ such that $\hat{x}(\mu^{-1}(0))=0$ for all $\mu^{-1}(0) \in E$.

Thus given any set $E \subset \Delta(A)$ there corresponds to it a closed set $h(k(E))$ in $\Delta(A)$ which contains E. This correspondence is a closure operation which defines a topology on $\Delta(A)$. The topology so defined is called the *hull-kernel topology* on $\Delta(A)$. It is, in general, a weaker topology than the Gelfand topology and need not be Hausdorff.

Theorem E.4.1. (Hoffman [1], 6.4, Katznelson [2], III.5.4, and Wang [2], p. 27). *Let A be a commutative Banach algebra. Then the following are equivalent:*

(i) *The hull kernel topology on $\Delta(A)$ is equivalent to the Gelfand topology.*

(ii) *The hull-kernel topology is Hausdorff.*

(iii) *If $E \subset \Delta(A)$ is closed in the Gelfand topology and $m_0 \in \Delta(A) \sim E$ then there is an $x \in A$ such that $\hat{x}(m_0) = 1$ and $\hat{x}(m) = 0$, $m \in E$.*

We say that a commutative Banach algebra A is *regular* if any one of the three equivalent assertions of the preceding theorem are valid. Part (iii) of the theorem says that if A is regular then $\hat{A}$ is a *regular function algebra*. An algebra B of functions in $C_0(S)$, S a locally compact Hausdorff space, is said to be *normal* if, given two disjoint closed subsets, E_1 and E_2 of S, there exists an $f \in B$ such that $f(t) = 0$, $t \in E_1$ and $f(t) = 1$, $t \in E_2$. If A is a regular commutative Banach algebra then $\hat{A}$ is normal (Naimark [1], p. 224).

If S is a locally compact Hausdorff space then $C_0(S)$ is a regular commutative Banach algebra and $\Delta[C_0(S)] = S$ (Dugundji [1], 6.4, Hoffman [1], pp. 19–20).

Finally, we observe that if A is a commutative Banach algebra and I is a closed ideal in A then $(A/I)^{\wedge}$ can be identified with the algebra of continuous functions on $h(I)$ obtained by restricting the elements of $\hat{A}$ to $h(I)$. Moreover the regular maximal ideal space of A/I can be identified with $h(I)$ (Hoffman [1], 6.6, Loomis [1], 20G).

E.5. Let A be a commutative Banach algebra. A net $\{x_\alpha\} \subset A$ is called an *approximate identity* if $\lim_\alpha \|x_\alpha x - x\| = 0$ for each $x \in A$. An approximate identity is *minimal* if $\lim_\alpha \|x_\alpha\| = 1$. *A admits factorization* if for every $x \in A$ there exists $y, z \in A$ such that $x = yz$. If A possesses a minimal approximate identity then it admits factorization (Cohen [1], Gulick, Liu and van Rooij [1_{II}]).

E.6. If A is a semi-simple commutative Banach algebra then the Gelfand transform is an isomorphism and we can consider A as an algebra of continuous functions in $C_0(\Delta(A))$. In this case we say that a subset $E \subset \Delta(A)$ is a *boundary for A* if

$$\sup_{m \in \Delta(A)} |\hat{x}(m)| = \sup_{m \in E} |\hat{x}(m)| \qquad (x \in A).$$

A minimal boundary is a boundary which is contained in all other boundaries. A minimal boundary need not exist for a given Banach algebra A (Bishop [1], p. 632, and Wang [2], p. 43). If a minimal boundary does exist for a semi-simple commutative Banach algebra A then it is called the *Bishop boundary* of A and denoted by $\rho\,\Delta(A)$. If a minimal closed boundary exists then it is called the *Šilov boundary* of A and denoted by $\partial\Delta(A)$. Clearly if the Bishop and Šilov boundaries exist they are unique, $\rho\Delta(A)\subset\partial\Delta(A)$ and the closure of $\rho\Delta(A)$ is $\partial\Delta(A)$.

Theorem E.6.1. (Hoffman [1], 7.1, Loomis [1], 24E, and Rickart [1], 3.3.1). *Let A be a semi-simple commutative Banach algebra. Then the Šilov boundary of A exists.*

If A is a semi-simple commutative Banach algebra which is either regular or self-adjoint then $\partial\Delta(A)=\Delta(A)$ (Hoffman [1], 7.3, and Rickart [1], 3.3.2).

Theorem E.6.2. (Bishop [1], p. 630, Wang [2], p. 39, and Wermer [2], 6.2). *Let A be a commutative Banach algebra with identity which is a supremum norm algebra. If $\Delta(A)$ is metrizable then the Bishop boundary of A exists.*

In this case $\rho\Delta(A)$ is a G_δ-set (Bishop [1], p. 633, and Wermer [2], 6.3).

Under certain conditions when the Bishop and Šilov boundaries exist one can represent the elements of $\hat{A}$ as certain integrals over these boundaries.

Theorem E.6.3. (Arens and Singer [1]). *Let A be a semi-simple commutative Banach algebra. Then for each $m_0\in\Delta(A)$ there exists a regular nonnegative Baire measure μ_0 defined on $\partial\Delta(A)$ such that $\mu_0(\partial\Delta(A))\leqq 1$ and*

$$\hat{x}(m_0)=\int_{\partial\Delta(A)}\hat{x}(m)\,d\mu_0(m)\qquad (x\in A).$$

For the Bishop boundary we have the following result.

Theorem E.6.4. (Bishop and de Leeuw [1], and Wermer [2], 6.5). *Let A be a commutative Banach algebra with identity which is a supremum norm algebra. If $\Delta(A)$ is metrizable then for each $m_0\in\Delta(A)$ there exists a regular nonnegative Baire measure μ_0 on $\Delta(A)$ such that $|\mu_0|(\Delta(A))=1$, $\mu_0(E)=0$ for each Baire set E for which $E\cap\rho\Delta(A)=\emptyset$ and*

$$\hat{x}(m_0)=\int_{\Delta(A)}\hat{x}(m)\,d\mu_0(m)\qquad (x\in A).$$

E.7. Let S be a locally compact Hausdorff space and consider the space $C_0(S)$. As we have remarked previously, $C_0(S)$ is a commutative Banach algebra under pointwise operations and the supremum norm. We state the Stone-Weierstrass approximation theorem.

Theorem E.7.1. [Stone-Weierstrass Theorem (Rickart [1], 3.2.12)]. *Let S be a locally compact Hausdorff space and suppose B is a subalgebra of $C_0(S)$ such that:*

(i) *If $s, t\in S$ then there exists $f\in B$ such that $f(s)\neq f(t)$.*
(ii) *If $s\in S$ then there exists $f\in B$ such that $f(s)\neq 0$.*

(iii) *If* $f \in B$ *then* $\bar{f} \in B$ *where* $\bar{f}(s) = \overline{f(s)}$, $s \in S$, *and the bar denotes complex conjugation.*

Then B *is supremum norm dense in* $C_0(S)$.

E.8. Let X be a Banach space. As indicated in D.3 the space $E(X, X) = E(X)$ of all continuous linear operators from X to X is a Banach space with the usual operator norm and linear operations on operators. If one defines a multiplication in $E(X)$ by operator composition then $E(X)$ is a Banach algebra with identity. In general, it is not a commutative Banach algebra (Naimark [1], I.4, and Rickart [1], Appendix A.1.1). An operator $T \in E(X)$ is *compact* if it maps bounded subsets of X into sets with compact closure. Denote the set of compact operators in $E(X)$ by $K(X)$. Then $K(X)$ is a norm closed two-sided ideal in $E(X)$ (Naimark [1], I.4.6, and Rickart [1], Appendix A.1.2).

Appendix F: Harmonic Analysis

F.1. Let G be a locally compact Abelian group and λ a Haar measure on G. For a given λ we shall denote the spaces $L_p(\lambda)$ by $L_p(G)$, $1 \leqq p \leqq \infty$. As indicated previously (D.9), these spaces are Banach spaces. Similarly the space $M(G)$ consisting of all regular complex valued Borel measures on G with finite total mass is a Banach space (D.9). If f, $g \in L_1(G)$ then the formula

$$f * g(t) = \int_G f(t s^{-1})\, g(s)\, d\lambda(s) \qquad (t \in G),$$

can be shown to define an element of $L_1(G)$ (Hewitt and Ross [1], 20.14). $L_1(G)$ with this *convolution* operation as multiplication is a commutative Banach algebra (Rudin [5], 1.1.7). Similarly, if one defines for μ, $\nu \in M(G)$ the *convolution* $\mu * \nu$ by

$$\mu * \nu(E) = \int_G \nu(E s^{-1})\, d\mu(s) = \int_G \mu(E s^{-1})\, d\nu(s)$$

then $M(G)$ is a commutative Banach algebra with convolution as multiplication (Rudin [5], 1.3.2). By the Radon-Nikodym Theorem (Theorem C.7.1) we see that $L_1(G)$ corresponds precisely to the subspace of $M(G)$ consisting of those measures which are absolutely continuous with respect to λ. Moreover $L_1(G)$ so considered is a closed ideal in $M(G)$ (Hewitt and Ross [1], 19.18, and Rudin [5], 1.3.4).

F.2. More generally it can be shown (Hewitt and Ross [1], 20.12) that given $\mu \in M(G)$ and $f \in L_p(G)$, $1 \leqq p \leqq \infty$, then

$$f * \mu(t) = \int_G f(t s^{-1})\, d\mu(s) \qquad (t \in G)$$

defines an element of $L_p(G)$, $1 \leqq p \leqq \infty$, and that

$$\|\mu * f\|_p \leqq \|\mu\| \, \|f\|_p.$$

If $\mu \in L_1(G)$ then clearly $\|\mu * f\|_p \leqq \|\mu\|_1 \|f\|_p$. Furthermore if $f \in L_p(G)$, $1 \leqq p < \infty$, and $\varepsilon > 0$ then there exists a nonnegative $g \in L_1(G)$ with $\|g\|_1 = 1$ such that $\|g * f - f\|_p < \varepsilon$ (Hewitt and Ross [1], 20.15).

Thus, if G is a compact Abelian group then $L_p(G) \subseteq L_1(G)$, $1 < p \leqq \infty$, and so each $L_p(G)$ is a closed ideal under convolution in $L_1(G)$. In particular then, $L_p(G)$, $1 \leqq p \leqq \infty$, is a commutative Banach algebra under convolution when G is compact, since by Hölder's inequality and the usual normalization of λ we have $\|f\|_1 \leqq \|f\|_p$, $f \in L_p(G)$, $1 \leqq p \leqq \infty$.

Evidently $L_\infty(G)$ for arbitrary locally compact Abelian groups is a commutative Banach algebra under pointwise multiplication.

F.3. Furthermore we note for locally compact Abelian G that if $f \in L_p(G)$, $g \in L_{p'}(G)$, $1 \leqq p \leqq \infty$, $1/p + 1/p' = 1$, then

$$f * g(t) = \int_G f(t s^{-1}) \, g(s) \, d\lambda(s) \qquad (t \in G)$$

exists and is a bounded uniformly continuous function, and if $1 < p < \infty$, then it belongs to $C_0(G)$ (Hewitt and Ross [1], 20.19, and Rudin [5], 1.1.6).

F.4. In a similar vein we observe that if $1 < p, r < \infty$, $1/p + 1/r > 1$ and q is such that $1/p + 1/r = 1 - 1/q$ then for $f \in L_p(G)$, $g \in L_r(G)$ we have $f * g \in L_q(G)$, and $\|f * g\|_q \leqq \|f\|_p \|g\|_r$ (Hewitt and Ross [1], 20.18).

F.5. The regular maximal ideal space of $L_1(G)$ for arbitrary locally compact Abelian groups or of $L_p(G)$, $1 \leqq p \leqq \infty$, for G compact is easy to describe. In both cases it can be shown that the multiplicative linear functionals correspond precisely to the continuous characters on G, that is, to the elements of $\hat{G}$, the dual group of G (Gaudry [6], II.2.3, Rickart [1], Appendix A.3.4, and Rudin [5], 1.2.2). Indeed, if μ is a multiplicative linear functional on one of these algebras then there exists a unique $\gamma \in \hat{G}$ such that

$$\langle f, \mu \rangle = \int_G (t^{-1}, \gamma) f(t) \, d\lambda(t).$$

We write the Gelfand transform of an element f in one of these algebras as $\hat{f}$ where

$$\hat{f}(\gamma) = \int_G (t^{-1}, \gamma) f(t) \, d\lambda(t) \qquad (\gamma \in \hat{G}).$$

In this case $\hat{f}$ is generally called the *Fourier transform* of f. Evidently $(f * g)^\wedge(\gamma) = \hat{f}(\gamma) \, \hat{g}(\gamma)$, $\gamma \in \hat{G}$, and $\|\hat{f}\|_\infty \leqq \|f\|_1$.

It should also be noted that the topology defined on $\hat{G}$ in (B.2) is equivalent to the Gelfand topology on $\hat{G}$ considered as the regular maximal ideal space (Rudin [5], 1.2.6).

F.6. The regular maximal ideal space $\Delta(M(G))$ of $M(G)$ is considerably more complicated than that of $L_1(G)$ and no simple characterization of it is generaly available. It is easily seen that if $\gamma \in \hat{G}$ then the formula

$$\hat{\mu}(\gamma) = \int_G (t^{-1}, \gamma)\, d\mu(t) \qquad (\mu \in M(G))$$

defines a multiplicative linear functional on $M(G)$. $\hat{\mu}$ is called the *Fourier-Stieltjes transform* of μ. Clearly $(\mu * \nu)\hat{}\,(\gamma) = \hat{\mu}(\gamma)\,\hat{\nu}(\gamma)$, $\gamma \in \hat{G}$.

Thus $\hat{G}$ can be considered as a subset of $\Delta(M(G))$. But if G is non-discrete then $\hat{G}$ is always a proper subset of $\Delta(M(G))$ (Hewitt and Ross [1], 23.28). Moreover, in this case, $\hat{G}$ is not dense in $\Delta(M(G))$ (Hewitt [2], p. 143, and Rickart [1], Appendix A.3.4). When G is discrete then $L_1(G) = M(G)$ and $\Delta(M(G)) = \hat{G}$. Indeed $L_1(G) = M(G)$ and $L_1(G)$ has a unit if and only if G is discrete (Rudin [5], 1.7.3).

F.7. Next we wish to list some of the elementary properties of the spaces $L_p(G)$, $1 \leqq p \leqq \infty$, and $M(G)$.

a) If G is a locally compact Abelian group, $1 \leqq p < \infty$ and $f \in L_p(G)$ then the mapping $s \to \tau_s f$ is uniformly continuous from G to $L_p(G)$ where $\tau_s f(t) = f(t s^{-1})$, $t \in G$ (Hewitt and Ross [1], 20.4, and Rudin [4], 1.1.5). If $f \in L_\infty(G)$ then $s \to \tau_s f$ is weak* continuous (Gaudry [6], I.3.1) and is norm continuous if and only if f is (equal locally almost everywhere to) a uniformly continuous function (Gaudry [6], II.1.5). If $\mu \in M(G)$ then the mapping $s \to \tau_s \mu$ is continuous if and only if $\mu \in L_1(G)$ (Gaudry [6], II.1.7, and Rudin [3]).

b) If G is a locally compact Abelian group then $L_1(G)$ is a semi-simple regular self-adjoint commutative Banach algebra with $f^+(t) = \overline{f(t^{-1})}$. The same is true for $L_p(G)$, $1 < p \leqq \infty$, when G is a compact Abelian group (Rickart [1], Appendices A.3.1 and A.3.4). $M(G)$ is a semi-simple commutative Banach algebra with identity which is self-adjoint if and only if G is discrete (Rudin [5], 1.7.3 and 5.3.4). Moreover $\hat{\mu}(\gamma) = 0$ for all $\gamma \in \hat{G}$ implies $\mu = 0$ (Rudin [5], 1.7.3).

c) Let $L_1(G)\hat{} = \{\hat{f} \mid f \in L_1(G)\}$, $M(G)\hat{} = \{\hat{\mu} \mid \mu \in M(G)\}$. Then $L_1(G)\hat{}$ is a dense subalgebra of $C_0(\hat{G})$ (Rudin [5], 1.2.4). $L_1(G)\hat{} = C_0(\hat{G})$ only if $\hat{G}$ is finite (Rudin [5], 4.6.8). Moreover, $\{f \mid \hat{f} \in C_c(\hat{G})\}$ is norm dense in $L_1(G)$ (Gaudry [6], II.7.1, Loomis [1], 37A, Rudin [5], 2.6.6). Each $\hat{\mu}$ is a uniformly continuous bounded function on G and $M(G)\hat{}$ is a proper subalgebra of $C(\hat{G})$ unless $\hat{G}$ is finite, in which case $M(G)\hat{} = C(\hat{G})$. Furthermore, there always exist $\varphi \in C_0(\hat{G})$ such that $\varphi \notin M(G)\hat{}$ unless $\hat{G}$

is finite (Edwards [1], and Gaudry [6], V.4.3), and there exist $\hat{\mu} \in M(G)\hat{}$ such that $\hat{\mu} \notin C_0(\hat{G})$ (Rudin [5], 5.6.10).

d) Although $L_1(G)$ does not have a unit unless G is discrete it does always possess an approximate identity. More generally for any locally compact Abelian group G and $1 \leqq p < \infty$, there exists a net $\{u_\alpha\} \subset L_1(G) \cap L_\infty(G)$ of nonnegative functions such that $\|u_\alpha\|_1 = 1$ and $\lim_\alpha \|u_\alpha * f - f\|_p = 0$ for each $f \in L_p(G)$ (Gaudry [6], II.1.4, Loomis [1], 31E, and Rudin [5], 1.1.8). If $f \in C_0(G)$ then $\lim_\alpha \|u_\alpha * f - f\|_\infty = 0$ and if $f \in L_\infty(G)$ then $\{u_\alpha * f\}$ converges in the weak* topology to f. It converges uniformly to f if and only if f is uniformly continuous (Gaudry [6], pp. 19–20). It is possible to choose $\{u_\alpha\}$ such that each u_α vanishes off of some fixed compact neighborhood of the identity in G. This shows that $L_1(G)$ has a minimal approximation identity, and that if G is compact then $L_p(G)$, $1 < p < \infty$, also has an approximate identity. However, in this case, the approximate identity can never be minimal unless G is finite (Gaudry [6], p. 65, and Yap [2]). For an arbitrary locally compact Abelian group it is always possible to construct a minimal approximate identity $\{u_\alpha\} \subset L_1(G)$ such that $\{\hat{u}_\alpha\} \subset C_c(\hat{G})$ (Gaudry [6], II.7.1, and Rudin [5], 2.6.6). Moreover, given a compact set $K \subset \hat{G}$, it is possible to construct a minimal approximate identity $\{u_\alpha\}$ such that $\{\hat{u}_\alpha\} \subset C_c(\hat{G})$, $\hat{u}_\alpha(\gamma) = 1$, $\gamma \in K$, and $\|u_\alpha\|_1 \leqq 2$ (Gaudry [6], II.7.3). When G is compact then the trigonometric polynomials, that is, the finite linear combinations of continuous characters, are norm dense in $L_p(G)$, $1 \leqq p < \infty$, and there exist approximate identities consisting of trigonometric polynomials (Gaudry [6], pp. 52–53, and Rudin [5], 1.5.2).

e) Let G be a locally compact Abelian group. Using Theorem C.4.3 one can show that for each compact $K \subset \hat{G}$ and $\varepsilon > 0$, given an open set U containing K there exists an $f \in L_1(G)$ such that $0 \leqq \hat{f}(\gamma) \leqq 1$, $\gamma \in \hat{G}$, $\hat{f}(\gamma) = 1$, $\gamma \in K$, $\hat{f}(\gamma) = 0$, $\gamma \notin U$, and $\|f\|_1 \leqq 1 + \varepsilon$ (Rudin [5], 2.6.1, 2.6.2 and 2.6.7). In particular, given any open set $U \subset \hat{G}$ with compact closure it is possible to find $f \in L_1(G)$ such that $\hat{f}(\gamma) = 1$, $\gamma \in U$.

F.8. Since $L_1(G)$ is a semi-simple commutative Banach algebra there corresponds precisely one $f \in L_1(G)$ to each $\hat{f} \in L_1(G)\hat{}$. Under certain conditions it is possible to write down an explicit formula for f in terms of $\hat{f}$.

Theorem F.8.1. [Fourier Inversion Formula (Gaudry [6], II.4.1, II.5.2 and Rudin [5], 1.5.1)]. *Let G be a locally compact Abelian group and λ a given Haar measure on G. Then it is possible to choose a Haar measure η on $\hat{G}$ in such a way that for every $f \in L_1(G)$ for which $\hat{f} \in L_1(\hat{G})$ we have, for almost all $t \in G$, that*

$$f(t) = \int_{\hat{G}} (t, \gamma)\, \hat{f}(\gamma)\, d\eta(\gamma).$$

It should be noted, when $\hat{f} \in L_1(\hat{G})$, that the equivalence class determined by f contains a continuous function. Thus, if we choose such a continuous representative for f, then the above inversion formula is valid for all $t \in G$.

The choice of the Haar measure η is consistent with the normalizations for Haar measure discussed in (C.4).

One can use the density of $L_1(G) \cap L_p(G)$ in $L_p(G)$, $1 \leqq p \leqq 2$, and the Fourier transform for $L_1(G)$ to define extensions of these transforms to all of $L_p(G)$, $1 \leqq p \leqq 2$. If G is compact then clearly the Fourier transform will suffice as $L_p(G) \subset L_1(G)$. If $1 < p \leqq 2$ and G is noncompact then the spaces involved are not Banach algebras. Nevertheless we can obtain the following results.

Theorem F.8.2. [Plancherel Theorem (Gaudry [6], II.4.2, Rudin [5], 1.6.1)]. *Let G be a locally compact Abelian group, λ a Haar measure on G and suppose that a Haar measure η on $\hat{G}$ has been chosen such that the Fourier Inversion Formula is valid. If $f \in L_1(G) \cap L_2(G)$ then $\hat{f} \in L_2(\hat{G})$ and $\|\hat{f}\|_2 = \|f\|_2$. $\{\hat{f} \mid f \in L_1(G) \cap L_2(G)\}$ is norm dense in $L_2(\hat{G})$ and the Fourier transform can be uniquely extended to a linear isometry of $L_2(G)$ onto $L_2(\hat{G})$.*

The extension of the Fourier transform so obtained is called the *Fourier-Plancherel* or *Plancherel transform.* We denote the Fourier-Plancherel transform of $f \in L_2(G)$ by $\hat{f}$. The context will generally be sufficient to indicate whether $\hat{f}$ denotes the Fourier or the Fourier-Plancherel transform.

A similar process can be carried out for $1 < p < 2$. Before we give this, however, we wish to state another useful result concerning the Fourier-Plancherel transform.

Theorem F.8.3. [Parseval's Formula (Gaudry [6], II.4.4, and Rudin [5], 1.6.2)]. *Let G be a locally compact Abelian group, λ a Haar measure on G and η a Haar measure on $\hat{G}$ chosen so that the Plancherel theorem is valid. If $f, g \in L_2(G)$ then*

$$\int_G f(t)\, g(t)\, d\lambda(t) = \int_G \hat{f}(\gamma)\, \hat{g}(\gamma^{-1})\, d\eta(\gamma).$$

With the aid of the Riesz-Thorin Convexity Theorem (Theorem D.11.2), one can uniquely extend the Fourier transform from the norm dense subspace of $L_p(G)$, $1 < p < 2$, consisting of the integrable simple functions and obtain a continuous transformation from $L_p(G)$ into $L_{p'}(\hat{G})$, $1/p + 1/p' = 1$.

Theorem F.8.4. [Hausdorff-Young Theorem (Edwards [14_{II}], 13.5, Gaudry [6], IV.4.1, and Weil [1], p. 117)]. *Let G be a locally compact Abelian group, λ a Haar measure on G and η a Haar measure on G such that the Plancherel Theorem is valid. If $f \in \mathcal{S}(\lambda)$ then $\hat{f} \in L_{p'}(\hat{G})$ and $\|\hat{f}\|_{p'} \leqq \|f\|_p$, $1 < p < 2$, $1/p + 1/p' = 1$. This mapping from $\mathcal{S}(\lambda) \subset L_p(G)$ into $L_{p'}(\hat{G})$, $1 < p < 2$, $1/p + 1/p' = 1$, can be uniquely extended to a linear mapping from*

$L_p(G)$ into $L_{p'}(\hat{G})$, again denoted by $f \to \hat{f}$, such that

$$\|\hat{f}\|_{p'} \leqq \|f\|_p \qquad (f \in L_p(G)).$$

The mapping so obtained will be called the *Hausdorff-Young transform.* The Hausdorff-Young transform is surjective only when G is finite (Edwards [14_{II}], 15.4.1, and Gaudry [6], IV.4.1).

F.9. The following result provides a characterization of those elements of $C(\hat{G})$ which belong to $M(G)^\wedge$. The proof in Rudin [5] is not complete as indicated in Gaudry [6], VII.4.

Theorem F.9.1. (Gaudry [6], VII.4.3, and Rudin [5], 1.9.1). *Let G be a locally compact Abelian group and φ be a function on $\hat{G}$. Then the following are equivalent:*

(i) *$\varphi \in M(G)^\wedge$ and $\|\varphi\|_\infty \leqq B$.*

(ii) *φ is continuous and there exists a constant $B > 0$ such that*

$$\left|\sum_{i=1}^{n} c_i\, \varphi(\gamma_i)\right| \leqq B \left\|\sum_{i=1}^{n} c_i(\cdot, \gamma)\right\|_\infty$$

for each positive integer n and all choices of $c_i \in C$ and $\gamma_i \in \hat{G}$, $i = 1, 2, \ldots, n$.

If $\varphi = \hat{\mu}$ then $\|\mu\|$ is the smallest constant B for which (ii) *holds.*

The step in the proof of this theorem in Rudin [5] which is apparently missing is the next result, which is of some independent interest for us.

Theorem F.9.2. (Gaudry [6], VII.4.2). *Let G be a locally compact Abelian group. If $\mu \in M(G)$ then*

$$\|\mu\| = \sup_f \int_G f(t^{-1})\, d\mu(t)$$

where f is any trigonometric polynomial for which $\|f\|_\infty \leqq 1$.

F.10. Next we wish to collect some additional facts about $L_1(G)^\wedge$ and $M(G)^\wedge$. We observe, since $L_1(G)$ and $M(G)$ are semi-simple, that $L_1(G)^\wedge$ and $M(G)^\wedge$ are commutative Banach algebras under pointwise multiplication and the norms $\|\hat{f}\| = \|f\|_1$, $\hat{f} \in L_1(G)^\wedge$, and $\|\hat{\mu}\| = \|\mu\|$, $\hat{\mu} \in M(G)^\wedge$ (E.2). For an arbitrary locally compact Abelian group G and $f \in L_1(G)$ we set

$$\check{f}(\gamma) = \hat{f}(\gamma^{-1}) = \int_G (t, \gamma)\, f(t)\, d\lambda(t) \qquad (\gamma \in \hat{G}).$$

a) If G is a locally compact Abelian group then $L_1(G)^\wedge = L_2(\hat{G}) * L_2(\hat{G}) = \{g \mid g = f_1 * f_2, f_1, f_2 \in L_2(\hat{G})\}$ (Rudin [5], 1.6.3).

b) If G is an infinite discrete Abelian group and $p > 2$ then there exists $\varphi \in L_p(\hat{G})$ such that $\varphi \notin M(G)^\wedge$ (Rudin [5], 7.8.5 and 7.8.6).

c) If G is an infinite compact group and $1<p<2$ then there exists $f\in C(G)$ such that $\hat{f}\notin L_p(\hat{G})$ (Zygmund [6], V.4.11).

d) If $G=\Gamma$, the multiplicative group of complex numbers with unit modulus, then $\hat{G}=Z$, the additive group of the integers. If $f\in L_2(\Gamma)$ then there exists a function φ on Z which takes only the values $+1$ and -1 and such that $\varphi\hat{f}\in L_p(\Gamma)\hat{}$, $1\leqq p<\infty$ (Edwards [14_{II}], 14.3.2).

F.11. Let G be a compact Abelian group and $\hat{G}$ its discrete dual group. A subset E of $\hat{G}$ is called a *Sidon set* if there exists a constant $B>0$ such that

$$\int_{\hat{G}} |\hat{f}(\gamma)|\, d\eta(\gamma)\leqq B\,\|f\|_\infty$$

for every trigonometric polynomial f on G such that $\hat{f}(\gamma)=0$, $\gamma\notin E$. Such a trigonometric polynomial will be called an *E-polynomial*. We summarize a few of the properties of Sidon sets.

a) If G is an infinite compact Abelian group then $\hat{G}$ contains infinite Sidon Sets (Rudin [5], 5.7.6).

b) If G is a compact Abelian group and E is a Sidon set in $\hat{G}$ then

$$\|f\|_2\leqq 2B\,\|f\|_1$$

for every E-polynomial f (Rudin [5], 5.7.7).

c) If G is a compact Abelian group and $E\subset\hat{G}$ then $f\in L_1(G)$ is called an *E-function* if $\hat{f}(\gamma)=0$, $\gamma\notin E$. If E is an infinite Sidon set and $f\in L_p(G)$, $1<p\leqq 2$, then $\chi_E\hat{f}\in L_2(\hat{G})$ (Rudin [5], 5.7.8).

d) Let G be an infinite compact Abelian group and $E\subset\hat{G}$ an infinite Sidon set. Suppose φ is a bounded function on $\hat{G}$ such that $\varphi(\gamma)=0$, $\gamma\notin E$. If $1<p\leqq 2$ then $\varphi\hat{f}\in L_p(G)\hat{}$ for each $f\in L_p(G)$. Indeed, from the previous observation we see that $\varphi\hat{f}=\varphi\chi_E\hat{f}\in L_2(\hat{G})$, and so by the Plancherel Theorem (Theorem F.8.2) there exists a unique $g\in L_2(G)\subset L_p(G)$ such that $\hat{g}=\varphi\hat{f}$.

e) Let $G=\Gamma$ and $\hat{G}=Z$. A sequence of positive integers $\{n_k\}$ is called a *Hadamard sequence* if there exists a real number $r>1$ such that $n_{k+1}/n_k\geqq r$, $k=1,2,\ldots$. Every Hadamard sequence, and even any finite union of Hadamard sequences, is a Sidon set in Z (Edwards [14_{II}], 15.2.4, and Rudin [5], 5.7.6). If $E=\{n_k\}$ is a Hadamard sequence and $f\in L_1(\Gamma)$ is an E-function then $f\in L_p(\Gamma)$, $1\leqq p<\infty$, (Rudin [5], 5.7.7, and Zygmund [6], V.8.20).

f) Let G be a compact connected Abelian group and suppose that G has been given an order (B.3). Let $\hat{G}_+$ denote the nonnegative elements in $\hat{G}$. $H_1(G)$ is the subset of $L_1(G)$ consisting of all f such that $\hat{f}(\gamma)=0$, $\gamma\notin\hat{G}_+$. Clearly $H_1(G)$ is a closed ideal in $L_1(G)$ and hence is a commutative Banach algebra. If $G=\Gamma$, $\hat{G}_+=Z_+$, the nonnegative integers, and

$\{n_k\} \subset Z_+$ is a Hadamard sequence then $\sum_{k=1}^{\infty} |\hat{f}(n_k)|^2 < \infty$ whenever $f \in H_1(\Gamma)$ (Rudin [5], 8.6). A more general result is valid (Rudin [5], 8.6), but this special case is adequate for our needs.

F.12. Again let $G = \Gamma$, and $\hat{G}_+ = Z_+$. We state the F. and M. Riesz Theorem and an extension of it.

Theorem F.12.1. [F. and M. Riesz Theorem (Rudin [5], 8.2.1)]. *If $G = \Gamma$ and $\mu \in M(\Gamma)$ is such that $\hat{\mu}(n) = 0$ for each $n \notin Z_+$ then μ is absolutely continuous with respect to Haar measure on Γ.*

Theorem F.12.2. (Rudin [4]). *Let $G = \Gamma$ and suppose $E \subset Z_+$ is a Sidon set. If $\mu \in M(\Gamma)$ is such that $\hat{\mu}(n) = 0$ for each $n \in Z_+ \sim E$ then μ is absolutely continuous with respect to Haar measure on Γ.*

F.13. When G is a compact Abelian group then we can consider $\hat{G}$ as belonging to $L_1(G)$, that is, the functions $(\cdot, \gamma)$, $\gamma \in \hat{G}$, belong to $L_1(G)$. In this case the continuous characters form a complete orthonormal family for the Hilbert space $L_2(G)$. We shall not need the full generality of this, the Peter-Weyl Theorem (Weil [1], 21), but only the orthonormality of the continuous characters.

Theorem F.13.1. (Rudin [5], 1.2.5). *Let G be a compact Abelian group and suppose the Haar measure λ on G is normalized so that $\lambda(G) = 1$. Then*

$$\int_G (t, \gamma)(t, \delta^{-1})\, d\lambda(t) = \begin{cases} 1, & \gamma = \delta \\ 0, & \gamma \neq \delta \end{cases} \qquad (\gamma, \delta \in \hat{G}).$$

Note that the orthogonality of the continuous characters holds even if λ is not normalized.

Bibliography

Akemann, C. A.

[1] Some mapping properties of the group algebras of a compact group. Pacific J. Math. **22**, 1–8 (1967).

Ambrose, W.

[1] Structure theorems for a special class of Banach algebras. Trans. Amer. Math. Soc. **57**, 364–386 (1945).

Andersen, T. B.

[1] On multipliers and order-bounded operators on C^*-algebras. Proc. Amer. Math. Soc. **25**, 896–899 (1970).

Arens, R., Singer, I.

[1] Function values as boundary integrals. Proc. Amer. Math. Soc. **5**, 735–745 (1954).

Bachelis, G. F.

[1] On the ideal of unconditionally convergent Fourier series in $L_p(G)$. Proc. Amer. Math. Soc. **27**, 309–312 (1971).

Bachelis, G. F., Rosenthal, H. P.

[1] On unconditionally converging series and biorthogonal systems in a Banach space.

Benedek, A., Calderón, A. P., Panzone, R.

[1] Convolution operators on Banach space valued functions. Proc. Nat. Acad. Sci. U.S.A. **48**, 356–365 (1962).

Benedek, A., Panzone, R.

[1] The spaces L^P with mixed norm. Duke Math. J. **28**, 301–324 (1968).

Beurling, A., Helson, H.

[1] Fourier-Stieltjes transforms with bounded powers. Math. Scand. **1**, 120–126 (1953).

Birtel, F. T.

[1] Banach algebras of multipliers. Duke Math. J. **28**, 203–211 (1961).

[2] Isomorphisms and isometric multipliers. Proc. Amer. Math. Soc. **13**, 204–210 (1962).

[3] On a commutative extension of a Banach algebra. Proc. Amer. Math. Soc. **13**, 815–822 (1962).

Bishop, E.

[1] A minimal boundary for function algebras. Pacific J. Math. **9**, 629–642 (1959).

Bishop, E., de Leeuw, K.

[1] The representation of linear functionals by measures on sets of extreme points. Ann. Inst. Fourier (Grenoble) **9**, 305–331 (1959).

Boas, R. P., Jr.
[1] Majorant problems for trigonometric series. J. Analyse Math. **10**, 253–271 (1962–63).

Bochner, S.
[1] Über Faktorfolgen für Fouriersche Reihen. Acta Sci. Math. (Szeged) **4**, 125–129 (1928–29).

Boehme, T. K.
[1] Concerning convolution on the half-line. Arch. Rational Mech. Anal. **10**, 220–228 (1962).
[2] Continuity and perfect operators. J. London Math. Soc. **39**, 355–358 (1964).

Bohr, H.
[1] Über die Bedeutung der Potenzreihen unendlich vieler Variablen in der Theorie der Dirichletchen Reihen $\Sigma\, a_n/n^s$. Nachr. Akad. Wiss. Göttingen Math.-Phys. Kl. **2**, 441–448 (1913).

Bourbaki, N.
[1] Éléments de mathématique. XIII. Livre VI. Intégration. Actualités Sci. et Ind. 1175. Paris: Hermann & Cie. 1952.

Brainerd, B., Edwards, R. E.
[1] Linear operators which commute with translations. I. Representation theorems; II. Applications of the representation theorems. J. Austral. Math. Soc. **6**, 289–327 (1966); **6**, 328–350 (1966).

Breuer, M., Cordes, H. O.
[1] On Banach algebras with σ-symbol. I; II. J. Math. Mech. **13**, 313–323 (1964); **14**, 299–314 (1965).

Busby, R. C.
[1] Some remarks on the structure space and extensions of C^*-algebras. J. Functional Analysis **1**, 370–377 (1967).
[2] Double centralizers and extensions of C^*-algebras. Trans. Amer. Math. Soc. **132**, 79–99 (1968).

Byrnes, J. S., Newman, D. J.
[1] Completeness preserving multipliers. Proc. Amer. Math. Soc. **21**, 445–450 (1969).

Calderón, A. P.
[1] On theorems of M. Riesz and Zygmund. Proc. Amer. Math. Soc. **1**, 533–535 (1950).
[2] Lebesgue spaces of differentiable functions and distributions. In: Proc. Sympos. Pure Math. IV, p. 33–49. Providence, R. I.: American Mathematical Society 1961.
[3] Singular integrals. Bull. Amer. Math. Soc. **72**, 427–465 (1966).

Calderón, A. P., Zygmund, A.
[1] On the existence of certain singular integrals. Acta Math. **88**, 85–139 (1952).
[2] On singular integrals. Amer. J. Math. **78**, 289–309 (1956).
[3] Singular integrals and periodic functions. Studia Math. **14**, 249–271 (1956).
[4] Algebras of certain singular operators. Amer. J. Math. **78**, 310–320 (1956).
See Benedek, A.

Caveny, D. J.
[1] Bounded Hadamard products of H^p functions. Duke Math. J. **33**, 389–394 (1966).
[2] Absolute convergence factors for H^p series. Canad. J. Math. **21**, 187–195 (1969).

Ching, W. M., Wong, J. S. W.
[1] Multipliers and H^*-algebras. Pacific J. Math. **22**, 387–396 (1967).

Cigler, J.
[1] Normed ideals in $L_1(G)$. Indag. Math. **31**, 273–282 (1969).

Coddington, E. A.
[1] Some Banach algebras. Proc. Amer. Math. Soc. **8**, 258–261 (1957).

Cohen, P. J.
[1] Factorization in group algebras. Duke Math. J. **26**, 199–205 (1959).
[2] On the homomorphisms of group algebras. Amer. J. Math. **82**, 213–226 (1960).

Coifman, R. R.
[1] Remarks on weak type inequalities for operators commuting with translations. Bull. Amer. Math. Soc. **74**, 710–714 (1968).

Comisky, C. V.
[1] Multipliers of Banach modules. Notices Amer. Math. Soc. **17**, 213 (1970). Doctoral dissertation, University of Oregon 1970. Eugene, Oregon: University of Oregon Library.

Cordes, H. O.
[1] The algebra of singular integral operators on R^n. J. Math. Mech. **14**, 1007–1032 (1965).
See Breuer, M.

Cotlar, M.
[1] A unified theory of Hilbert transforms and ergodic theorems. Rev. Math. Cuyana **1**, 105–167 (1955).

Curtis, P. C., Jr., Figà-Talamanca, A.
[1] Factorization theorems for Banach algebras. In: Function algebras, p. 169–185. Chicago, Ill.: Scott, Foresman & Company 1966.

Dauns, J.
[1] Multiplier rings and primitive ideals. Trans. Amer. Math. Soc. **145**, 125–158 (1969).

de Leeuw, K.
[1] On L_p multipliers. Ann. of Math. **81**, 364–379 (1965).
See Bishop, E.

Devinatz, A., Hirschman, I. I., Jr.
[1] The spectra of multiplier transforms on l^p. Amer. J. Math. **80**, 829–842 (1958).

De Vore, R.
[1] Multipliers of uniform convergence. Enseignement Math. **14**, 175–188 (1968).

Doss, R.
[1] On the multiplicators of some classes of Fourier transforms. Proc. London Math. Soc. (2) **50**, 169–195 (1949).
[2] Some inclusions in multipliers. Pacific J. Math. **32**, 643–646 (1970).

Dugundji, J.
[1] Topology. Boston, Mass.: Allyn & Bacon, Inc. 1966.

Dunford, N., Schwartz, J.
[1] Linear operators. Part I: General theory. New York, N.Y.: Interscience Publishers, Inc. 1958.

Duren, P. L.
[1] On the multipliers of H^p spaces. Proc. Amer. Math. Soc. **22**, 24–27 (1969).

Duren, P. L., Romberg, B. W., Shields, A. L.
[1] Linear functionals on H^p spaces with $0<p<1$. J. Reine Angew. Math. **238**, 32–60 (1969).

Duren, P. L., Shields, A. L.
[1] Coefficient multipliers of H^p and B^p spaces. Pacific J. Math. **32**, 69–78 (1970).
[2] Properties of H^p $(0<p<1)$ and its containing Banach space. Trans. Amer. Math. Soc. **141**, 255–262 (1969).

Edwards, R. E.

[1] On functions which are Fourier transforms. Proc. Amer. Math. Soc. **5**, 71–78 (1954).

[2] On factor functions. Pacific J. Math. **5**, 367–378 (1955).

[3] Representation theorems for certain functional operators. Pacific J. Math. **7**, 1333–1339 (1957).

[4] The stability of weighted Lebesgue spaces. Trans. Amer. Math. Soc. **93**, 369–394 (1959).

[5] Convolutions as bilinear and linear operators. Canad. J. Math. **16**, 274–285 (1964).

[6] Endomorphisms of function spaces which leave stable all translation invariant manifolds. Pacific J. Math. **14**, 31–47 (1964).

[7] Translates of L^∞ functions and of bounded measures. J. Austral. Math. Soc. **4**, 403–409 (1964).

[8] Approximation by convolutions. Pacific J. Math. **15**, 85–95 (1965).

[9] Bipositive and isometric isomorphisms of same convolution algebras. Canad. J. Math. **17**, 839–846 (1965).

[10] Changing signs of Fourier coefficients. Pacific J. Math. **15**, 463–475 (1965).

[11] Functional analysis: Theory and applications. New York, N.Y.: Holt, Rinehart & Winston, Inc. 1965.

[12] Operators commuting with translations. Pacific J. Math. **16**, 259–265 (1966).

[13] Supports and singular supports of pseudomeasures. J. Austral. Math. Soc. **6**, 65–75 (1966).

[14] Fourier series: A modern introduction. I; II. New York, N.Y.: Holt, Rinehart & Winston, Inc. 1967.

[15] A class of multipliers. J. Austral. Math. Soc. **8**, 584–590 (1968).

Edwards, R. E., Price, J. F.

[1] A naively constructive approach to boundedness principles, with applications to harmonic analysis.

See Brainerd, B.

Eymard, P.

[1] L'algèbre de Fourier d'un groupe localement compact. C. R. Acad. Sci. Paris Sér. A **256**, 1429–1431 (1963).

[2] L'algèbre de Fourier d'un groupe localement compact. Bull. Soc. Math. France **92**, 181–236 (1964).

[3] Algèbres A_p et convoluteurs de L^p. Séminaire Bourbaki, No. 367 (1969/1970).

Fekete, M.

[1] Über Faktorenfolgen, welche die „Klasse" einer Fourierschen Reihe unverändert lassen. Acta Sci. Math. (Szeged) **1**, 148–166 (1922–23).

Figà-Talamanca, A.

[1] Multipliers of p-integrable functions. Bull. Amer. Math. Soc. **70**, 666–669 (1964).

[2] On the subspace of L^p invariant under multiplication of transforms of bounded continuous functions. Rend. Sem. Mat. Univ. Padova **35**, 176–189 (1965).

[3] Translation invariant operators in L^p. Duke Math. J. **32**, 495–502 (1965).

Figà-Talamanca, A., Gaudry, G. I.

[1] Density and representation theorems for multipliers of type (p, q). J. Austral. Math. Soc. **7**, 1–6 (1967).

[2] Multipliers and sets of uniqueness of L^p. Michigan Math. J. **17**, 179–191 (1970).

[3] Multipliers of L^p which vanish at infinity.

[4] Extensions of multipliers. Boll. Un. Mat. Ital. **3**, 1003–1014 (1970).

Figà-Talamanca, A., Rider, D.

[1] A theorem of Littlewood and lacunary series for compact groups. Pacific J. Math. **16**, 505–514 (1966).

[2] A theorem on random Fourier series on noncommutative groups. Pacific J. Math. **21**, 487–492 (1967).
See Curtis, P. C., Jr.

Flanders, M. C.
[1] Ideal C^*-algebras. Doctoral dissertation, Tulane University 1968. New Orleans, Louisiana: Tulane University Library.

Foias, C.
[1] On a commutative extension of a commutative Banach algebra. Pacific J. Math. **8**, 407–410 (1958).

Forelli, F.
[1] Homomorphisms of ideals in group algebras. Illinois J. Math. **9**, 410–417 (1965).

Fournier, J.
[1] Extensions of a Fourier multiplier theorem of Paley. Pacific J. Math. **30**, 415–432 (1969).

Gaudry, G. I.
[1] Quasimeasures and operators commuting with convolution. Pacific J. Math. **18**, 461–476 (1966).
[2] Multipliers of type (p, q). Pacific J. Math. **18**, 477–488 (1966).
[3] Quasimeasures and multiplier problems. Doctoral dissertation. Australian National University, 1966. Canberra, Australia.
[4] Multipliers of weighted Lebesgue and measure spaces. Proc. London Math. Soc. **19**, 327–340 (1969).
[5] H^p multipliers and inequalities of Hardy and Littlewood. J. Austral. Math. Soc. **10**, 23–32 (1969).
[6] Topics in harmonic analysis. Lecture notes, Department of Mathematics, Yale University, New Haven, Ct. 1969.
[7] Isomorphisms of multiplier algebras. Canad. J. Math. **20**, 1165–1172 (1968).
[8] Some results on multipliers.
[9] Bad behavior and inclusion results for multipliers of type (p, q). Pacific J. Math. **35**, 83–94 (1970).
See Figà-Talamanca, A.

Gelfand, I., Raikov, D., Šilov, G.
[1] Commutative normed rings. Amer. Math. Soc. Transl. (2) **5**, 115–220 (1957).

Gilbert, J. E.
[1] Convolution operators on $L^p(G)$ and properties of locally compact groups. Pacific J. Math. **24**, 257–268 (1968).

Glicksberg, I.
[1] Homomorphisms of certain measure algebras. Pacific J. Math. **10**, 167–191 (1960).

Goes, G.
[1] Komplementäre Fourier Koeffizientenräume und Multiplikatoren. Math. Ann. **137**, 371–384 (1959).

Gohberg, I. C.
[1] On the theory of multidimensional singular integral equations. Soviet Math. Dokl. **1**, 960–963 (1960).

Greenleaf, F. P.
[1] Norm decreasing homomorphisms of group algebras. Bull. Amer. Math. Soc. **70**, 536–539 (1964).

[2] Norm decreasing homomorphisms of group algebras. Pacific J. Math. **15**, 1187–1219 (1965).
[3] Characterization of group algebras in terms of their translation operators. Pacific J. Math. **18**, 243–276 (1966).
[4] Closed subalgebras of group algebras which are group algebras. In: Function algebras, p. 276–281. Chicago, Illinois: Scott, Foresman & Company 1966.

Grothendieck, A.
[1] Résultats nouveaux dans la théorie des opérations linéaires I; II. C. R. Acad. Sci. Paris Sér. A **239**, 557–579 (1954); C. R. Acad Sci. Paris Sér. A **239**, 607–609 (1954).

Gulick, S. L., Liu, T. S., van Rooij, A. C. M.
[1] Group algebra modules I; II; III; IV; V. Canad. J. Math. **19**, 133–150 (1967); **19**, 151–173 (1967); — Trans. Amer. Math. Soc. **152**, 561–580 (1971); **152**, 581–596 (1971).

Guy, D. L.
[1] Hankel multiplier transformations and weighted p-norms. Trans. Amer. Math. Soc. **95**, 137–189 (1960).

Hahn, L. S.
[1] On multipliers of p-integrable functions. Trans. Amer. Math. Soc. **128**, 321–335 (1967).
[2] A theorem on multipliers of type (p, q). Proc. Amer. Math. Soc. **21**, 493–496 (1969).

Halmos, P.
[1] Measure theory. Princeton, N. J.: D. van Nostrand Company, Inc. 1950.

Hardy, G. H.
[1] Remarks on three recent notes in the Journal. J. London Math. Soc. **3**, 166–169 (1928).

Hardy, G. H., Littlewood, J. E.
[1] Same properties of fractional integrals. I; II. Math. Z. **27**, 565–606 (1928); **34**, 403–439 (1931–32).
[2] Some theorems on Fourier series and Fourier power series. Duke Math. J. **2**, 354–381 (1936).
[3] Notes on the theory of series (XX): Generalizations of a theorem of Paley. Quart. J. Math. Oxford Ser. **8**, 161–171 (1937).
[4] Theorems concerning mean values of analytic or harmonic functions. Quart. J. Math. Oxford Ser. **12**, 221–256 (1942).

Harte, R. E.
[1] Tensor products of normed modules.

Hedlund, J. H.
[1] Multipliers of H^1 and Hankel matrices. Proc. Amer. Math. Soc. **22**, 20–23 (1969).
[2] Multipliers of H^p spaces. J. Math. Mech. **18**, 1067–1074 (1969).

Helgason, S.
[1] Multipliers of Banach algebras. Ann. of Math. **64**, 240–254 (1956).
[2] Topologies of group algebras and a theorem of Littlewood. Trans. Amer. Math. Soc. **86**, 269–283 (1957).
[3] Lacunary Fourier series on noncommutative groups. Proc. Amer. Math. Soc. **9**, 782–790 (1958).

Helson, H.
[1] Isomorphisms of abelian group algebras. Ark. Mat. **2**, 475–487 (1953).
[2] Conjugate series and a theorem of Paley. Pacific J. Math. **8**, 437–446 (1958).
See Beurling, A.

Herz, C.
[1] On the mean inversion of Fourier and Hankel transforms. Proc. Nat. Acad. Sci. U.S.A. **40**, 996–999 (1954).
[2] Fourier transforms related to convex sets. Ann. of Math. **75**, 81–92 (1962).
[3] Remarques sur la Note precédente de M. Varopoulos. C. R. Acad. Sci. Paris Sér. A **260**, 6001–6004 (1965).

Hewitt, E.
[1] The asymmetry of certain algebras of Fourier-Stieltjes transforms. Michigan Math. J. **5**, 149–158 (1958).
[2] A survey of abstract harmonic analysis. In: Some aspects of analysis and probability, p. 105–168. Surveys in applied mathematics, vol. 4. New York, N.Y.: John Wiley & Sons, Inc. 1958.
[3] The ranges of certain convolution operators. Math. Scand. **15**, 147–155 (1964).

Hewitt, E., Ross, K. A.
[1] Abstract harmonic analysis. I. Berlin-Göttingen-Heidelberg: Springer 1963.
[2] Abstract harmonic analysis. II. Berlin-Heidelberg-New York: Springer 1970.

Hille, E.
[1] On functions of bounded deviation. Proc. London Math. Soc. (2) **31**, 165–173 (1930).

Hille, E., Phillips, R. S.
[1] Functional analysis and semi-groups. Amer. Math. Soc. Colloquium Publ. XXXI. Providence, R. I.: American Mathematical Society 1957.

Hille, E., Tamarkin, J. D.
[1] On the summability of Fourier series. I; II. Ann. of Math. (2) **34**, 329–348 (1933); (2) **34**, 602–605 (1933).

Hirschman, I. I., Jr.
[1] The decomposition of Walsh and Fourier series. Mem. Amer. Math. Soc. **15** (1955).
[2] On multiplier transformations. Duke Math. J. **26**, 221–242 (1959).
See Devinatz, A.

Hörmander, L.
[1] Estimates for translation invariant operators in L^p spaces. Acta Math. **104**, 93–140 (1960).

Hoffman, K.
[1] Fundamentals of Banach algebras. Monografias Mathemáticas da Universidade do Paraná. Curitiba, Brazil: Instituto de Matemática da Universidade do Paraná 1962.

Husain, T.
[1] Introduction to topological groups. Philadelphia, Pa.: W. B. Saunders Company 1966.

Igari, S.
[1] On the decomposition theorems of Fourier transforms with weighted norms. Tôhoku Math. J. **15**, 6–36 (1963).
[2] Functions of L^p-multipliers. Tôhoku Math. J. **21**, 304–320 (1969).

Iltis, R.
[1] Harmonic analysis on compact topological groups. Doctoral dissertation, University of Oregon 1966. Eugene, Oregon: University of Oregon Library.

Jodeit, M.
[1] Restrictions and extensions of Fourier multipliers. Studia Math. **34**, 215–226 (1970).

Johnson, B. E.
[1] An introduction to the theory of centralizers. Proc. London Math. Soc. **14**, 299–320 (1964).
[2] Centralisers on certain topological algebras. J. London Math. Soc. **39**, 603–614 (1964).
[3] Isometric isomorphisms of measure algebras. Proc. Amer. Math. Soc. **15**, 186–189 (1964).
[4] Continuity of centralizers on Banach algebras. J. London Math. Soc. **41**, 639–640 (1966).
[5] Continuity of transformations which leave invariant certain translation invariant subspaces. Pacific J. Math. **20**, 223–230 (1967).

Jones, B. F., Jr.
[1] Singular integrals and parabolic equations. Bull. Amer. Math. Soc. **69**, 501–503 (1963).
[2] A class of singular integrals. Amer. J. Math. **86**, 441–462 (1964).

Kaczmarz, S.
[1] On some classes of Fourier series. J. London Math. Soc. **8**, 39–46 (1933).

Kaczmarz, S., Marcinkiewicz, J.
[1] Sur les multiplicateurs des séries orthogonales. Studia Math. **7**, 73–81 (1938).

Kaczmarz, S., Steinhaus, H.
[1] Theorie der Orthogonalreihen. Monografje Matematyczne, tom VI. Warszawa: 1935. Reprinted by Chelsea Publ. Co., New York, 1951.

Kadec, M., Pelczynski, A.
[1] Basic sequences, biorthogonal sequences and norming sets in Banach and Fréchet spaces. Studia Math. **25**, 297–323 (1965) [Russian].

Kahane, J. P.
[1] Transformées de Fourier des fonctions sommables. Proc. Int. Congr. Mathematicians 1962, p. 114–131. Djursholm, Sweden: Institut Mittag-Leffler 1963.

Karamata, J.
[1] Suite de fonctionelles linéaires et facteurs de convergence des séries de Fourier. J. Math. Pures Appl. **35**, 87–95 (1956).
[2] Sur les facteurs de convergence uniforme des séries de Fourier. Rev. Fac. Sci. Univ. Istanbul Ser. A **22**, 35–43 (1957).

Karamata, J., Tomić, M.
[1] Sur la sommation des séries de Fourier des fonctions continues. Acad. Serbe. Sci. Publ. Inst. Math. **8**, 123–138 (1955).

Katayama, M.
[1] Fourier Series. VII. Uniform convergence factors of Fourier series. J. Fac. Sci. Hokkaido Univ. Ser. I **13**, 121–129 (1957).

Katznelson, Y.
[1] Sets of uniqueness for some classes of trigonometric series. Bull. Amer. Math. Soc. **70**, 722–723 (1964).
[2] An introduction to harmonic analysis. New York, N.Y.: John Wiley & Sons, Inc. 1968.

Kawada, Y.
[1] On the group ring of a topological group. Math. Japon. **1**, 1–5 (1948).

Kelley, J. L.
[1] General topology. Princeton, N.J.: D. van Nostrand Company, Inc. 1955.

Kelley, J. L., Namioka, I., and coauthors
[1] Linear topological spaces. Princeton, N.J.: D. van Nostrand Company, Inc. 1963.

Kellogg, C. N.
[1] Centralizers and H^*-algebras. Pacific J. Math. **17**, 121–129 (1966).

Kitchen, J. W., Jr.
[1] Normed modules and almost periodicity. Monatsh. Math. **70**, 233–243 (1966).
[2] The almost periodic measures on a compact Abelian group. Monatsh. Math. **72**, 217–219 (1968).

König, H., Meixner, J.
[1] Lineare Systeme und lineare Transformationen. Math. Nachr. **19**, 265–322 (1958).

Kolmogorov, A. N.
[1] Sur les fonctions harmoniques conjuguées et les séries de Fourier. Fund. Math. **7**, 23–28 (1925).

Koosis, P.
[1] Sur un théorème de Paul Cohen. C.R. Acad. Sci. Paris Sér. A **259**, 1380–1382 (1964).

Krée, P.
[1] Sur les multiplicateurs dans $\mathscr{F} L^p$ avec poids. C.R. Acad. Sci. Paris Sér. A **258**, 1692–1695 (1964).
[2] Sur les multiplicateurs dans $\mathscr{F} L^p$. C.R. Acad. Sci. Paris Sér. A **260**, 4400–4403 (1965).

LaDuke, J.
[1] On a certain generalization of l_p spaces.

Lai, H. C.
[1] On some properties of $A^p(G)$-algebras. Proc. Japan Acad. **45**, 572–576 (1969).
[2] On the category of $L^1(G) \cap L^p(G)$ in $A^q(G)$. Proc. Japan Acad. **45**, 577–581 (1969).
[3] Remark on the $A^p(G)$-algebras. Proc. Japan Acad. **46**, 58–63 (1970).

Larsen, R.
[1] Closed ideals in Banach algebras with Gelfand transforms in $L_p(\mu)$. Rev. Roumaine Math. Pures Appl. **14**, 1295–1302 (1969).
[2] The Multiplier Problem. Lecture Notes in Mathematics, vol. 105. Berlin-Heidelberg-New York: Springer 1969.
[3] The multipliers for functions with Fourier transforms in L_p. Math. Scand. **28**, 1–11 (1971).

Larsen, R., Liu, T. S., Wang, J. K.
[1] On functions with Fourier transforms in L_p. Michigan Math. J. **11**, 369–378 (1964).

Lax, P. D.
[1] The L_2-operators of Mihlin, Calderón-Zygmund.

Leptin, H.
[1] On locally compact groups with invariant means. Proc. Amer. Math. Soc. **19**, 489–494 (1968).

Littlewood, J. E., Paley, R. E. A. C.
[1] Theorems on Fourier series and power series. I; II; III. J. London Math Soc. **6**, 230–233 (1931); – Proc. London Math. Soc. (2) **42**, 52–89 (1937); (2) **43**, 105–126 (1937).
See Hardy, G. H.

Littman, W.
[1] Multipliers in L^p and interpolation. Bull. Amer. Math. Soc. **71**, 764–766 (1965).

Littman, W., McCarthy, C., Rivière, N. M.
[1] L^p-multiplier theorems. Studia Math. **30**, 193–217 (1968).
[2] The non-existence of L^p estimates for certain translation-invariant operators. Studia Math. **30**, 219–229 (1968).

Liu, T. S., van Rooij, A. C. M.
[1] Sums and intersections of normed linear spaces. Math. Nachr. **42**, 29–42 (1969).

Liu, T. S., van Rooij, A. C. M., Wang, J. K.
[1] Characters of group algebra modules.

Liu, T. S., Wang, J. K.
[1] Sums and intersections of Lebesgue spaces. Math. Scand. **23**, 241–251 (1968).
See Gulick, S. L., Larsen, R.

Lizorkin, P. I.
[1] (L_p, L_q)-multipliers of Fourier integrals. Soviet Math. Dokl. **4**, 1420–1424 (1963).
[2] Generalized Liouville differentiation and the method of multipliers in the theory of imbeddings of function classes. Mat. Zametki **4**, 467–482 (1968) [Russian].

Loomis, L.
[1] An introduction to abstract harmonic analysis. Princeton, N.J.: D. van Nostrand Company, Inc. 1952.

Mackey, G.
[1] Functions on locally compact groups. Bull. Amer. Math. Soc. **56**, 385–412 (1950).

Marcinkiewicz, J.
[1] Sur les multiplicateurs des séries de Fourier. Studia Math. **8**, 78–91 (1939).
See Kaczmarz, S.

Martin, J. C., Yap, L. Y. H.
[1] The algebra of functions with Fourier transforms in L^p. Proc. Amer. Math. Soc. **24**, 217–219 (1970).

Máté, L.
[1] On the factor theory of commutative Banach algebras. Magyar Tud. Akad. Mat. Kutato Int. Köze **9**, 349–364 (1964).
[2] Embedding multiplier operators of a Banach algebra B into its second conjugate space B^{**}. Bull. Acad. Polon. Sci. Sér. Sci. Math. Astronom. Phys. **13**, 809–812 (1965).
[3] Multiplier operators and quotient algebras. Bull. Acad. Polon. Sci. Sér. Sci. Math. Astronom. Phys. **13**, 523–526 (1965).
[4] Some abstract results concerning multiplier algebras. Rev. Roumaine Math. Pures. Appl. **10**, 261–266 (1965).
[5] The Arens product and multiplier operators. Studia Math. **28**, 227–234 (1967).
[6] On the connection between Helson sets and approximation by convolutions.
[7] On the connections between approximation by convolution and the multiplier problem. To appear in Studia Sci. Math. Hungar.
[8] On a dual space representation of module homomorphisms.

Mazurkiewicz, St.
[1] O sumowalności szeregów kształtu $\sum_{n=1}^{\infty} a_n u_n$. C.R. Soc. Sci. Varsovie **8**, 649–655 (1915).

McCarthy, C.
See Littman, W.

McGiveney, R.J., Ruckle, W.
[1] Multiplier algebras of biorthogonal systems. Pacific J. Math. **29**, 375–387 (1969).

Meixner, J.
See König, H.

Merlo, J.C.
[1] Some remarks on multipliers and their applications to singular kernels. Rev. Un. Mat. Argentina **20**, 210–230 (1962). [Spanish].

Meyer, Y.
[1] Prolongement des multiplicateurs d'idéaux fermés de $L^1(R^n)$. C.R. Acad. Sci. Paris Sér. A **262**, 744–745 (1966).
[2] Multiplicateurs des coefficients de Fourier des fonctions intégrables analytiques. C.R. Acad. Sci. Paris Sér. A **263**, 385–387 (1966).
[3] Endomorphisms des idéaux fermés de $L_1(G)$, classes de Hardy, et séries de Fourier lacunaires. Ann. Sci. École Norm. Sup. (4) **1**, 499–580 (1968).

Mihlin, S.G.
[1] On the multipliers of Fourier integrals. Dokl. Akad. Nauk SSSR, N.S. **109**, 701–703 (1956) [Russian].
[2] On the theory of multi-dimensional singular integral equations. Vestnik Leningrad. Univ. Ser. Mat. Mech. Astr. **11**, 3–24 (1956) [Russian].
[3] Singular integrals in L_p spaces. Dokl. Akad. Nauk SSSR, N.S. **117**, 28–31 (1957) [Russian].
[4] Fourier integrals and multiple singular integrals. Vestnik Leningrad. Univ. Ser. Mat. Mech. Astr. **12**, 143–155 (1957) [Russian].

Muckenhoupt, B.
[1] On certain singular integrals. Pacific J. Math. **10**, 239–261 (1960).

Nachbin, L.
[1] The Haar integral. Princeton, N.J.: D. van Nostrand Company, Inc. 1965.

Naimark, M.A.
[1] Normed rings. Groningen, The Netherlands: P. Noordhoff, Ltd. 1964.

Namioka, I.
See Kelley, J.L.

Newman, D.J.
See Byrnes, J.S.

Orlicz, W.
[1] Beiträge zur Theorie der Orthogonalentwicklungen. Studia Math. **1**, 1–39 (1929).

Paley, R.E.A.C.
[1] A note on power series. J. London Math. Soc. **7**, 122–130 (1932).
See Littlewood, J.E.

Paneyakh, B.P.
See Volevich, L.R.

Panzone, R.
See Benedek, A.

Parrott, S. K.
[1] Isometric multipliers. Pacific J. Math. **25**, 159–166 (1968).

Peetre, J.
[1] Espaces d'interpolation et théorème de Soboleff. Ann. Inst. Fourier (Grenoble) **16**, 279–317 (1966).
[2] On convolution operators leaving $L^{p,\lambda}$ spaces invariant. Ann. Mat. Pura Appl. Ser. IV **72**, 295–304 (1966).
[3] Applications de la théorie des espaces d'interpolation dans l'analyse harmonique. Ricerche Mat. **15**, 1–34 (1966).

Pelczynski, A.
See Kadec, M.

Peyrière, J., Spector, R.
[1] Sur les multiplicateurs radiaux $\mathscr{F} L^p(G)$ pour un groupe abélian localement compact totalement discontinu. C.R. Acad. Sci. Paris Sér. A **269**, 973–974 (1969).

Phillips, R. S.
See Hille, E.

Pigno, L.
[1] A multiplier theorem. Pacific J. Math. **34**, 755–758 (1970).
[2] On multipliers of Fourier transforms.
[3] Some multiplier problems in Fourier analysis on groups.

Pontryagin, L.
[1] Topological groups. Princeton, N.J.: Princeton University Press 1946.

Powell, J.
[1] Multipliers on locally convex algebras. To appear in Studia Math.

Price, J. F.
[1] Multipliers between some spaces of distributions. J. Austral. Math. Soc. **9**, 415–423 (1969).
[2] Some strict inclusions between spaces of L^p multipliers.
[3] (L^p, L^q)-multiplier problems. Doctoral dissertation. Australian National University 1970. Canberra, Australia.
See Edwards, R. E.

Raikov, D.
See Gelfand, I.

Reid, G. A.
[1] A generalization of W^*-algebras. Pacific J. Math. **15**, 1019–1026 (1965).

Reiter, H.
[1] Subalgebras of $L^1(G)$. Indag. Math. **27**, 691–696 (1965).
[2] Classical harmonic analysis and locally compact groups. Oxford: Oxford University Press 1968.

Rickart, C. E.
[1] General theory of Banach algebras. Princeton, N.J.: D. van Nostrand Company, Inc. 1960.

Rider, D.
[1] A relation between a theorem of Bohr and Sidon sets. Bull. Amer. Math. Soc. **72**, 558–561 (1966).
See Figà-Talamanca, A.

Rieffel, M. A.
[1] A characterization of commutative group algebras and measure algebras. Trans. Amer. Math. Soc. **116**, 32–65 (1965).
[2] Induced Banach representations of Banach algebras and locally compact groups. J. Functional Analysis **1**, 443–491 (1967).
[3] On the continuity of certain intertwining operators, centralizers and positive linear functionals. Proc. Amer. Math. Soc. **20**, 455–457 (1969).
[4] Multipliers and tensor products of L^p spaces of locally compact groups. Studia Math. **33**, 71–82 (1969).

Riesz, M.
[1] Sur les maxima des formes bilinéaires et sur les fonctionelles linéaires. Acta Math. **49**, 465–497 (1926).
[2] Sur les fonctions conjuguées. Math. Z. **27**, 218–244 (1927).

Rigelhof, R. P.
[1] Norm decreasing homomorphisms of measure algebras. Trans. Amer. Math. Soc. **136**, 361–371 (1969).

Rivière, N. M.
[1] Vector valued multipliers and applications. Bull. Amer. Math. Soc. **74**, 946–948 (1968).

Riviére, N. M., Sagher, Y.
[1] Multipliers of trigonometric series and pointwise convergence. Trans. Amer. Math. Soc. **140**, 301–308 (1969).
See Littman, W.

Romberg, B. W.
See Duren, P. L.

Rosenthal, H. P.
[1] Projections onto translation invariant subspaces of $L^p(G)$. Mem. Amer. Math. Soc. **63**, (1966).
See Bachelis, G. F.

Ross, K. A.
See Hewitt, E.

Rowlands, K.
[1] Some new characterizations of perfect operators. J. London Math. Soc. **44**, 531–541 (1969).

Royden, H.
[1] Real analysis, 2nd edit. New York, N. Y.: The Macmillan Company 1968.

Ruckle, W.
See McGiveney, R. J.

Rudin, W.
[1] Remarks on a theorem of Paley. J. London Math. Soc. **32**, 307–311 (1957).
[2] Representation of functions by convolutions. J. Math. Mech. **7**, 103–116 (1958).
[3] Measure algebras on Abelian groups. Bull. Amer. Math. Soc. **65**, 227–247 (1959).
[4] Trigonometric series with gaps. J. Math. Mech. **9**, 203–228 (1960).
[5] Fourier analysis on groups. New York, N. Y.: Interscience Publishers, Inc. 1962.

[6] Ideals with small automorphisms. Bull. Amer. Math. Soc. **72**, 339–341 (1966).
[7] Real and complex analysis. New York, N.Y.: McGraw-Hill Book Co. 1966.

Ryan, R.
[1] Fourier transforms of certain classes of integrable functions. Trans. Amer. Math. Soc. **105**, 102–111 (1962).

Saeki, S.
[1] Translation invariant operators on groups. Tôhoku Math. J. **22**, 409–419 (1970).

Sagher, Y.
See Rivière, N. M.

Sakai, S.
[1] Weakly compact operators on operator algebras. Pacific J. Math. **14**, 659–664 (1964).

Salem, R.
[1] Sur les transformations des séries de Fourier. Fund. Math. **33**, 108–114 (1939).

Saworotnow, P. P.
[1] Trace-class and centralizers of an H^*-algebra. Proc. Amer. Math. Soc. **26**, 101–104 (1970).

Schwartz, J.
[1] A remark on inequalities of Calderón-Zygmund type for vector valued functions. Comm. Pure Appl. Math. **14**, 785–799 (1961).
See Dunford, N.

Schwartz, L.
[1] Sur les multiplicateurs de $\mathscr{F}L^p$. Kungl. Fysiogr. Sällsk. Lund. Förh. **22**, 124–128 (1953).
[2] Théorie des distributions. Actualités Sci. et Ind. 1245. Paris: Hermann & Cie. 1959.

Shields, A. L.
See Duren, P. L.

Sidon, S.
[1] Reihentheoretische Sätze und ihre Anwendungen in der Theorie der Fourierschen Reihen. Math. Z. **10**, 121–127 (1921).
[2] Ein Satz über die Fourierschen Reihen stetiger Funktionen. Math. Z. **34**, 485–486 (1932).

Šilov, G.
See Gelfand, I.

Singer, I.
See Arens, R.

Skvortsova, M. G.
[1] Fourier series multipliers. Siberian Math. J. **10**, 97–135 (1969).

Sobolev, S.
[1] On a theorem in functional analysis. Dokl. Akad. Nauk SSSR **20**, 5 (1938) [Russian].

Spector, R.
See Peyrière, J.

Stein, E. M.
[1] On the functions of Littlewood-Paley, Lusin and Marcinkiewicz. Trans. Amer. Math. Soc. **88**, 430–466 (1958).

Waterman, D.
[1] On functions analytic in a half-plane. Trans. Amer. Math. Soc. **81**, 167–194 (1956).
[2] On an integral of Marcinkiewicz. Trans. Amer. Math. Soc. **91**, 129–138 (1959).

Weil, A.
[1] L'Intégration dans les groupes topologiques et ses applications. Actualités Sci. et Ind., 869–1145. Paris: Hermann & Cie. 1951.

[1] Multipliers of ideals in function algebras. Duke Math. J. 31, 703–709 (1964).
[2] Restrictions of Fourier-Stieltjes transforms. Proc. Amer. Math. Soc. **15**, 243–246 (1964).
[3] Some results concerning multipliers of H^p.

Wendel, J. G.
[1] On isometric isomorphism of group algebras. Pacific J. Math. **1**, 305–311 (1951).
[2] Left centralizers and isomorphisms of group algebras. Pacific J. Math. **2**, 251–261 (1952).

Wermer, J.
[1] Ideals in a class of commutative Banach algebras. Duke Math. J. **20**, 273–278 (1953).
[2] Banach algebras and analytic functions. In: Advances in mathematics, vol. 1, fasc. 1. New York, N.Y.: Academic Press, Inc. 1961.

Weston, J. D.
[1] An extension of the Laplace-transform calculus. Rend. Circ. Mat. Palermo **6**, 325–333 (1957).
[2] Characterizations of Laplace transforms and perfect operators. Arch. Rational Mech. Anal. **3**, 348–354 (1959).
[3] Operational calculus and generalized functions. Proc. Roy. Soc. Ser. A **250**, 460–471 (1959).
[4] On the representation of operators by convolution integrals. Pacific J. Math. **10**, 1453–1486 (1960).
[5] Positive perfect operators. Proc. London Math. Soc. **10**, 545–565 (1960).

Widom, H.
[1] On the spectrum of a Toeplitz operator. Pacific J. Math. **14**, 365–375 (1964).
[2] Toeplitz operators on H_p. Pacific J. Math. **19**, 573–582 (1966).

Wong, J. S. W.
See Ching, W. M.

Wong, P. K.
See Tomiuk, B. J.

Yap, L.Y.H.

[1] Ideals in subalgebras of the group algebras. Studia Math. **35**, 165–175 (1970).

[2] On the impossibility of representing certain functions by convolution. Math. Scand. **26**, 132–140 (1970).

See Martin, J.C.

Young, W.H.

[1] On the Fourier series of bounded functions. Proc. London Math. Soc. (2) **12**, 41–70 (1913).

[2] On Fourier series and functions of bounded variation. Proc. Roy. Soc. Ser. A **88**, 561–568 (1913).

Zygmund, A.

[1] Sur un théorème de M. Fekete. Bull. Int. Acad. Polon. Sci. Lett. Ser. A. Math. Cracowie, No. 6A, 343–347 (1927).

[2] On the convergence and summability of power series on the circle of convergence. I; II. Fund. Math. **30**, 170–196 (1928);—Proc. London Math. Soc. **47**, 326–350 (1941).

[3] On a theorem of Marcinkiewicz concerning interpolation of operations. J. Math. Pures Appl. **35**, 223–248 (1956).

[4] On the Littlewood-Paley function $g^*(\theta)$. Proc. Nat. Acad. Sci. U.S.A. **42**, 208–212 (1956).

[5] On the preservation of classes of functions. J. Math. Mech. **8**, 889–895 (1959).

[6] Trigonometrical series. 2nd edit. New York, N.Y.: Cambridge University Press 1968.

See Calderón, A.P., Stein, E.M., Weiss, M.

Author and Subject Index

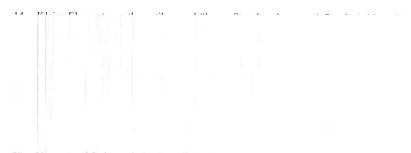

141. Mathematische Hilfsmittel des Ingenieurs. Herausgeber: Sauer/Szabó. 3. Teil. DM 98,—; US $ 27.00
142. Mathematische Hilfsmittel des Ingenieurs. Herausgeber: Sauer/Szabó. 4. Teil. DM 124,—; US $ 34.10
143. Schur/Grunsky: Vorlesungen über Invariantentheorie. DM 32,—; US $ 8.80
144. Weil: Basic Number Theory. DM 48,—; US $ 12.00
145. Butzer/Berens: Semi-Groups of Operators and Approximation. DM 56,—; US $ 14.00
146. Treves: Locally Convex Spaces and Linear Differential Equations. DM 36,—; US $ 9.90
147. Lamotke: Semisimpliziale algebraische Topologie. DM 48,—; US $ 13.20
148. Chandrasekharan: Introduction to Analytic Number Theory. DM 28,—; US $ 7.00
149. Sario/Oikawa: Capacity Functions. DM 96,—; US $ 24.00
150. Iosifescu/Theodorescu: Random Processes and Learning. DM 68,—; US $ 18.70
151. Mandl: Analytical Treatment of One-dimensional Markov Processes. DM 36,—; US $ 9.80
152. Hewitt/Ross: Abstract Harmonic Analysis. Voll. II. Structure and Analysis for Compact Groups. Analysis on Locally Compact Abelian Groups. DM 140,—; US $ 38.50
153. Federer: Geometric Measure Theory. DM 118,—; US $ 29.50
154. Singer: Bases in Banach Spaces I. DM 112,—; US $ 30.80
155. Müller: Foundations of the Mathematical Theory of Electromagnetic Waves. DM 58,—; US $ 16.00
156. van der Waerden: Mathematical Statistics. DM 68,—; US $ 18.70
157. Prohorov/Rozanov: Probability Theory. DM 68,—; US $ 18.70
158. Constantinescu/Cornea: Potential Theory on Harmonic Spaces. In preparation
159. Köthe: Topological Vector Spaces I. DM 78,—; US $ 21.50
160. Agrest/Maksimov: Theory of Incomplete Cylindrical Functions and their Applications. In preparation
161. Bhatia/Szegö: Stability Theory of Dynamical Systems. DM 58,—; US $ 16.00
162. Nevanlinna: Analytic Functions. DM 76,—; US $ 20.90
163. Stoer/Witzgall: Convexity and Optimization in Finite Dimensions I. DM 54,—; US $ 14.90
164. Sario/Nakai: Classification Theory of Riemann Surfaces. DM 98,—; US $ 27.00
165. Mitrinović: Analytic Inequalities. DM 88,—; US $ 26.00
166. Grothendieck/Dieudonné: Eléments de Géométrie Algébrique I. DM 84,—; US $ 23.10
167. Chandrasekharan: Arithmetical Functions. DM 58,—; US $ 16.00
168. Palamodov: Linear Differential Operators with Constant Coefficients. DM 98,—; US $ 27.00
169. Rademacher: Topics in Analytic Number Theory. In preparation
170. Lions: Optimal Control of Systems Governed by Partial Differential Equations. DM 78,—; US $ 21.50
171. Singer: Best Approximation in Normed Linear Spaces by Elements of Linear Subspaces. DM 60,—; US $ 16.50
172. Bühlmann: Mathematical Methods in Risk Theory. DM 52,—; US $ 14.30
173. F. Maeda/S. Maeda: Theory of Symmetric Lattices. DM 48,—; US $ 13.20
174. Stiefel/Scheifele: Linear and Regular Celestial Mechanics. DM 68,—; US $ 18.70
175. Larsen: An Introduction to the Theory of Multipliers. DM 84,—; US $ 23.10
176. Grauert/Remmert: Analytische Stellenalgebren. DM 64,—; US $ 17.60
177. Flügge: Practical Quantum Mechanics I. DM 70,—; US $ 19.30
178. Flügge: Practical Quantum Mechanics II. DM 60,—; US $ 16.50
179. Giraud: Cohomologie non abélienne. En préparation